Richi Nayak, Nikhil Ichalkaranje, and Lakhmi C. Jain (Eds.)

Evolution of the Web in Artificial Intelligence Environments

T0180971

Studies in Computational Intelligence, Volume 130

Editor-in-Chief

Prof. Janusz Kacprzyk
Systems Research Institute
Polish Academy of Sciences
ul. Newelska 6
01-447 Warsaw
Poland
E-mail: kacprzyk@ibspan.waw.pl

Further volumes of this series can be found on our homepage: springer.com

Vol. 113. Gemma Bel-Enguix, M. Dolores Jiménez-López and Carlos Martín-Vide (Eds.)
New Developments in Formal Languages and Applications, 2008
ISBN 978-3-540-78290-2

Vol. 114. Christian Blum, Maria José Blesa Aguilera, Andrea Roli and Michael Sampels (Eds.)
Hybrid Metaheuristics, 2008
ISBN 978-3-540-78294-0

Vol. 115. John Fulcher and Lakhmi C. Jain (Eds.)
Computational Intelligence: A Compendium, 2008
ISBN 978-3-540-78292-6

Vol. 116. Ying Liu, Aixin Sun, Han Tong Loh, Wen Feng Lu and Ee-Peng Lim (Eds.)
Advances of Computational Intelligence in Industrial Systems, 2008
ISBN 978-3-540-78296-4

Vol. 117. Da Ruan, Frank Hardeman and Klaas van der Meer (Eds.)
Intelligent Decision and Policy Making Support Systems, 2008
ISBN 978-3-540-78306-0

Vol. 118. Tsau Young Lin, Ying Xie, Anita Wasilewska and Churn-Jung Liau (Eds.)
Data Mining: Foundations and Practice, 2008
ISBN 978-3-540-78487-6

Vol. 119. Slawomir Wiak, Andrzej Krawczyk and Ivo Dolezel (Eds.)
Intelligent Computer Techniques in Applied Electromagnetics, 2008
ISBN 978-3-540-78489-0

Vol. 120. George A. Tsihrintzis and Lakhmi C. Jain (Eds.)
Multimedia Interactive Services in Intelligent Environments, 2008
ISBN 978-3-540-78491-3

Vol. 121. Nadia Nedjah, Leandro dos Santos Coelho and Luiza de Macedo Mourelle (Eds.)
Quantum Inspired Intelligent Systems, 2008
ISBN 978-3-540-78531-6

Vol. 122. Tomasz G. Smolinski, Mariofanna G. Milanova and Aboul-Ella Hassanien (Eds.)
Applications of Computational Intelligence in Biology, 2008
ISBN 978-3-540-78533-0

Vol. 123. Shuichi Iwata, Yukio Ohsawa, Shusaku Tsumoto, Ning Zhong, Yong Shi and Lorenzo Magnani (Eds.)
Communications and Discoveries from Multidisciplinary Data, 2008
ISBN 978-3-540-78732-7

Vol. 124. Ricardo Zavala Yoe
Modelling and Control of Dynamical Systems: Numerical Implementation in a Behavioral Framework, 2008
ISBN 978-3-540-78734-1

Vol. 125. Larry Bull, Bernadó-Mansilla Ester and John Holmes (Eds.)
Learning Classifier Systems in Data Mining, 2008
ISBN 978-3-540-78978-9

Vol. 126. Oleg Okun and Giorgio Valentini (Eds.)
Supervised and Unsupervised Ensemble Methods and their Applications, 2008
ISBN 978-3-540-78980-2

Vol. 127. Régie Gras, Einoshin Suzuki, Fabrice Guillet and Filippo Spagnolo (Eds.)
Statistical Implicative Analysis, 2008
ISBN 978-3-540-78982-6

Vol. 128. Fatos Xhafa and Ajith Abraham (Eds.)
Metaheuristics for Scheduling in Industrial and Manufacturing Applications, 2008
ISBN 978-3-540-78984-0

Vol. 129. Natalio Krasnogor, Giuseppe Nicosia, Mario Pavone and David Pelta (Eds.)
Nature Inspired Cooperative Strategies for Optimization (NICSO 2007), 2008
ISBN 978-3-540-78986-4

Vol. 130. Richi Nayak, Nikhil Ichalkaranje and Lakhmi C. Jain (Eds.)
Evolution of the Web in Artificial Intelligence Environments, 2008
ISBN 978-3-540-79139-3

Richi Nayak
Nikhil Ichalkaranje
Lakhmi C. Jain
(Eds.)

Evolution of the Web in Artificial Intelligence Environments

With 71 Figures and 14 Tables

 Springer

Dr. Richi Nayak
School of Information Systems
Queensland University of Technology
Brisbane, QLD 4001
Australia

Dr. Nikhil Ichalkaranje
School of Electrical and Information Engineering
University of South Australia
Adelaide
South Australia
Australia

Professor Dr. Lakhmi C. Jain
School of Electrical and Information Engineering
University of South Australia
Adelaide
South Australia
Australia
Email: Lakhmi.jain@unisa.edu.au

ISBN 978-3-642-09802-4   e-ISBN 978-3-540-79140-9

DOI 10.1007/978-3-540-79140-9

Studies in Computational Intelligence   ISSN 1860949X

Cover Design: Deblik, Berlin, Germany.

Printed in acid-free paper

9 8 7 6 5 4 3 2 1
springer.com

Preface

The Web has revolutionized the way we seek information on all aspects of education, entertainment, business, health and so on. The Web has evolved into a publishing medium, global electronic market and increasingly, a platform for conducting electronic commerce. A part of this success can be attributed to the tremendous advances made in the Artificial Intelligence field. The popularity of the Web has opened many opportunities to develop smart Web-based systems using artificial intelligence techniques.

There exist numerous Web technology and applications that can benefit with the application of artificial intelligence techniques. It is not possible to cover them all in one book with a required degree of quality, depth and width. We present this book to discuss some important Web developments by using artificial intelligence techniques in the areas of Web personalisation, semantic Web and Web services.

The primary readers of this book are undergraduate/postgraduate students, researchers and practitioners in information technology and computer science related areas.

The success of this book is largely due to the collective efforts of a great team consisting of authors and reviewers. We are grateful to them for their vision and wonderful support. The final quality of selected papers reflects their efforts.

Finally we would like to thank the Queensland University of Technology, Brisbane Australia and University of South Australia, Adelaide Australia for providing us the resources and time to undertake this task. We extend our sincere thanks to Scientific Publishing Services Pvt. Ltd., for the editorial support.

Richi Nayak, Australia
Lakhmi C. Jain, Australia
Nikhil Ichalkaranje, Australia

Contents

1

An Introduction to the Evolution of the Web in an Artificial Intelligence Environment

R. Nayak[1] and L.C. Jain[2]

[1] Faculty of Information technology
Queensland University of Technology
Brisbane, Australia
r.nayak@qut.edu.au
[2] School of Electrical and Information Engineering
University of South Australia
Adelaide, Australia
Lakhmi.Jain@unisa.edu.au

Abstract. This chapter provides a brief introduction to the developments of the Artificial Intelligence field in the evolution of the Web covered in this book. This chapter presents the organization of this book. A comprehensive list of resources is also provided in the end to help the reader in gaining deep insights of the field.

1.1 Introduction

The digital revolution and the overwhelming acceptance of the rapidly growing World Wide Web as a primary medium for content delivery and business transactions have created many unique opportunities and challenges for researchers in the Artificial Intelligence area. The phenomenal growth of the Web has caused and resulted into developing many smart Web applications using artificial intelligence techniques. These applications range from the simple access to information in the Web to the advanced articulation of automatic tasks on the Web.

The material in this book is designed to be drawn from the different Web application areas so as to provide a good overview of the important contributions in the development of the Web applications in the artificial intelligence environment. This book predominately presents three research areas: Web personalisation, semantic Web and Web services. Information is growing at an ever increasing rate. This leaves web users drowning in an apparent endless sea of information. Web Personalisation is a way to deal with this by filtering the information in the background for the user by using implicit and explicit data relating, generally, to the user. Personalisation also helps in developing the context to which the user associates to information as well as aiding users in refining exactly what it is that they are after as users often find it hard to express what it is they are looking for in terms of a search query.

The Semantic web is the second generation Web enriched by machine process able information which supports the user in their tasks. The Semantic Web is a vision of Tim Berners-Lee of having data on the web defined and linked in such a way that it can be used by machines - not just for display purposes, but for using it in automation, integration, and reuse across applications. The development of the Semantic Web is a fast moving process that has raised a lot of research challenges. A significant task in

R. Nayak et al. (Eds.): Evolution of the Web in Artificial Intel. Environ., SCI 130, pp. 1–15, 2008.
springerlink.com   © Springer-Verlag Berlin Heidelberg 2008

semantic web is to deal with the semantic interoperability problem and to learn the Ontology from the Web. The semantic Web with artificial intelligence techniques would be able to address the dilemma that data on the Web is mostly unstructured and understood only by humans, but data on the Web is huge, and the need of machines is inevitable if the Web is to continue to grow at its current speed.

Web service provides a common platform for interacting with and for integrating the available services on the Web. As databases, data, archives and algorithms are located at different sites, Web services are essential for passing the algorithms or queries to multiple sites, and combining the responses from these sites to present a single interface to the user. Web services technology represents the next generation of distributed computing. Web service technology is still in its infancy. Artificial intelligence techniques will play an important role in the emerging technologies in Web service composition, discovery and matching, and personalisation.

1.2 Chapters Included in the Book

This book is structured as an edited book from prominent researchers in the field. The book includes eleven chapters. The subsequent chapter introduces various Web applications that are evolved with artificial intelligence techniques to improve the Web functionality. The remaining chapters will provide a comprehensive treatment of these Web applications.

Chapter two, Innovations in Web applications with Artificial Intelligence Technology by Richi Nayak and Lakhmi C. Jain, presents a broad overview of the main Web application topics discussed in this book. These applications are Web personalisation, Web services and semantic Web. Each Web application is represented with its key research content and the future directions of research.

Chapter three, Modelling Interests of Web users for recommendation by Daniela Godoy and Analia Amandi addresses the problem of modelling the information preferences of Web users and adapt them to changes in the user interests over time for Web personalisation. This chapter presents a novel user profiling technique that enables agents to build user profiles by incrementally learning user interests for Web based tasks.

Chapter four, Context-aware Web Content adaptation for mobile User Agents by Timo Laakko presents adaptation approaches and techniques to personalise the Web content to mobile devices of the users. This chapter describes an adaptation proxy system in the context of current delivery with its typical use-cases.

Chapter five, Knowledge on the Web: Models, Algorithms, and an Effective Framework based on Web Services by Alfredo Cuzzocrea proposes models and algorithms for efficiently representing, managing and personalizing data and services in distributed knowledge environment. This chapter adopts the On-Line Analytical Process (OLAP) technology to build the user profile to personalise Web services.

Chapter six, Model-based Data Engineering for Web Services by Andreas Tolk and Saikou Y. Diallo discusses the applicability of Model-based Data Engineering (MBDE) process in composing Web services. The MBDE process represents the information exchange requirements, contexts, and constraints of Web services in meta-data.

Chapter seven, SAM: Semantic Advanced Matchmaker by Erdem Savas Ilhan and Ayse Basar Bener addresses the problem of service matchmaking in the problem of Web service discovery and composition. This chapter presents an innovative approach of efficient matchmaking based on bipartite graphs to find the semantically matching services to the user query.

Chapter eight, Clarifying the Meta by Vladan Devedzic, Dragan Gasevic and Dragan Djuric presents a survey of some of the most interesting cases of using the concept of *meta* in computing. The chapter attempts to establish that the meta can imply much more than understanding some concepts in a specific domain.

Chapter nine, Emergent Semantics in Distributed Knowledge Management by Carola Aiello, Tiziana Catarci, Paolo Ceravolo, Ernesto Damiani, Monica Scannapieco, and Marco Viviani presents the principles, research area and current state of the art of the Emergent Semantics as a solution to the semantic interoperability problem.

Chapter ten, Web-based Bayesian Intelligent Tutoring Systems by Cory J. Butz, S. Hua and R.B. Maguire presents a Web-based intelligent tutoring system for computer programming. The decision making process conducted in the intelligent tutoring system is guided by Bayesian networks.

Final chapter, A Survey of Web-based Collective Decision Making Systems by Jennifer H. Watkins and Marko A. Rodriguez presents a taxonomy that defines a Web-based collective decision making system such as recommender system, folksonmy, document ranking, etc by the unique combination of their features. The taxonomy is further used to highlight the similarities between the systems.

1.3 Conclusion

In this chapter, we discussed the broad areas of the Web developments with the artificial intelligence techniques. We also discussed a broad overview of the different chapters included in this book. The subsequent chapter introduces the need of developing smart Web applications and systems with the use of artificial intelligence. The remaining chapters introduce specific problems and comprehensively discuss the advancement in the respective field.

1.4 Resources

Following is a sample of resources on Web and Artificial Intelligence.

1.4.1 Journals

- IEEE Internet Computing, IEEE Press, USA
  www.computer.org/internet/
- IEEE Intelligent Systems, IEEE Press, USA
  www.computer.org/intelligent/
- AI Magazine, USA
  www.aaai.org/

- Web Intelligence and Agent systems, IOS Press, The Netherlands.
- International Journal of Knowledge-Based intelligent Engineering systems, IOS Press, The Netherlands.
  http://www.kesinternational.org/journal/
- International Journal of Knowledge and Web Intelligence (IJKWI), Inder-Science
- International Journal of Hybrid Intelligent Systems, IOS Press, The Netherlands.
- Intelligent Decision Technologies: An International Journal, IOS Press, The Netherlands.

1.4.2 Special Issue of Journals

- Wade, V.P. and Ashman, H., Evolving the Infrastructure for Technology Enhanced Distance Learning, IEEE Internet Computing, Volume 11, Number 3, 2007.
- Tu, S. and Abdelguerfi, M., Web Services for GIS, IEEE Internet Computing, Volume 10, Number 5, 2006.
- Cahill, V. and Clarke, S., Roaming: Technology for a Connected Society, IEEE Internet Computing, Volume 11, Number 2, 2007.
- Ghosh, A., Seiffert, U. and **Jain, L.C.**, Evolutionary Computation in Bioinformatics, Journal of Intelligent and Fuzzy Systems, IOS Press, The Netherlands, Volume 18, Number 6, 2007.
- Abraham, A., Smith, K., Jain, R. and **Jain, L.C.**, Network and Information Security: A Computational Intelligence Approach, Journal of Network and Computer Applications, Elsevier Publishers, Volume 30, Issue 1, 2007.
- Palade, V. and Jain, L.C., Practical Applications of Neural Networks, Journal of "Neural Computing and Applications", Springer, Germany, Volume 14, No. 2, 2005.
- Abraham, A. and Jain, L.C., Computational Intelligence on the Internet, Journal of Network and Computer Applications, Elsevier Publishers, Volume 28, Number 2, 2005.
- Abraham, A., Thomas, J., Sanyal, S. and Jain, L.C., Information Assurance and Security, Journal of Universal Computer Science, Volume 11, Issue 1, 2005.
- Abraham, A. and Jain, L.C., Optimal Knowledge Mining, Journal of Fuzzy Optimization and Decision Making, Kluwer Academic Publishers, Volume 3, Number 2, 2004.
- Palade, V., Ghaoui, C. and Jain, L.C., Intelligent Instructional Environments, Journal of Interactive Technology and Smart Education, Troubador Publishing Ltd, UK, Volume 1, Issue 3, August 2004.
- Palade, V., Ghaoui, C. and Jain, L.C., Engineering Applications of Computational Intelligence, Journal of Intelligent and Fuzzy systems, IOS Press, Volume 15, Number 3, 2004.

- Alahakoon, D., Abraham, A. and Jain, L.C., Neural Networks for Enhanced Intelligence, Neural Computing and Applications, Springer, UK, Volume 13, No. 2, June 2004.
- Abraham, A., Jonkar, I., Barakova, E., Jain, R. and Jain, L.C., Special issue on Hybrid Neurocomputing, Neurocomputing, Elsevier, The Netherlands, Volume 13, No. 2, June 2004.
- Abraham, A. and Jain, L.C., Knowledge Engineering in an Intelligent Environment, Journal of Intellgent and Fuzzy Systems, The IOS Press, The Netherlands, Volume 14, Number 3, 2003.
- Jain, L.C., Fusion of Neural Nets, Fuzzy Systems and Genetic Algorithms in Industrial Applications, IEEE Transactions on Industrial Electronics, USA, December 1999.
- De Silva, C. and Jain, L.C., Intelligent Electronic Systems, Engineering Applications of Artificial Intelligence, an international journal, USA, January 1998.
- Jain, L.C., Intelligent Systems: Design and Applications - 2, Journal of Network and Computer Applications (An International Journal published by Academic Press, England). **Vol. 2**, April 1996.
- Jain, L.C., Intelligent Systems: Design and Applications - 1, Journal of Network and Computer Applications (An International Journal published by Academic Press, England). **Vol.1**, January, 1996.

1.4.3 Conferences

- IEEE/WIC/ACM International Conferences on Web Intelligence
- AAAI Conference on Artificial Intelligence
  www.aaai.org/aaai08.php
- KES International Conference Series
  www.kesinternational.org/
- European Conferences on Artificial Intelligence (ECAI)
- Australian World Wide Web Conferences
  http://ausweb.scu.edu.au

1.4.4 Conference Proceedings

- Apolloni, B., Howlett, R.J. and Jain, L.C. (Editors), Knowledge-Based Intelligent Information and Engineering Systems, Lecture Notes in Artificial Intelligence, **Volume 1, LNAI 4692**, KES 2007, Springer-Verlag, Germany, 2007.
- Apolloni, B.,Howlett, R.J.and Jain, L.C. (Editors), Knowledge-Based Intelligent Information and Engineering Systems, Lecture Notes in Artificial Intelligence, **Volume 2, LNAI 4693,** , KES 2007, Springer-Verlag, Germany, 2007.
- Apolloni, B.,Howlett, R.J.and Jain, L.C. (Editors), Knowledge-Based Intelligent Information and Engineering Systems, Lecture Notes in Artificial Intelligence, **Volume 3, LNAI 4694,** KES 2007, Springer-Verlag, Germany, 2007.

- Nguyen, N.T., Grzech, A., Howlett, R.J. and Jain, L.C., Agents and Multi-Agents Systems: Technologies and Applications, Lecture Notes in artificial Intelligence, **LNAI 4696,** Springer-Verlag, Germany, 2007.
- Howlett, R.P., Gabrys, B. and Jain, L.C. (Editors), Knowledge-Based Intelligent Information and Engineering Systems, Lecture Notes in Artificial Intelligence, KES 2006, Springer-Verlag, Germany, Vol. **4251**, 2006.
- Howlett, R.P., Gabrys, B. and Jain, L.C. (Editors), Knowledge-Based Intelligent Information and Engineering Systems, Lecture Notes in Artificial Intelligence, KES 2006, Springer-Verlag, Germany, Vol. **4252**, 2006.
- Howlett, R.P., Gabrys, B. and Jain, L.C. (Editors), Knowledge-Based Intelligent Information and Engineering Systems, Lecture Notes in Artificial Intelligence, KES 2006, Springer-Verlag, Germany, Vol. **4253**, 2006.
- Liao, B.-H., Pan, J.-S., Jain, L.C., Liao, M., Noda, H. and Ho, A.T.S., Intelligent Information Hiding and Multimedia Signal Processing, IEEE Computer Society Press, USA, 2007. ISBN: 0-7695-2994-1.
- Khosla, R., Howlett, R.P., and Jain, L.C. (Editors), Knowledge-Based Intelligent Information and Engineering Systems, Lecture Notes in Artificial Intelligence, KES 2005, Springer-Verlag, Germany, Vol. **3682**, 2005.
- Skowron, A., Barthes, P., Jain, L.C., Sun, R.,Mahoudeaux, P., Liu, J. and Zhong, N.(Editors), Proceedings of the 2005 IEEE/WIC/ACM International Conference on Intelligent Agent Technology, Compiegne, France, IEEE Computer Society Press, USA, 2005.
- Khosla, R., Howlett, R.P., and Jain, L.C. (Editors), Knowledge-Based Intelligent Information and Engineering Systems, Lecture Notes in Artificial Intelligence, KES 2005, Springer-Verlag, Germany, Vol. **3683**, 2005.
- Khosla, R., Howlett, R.P., and Jain, L.C. (Editors), Knowledge-Based Intelligent Information and Engineering Systems, Lecture Notes in Artificial Intelligence, KES 2005, Springer-Verlag, Germany, Vol. **3684**, 2005.
- Khosla, R., Howlett, R.P., and Jain, L.C. (Editors), Knowledge-Based Intelligent Information and Engineering Systems, Lecture Notes in Artificial Intelligence, KES 2005, Springer-Verlag, Germany, Vol. **3685**, 2005.
- Negoita, M., Howlett, R.P., and Jain, L.C. (Editors), Knowledge-Based Intelligent Engineering Systems, KES 2004, Lecture Notes in Artificial Intelligence, Vol. **3213**, Springer, 2004.
- Negoita, M., Howlett, R.P., and Jain, L.C. (Editors), Knowledge-Based Intelligent Engineering Systems, KES 2004, Lecture Notes in Artificial Intelligence, Vol. **3214**, Springer, 2004.
- Negoita, M., Howlett, R.P., and Jain, L.C. (Editors), Knowledge-Based Intelligent Engineering Systems, KES 2004, Lecture Notes in Artificial Intelligence, Vol. **3215**, Springer, 2004.
- Murase, K., Jain, L.C., Sekiyama, K. and Asakura, T. (Editors), Proceedings of the Fourth International Symposium on Human and Artificial Intelligence Systems, University of Fukui, Japan, 2004.
- Palade, V., Howlett, R.P., and Jain, L.C. (Editors), Knowledge-Based Intelligent Engineering Systems, Lecture Notes in Artificial Intelligence, Vol. **2773**, Springer, 2003.

- Palade, V., Howlett, R.P., and Jain, L.C. (Editors), Knowledge-Based Intelligent Engineering Systems, Lecture Notes in Artificial Intelligence, Vol. **2774**, Springer, 2003.
- Damiani, E., Howlett, R.P., Jain, L.C. and Ichalkaranje, N. (Editors), Proceedings of the Fifth International Conference on Knowledge-Based Intelligent Engineering Systems, **Volume 1**, IOS Press, The Netherlands, 2002.
- Damiani, E., Howlett, R.P., Jain, L.C. and Ichalkaranje, N. (Editors), Proceedings of the Fifth International Conference on Knowledge-Based Intelligent Engineering Systems, **Volume 2**, IOS Press, The Netherlands, 2002.
- Baba, N., Jain, L.C. and Howlett, R.P. (Editors), Proceedings of the Fifth International Conference on Knowledge-Based Intelligent Engineering Systems (KES'2001), **Volume 1**, IOS Press, The Netherlands, 2001.
- Baba, N., Jain, L.C. and Howlett, R.P. (Editors), Proceedings of the Fifth International Conference on Knowledge-Based Intelligent Engineering Systems (KES'2001), **Volume 2**, IOS Press, The Netherlands, 2001.
- Howlett, R.P. and Jain, L.C.(Editors), Proceedings of the Fourth International Conference on Knowledge-Based Intelligent Engineering Systems, IEEE Press, USA, 2000. **Volume 1**.
- Howlett, R.P. and Jain, L.C.(Editors), Proceedings of the Fourth International Conference on Knowledge-Based Intelligent Engineering Systems, IEEE Press, USA, 2000. **Volume 2**.
- Jain, L.C.(Editor), Proceedings of the Third International Conference on Knowledge-Based Intelligent Engineering Systems, IEEE Press, USA, 1999.
- Jain, L.C. and Jain, R.K. (Editors), Proceedings of the Second International Conference on Knowledge-Based Intelligent Engineering Systems, **Volume 1**, IEEE Press, USA, 1998.
- Jain, L.C. and Jain, R.K. (Editors), Proceedings of the Second International Conference on Knowledge-Based Intelligent Engineering Systems, **Volume 2**, IEEE Press, USA, 1998.
- Jain, L.C. and Jain, R.K. (Editors), Proceedings of the Second International Conference on Knowledge-Based Intelligent Engineering Systems, **Volume 3**, IEEE Press, USA, 1998.
- Jain, L.C. (Editor), Proceedings of the First International Conference on Knowledge-Based Intelligent Engineering Systems, **Volume 1**, IEEE Press, USA, 1997.
- Jain, L.C. (Editor), Proceedings of the First International Conference on Knowledge-Based Intelligent Engineering Systems, **Volume 2**, IEEE Press, USA, 1997.
- Narasimhan, V.L., and Jain, L.C. (Editors), The Proceedings of the Australian and New Zealand Conference on Intelligent Information Systems, IEEE Press, USA, 1996.

1.4.5 Book Series

- Advanced Intelligence and Knowledge Processing, Springer-Verlag, Germany www.springer.com/series/4738

- Computational Intelligence and its Applications Series, Idea group Publishing, USA.
- The CRC Press International Series on Computational Intelligence, The CRC Press, USA.
- Advanced Information Processing, Springer-Verlag, Germany.
- Knowledge-Based Intelligent Engineering Systems Series, IOS Press, The Netherlands.
- International series on Natural and artificial Intelligence, AKI. http://www.innoknowledge.com

1.4.6 Books

- Henninger, M., The Hidden Web, Second Edition, University of New South Wales Press Ltd, Australia, 2008.
- Jain, L.C., Sato, M., Virvou, M., Tsihrintzis, G., Balas, V. and Abeynayake, C. (Editors), Computational Intelligence Paradigms: Volume 1 – Innovative Applications, Springer-Verlag, 2008.
- Phillips-Wren, G., et al. (Editors), Intelligent Decision Making-An AI-Based Approach, Springer-Verlag, 2008.
- Fulcher, J. and Jain, L.C., Computational Intelligence: A Compendium, Springer-Verlag, 2008.
- Sordo, M., Vaidya, S. and Jain, L.C.(Editors), Advanced Computational Intelligence Paradigms in **Healthcare 3**, Springer-Verlag, 2008.
- Virvou, M. And Jain, L.C.(Editors), Intelligent Interactive Systems in Knowledge-Based Environments, Springer-Verlag, 2008.
- Jarvis, J., Ronnquist, R, Jarvis, D. and Jain, L.C., Holonic Execution: A BDI Approach, Springer-Verlag, 2008.
- Tsihrintzis, G. and Jain, L.C., Multimedia Services in Intelligent Environments, Springer-Verlag, 2008.
- Jain, L.C., Srinivasan, D. and Palade, V. (Editors), Advances in Evolutionary Computing for system Design, Springer-Verlag, 2007.
- Baba, N., et al. (Editors), Advanced Intelligent Paradigms in Computer Games, Springer-Verlag, 2007.
- Chahl, J.S., Jain, L.C., Mizutani, A. and Sato-Ilic, M. (Editors), Innovations in Intelligent Machines 1, Springer-Verlag, 2007.
- Jain, L.C, Tedman, R. and Tedman, D. (Editors), Evolution of Teaching and Learning in Intelligent Environment, Springer-Verlag, 2007.
- Zharkova, V. and Jain, L.C. (Editors), Artificial Intelligence in Recognition and Classification of Astrophysical and Medical Images, Springer-Verlag, 2007.
- Pan, J-S. et al. (Editors), Intelligent Multimedia Data Hiding, Springer-Verlag, 2007.
- Yoshida, H., et al. (Editors), Advanced Computational Intelligence Paradigms in **Healthcare 1**, Springer-Verlag, 2007.
- Vaidya, S., et al. (Editors), Advanced Computational Intelligence Paradigms in **Healthcare 2**, Springer-Verlag, 2007.

- Sato, M. and Jain, L.C., Innovations in Fuzzy Clustering, Springer-Verlag, 2006.
- Patnaik, S. et al. (Editors), Innovations in Robot Mobility and Control, Springer-Verlag, 2006.
- Apolloni, B., et al. (Editors), Machine Learning and Robot Perception, Springer-Verlag, 2006.
- Palade, V., et al. (Editors), Computational Intelligence in Fault Diagnosis, Springer-Verlag, 2006.
- Holmes, D. and Jain, L.C. (Editors), Innovations in Machine Learning, Springer-Verlag, 2006.
- Seiffert, U., et al. (Editors), Bioinformatics Using Computational Intelligence Paradigms, Springer-Verlag, ISBN 3-54022-901-9, 2005.
- Khosla, R., Ichalkaranje, N. and Jain, L.C.(Editors), Design of Intelligent Multi-Agent Systems, Springer-Verlag, Germany, 2005.
- Ghosh, A. and Jain, L.C.(Editors), Evolutionary Computation in Data Mining, Springer-Verlag, Germany, 2005.
- Phillips-Wren, G. and Jain, L.C.(Editors), Intelligent Decision Support Systems in Agent-Mediated Environments, IOS Press, The Netherlands, 2005.
- Silvermann, B. et al. (Editors), Intelligent Paradigms in Healthcare Enterprises, Springer-Verlag, Germany, 2005.
- Ghaoui, C. et al. (Editors), Knowledge-Based Virtual Education, Springer-Verlag, Germany, 2005.
- Abraham, A., et al. (Editors), Evolutionary Multiobjective Optimization, Springer-Verlag, London, 2005
- Pal, N. and Jain, L.C.(Editors), Advanced Techniques in Knowledge Discovery and Data Mining, Springer-Verlag, London, 2005
- Chen, S-H. et al. (Editors), Computational Economics: A Perspective from Computational Intelligence, Idea Group Publishing, 2005
- Abraham, A. et al., (Editors), Innovations in Intelligent Systems, Springer-Verlag, Germany, 2004.
- Nikravesh, M., et al. (Editors), Enhancing the power of Internet, Springer-Verlag, Germany, 2004.
- Tonfoni, G. and Jain, L.C., Visualizing Document Processing, Mouton De Gruyter, Germany, 2004.
- Fulcher, J. and Jain, L.C.(Editors), Applied Intelligent Systems, Springer-Verlag, Germany, 2004.
- Damiani, E. et al. (Editors), Soft Computing in Software Engineering, Springer-Verlag, Germany, 2004.
- Resconi, G. and Jain, L.C., Intelligent Agents: Theory and Applications, Springer-Verlag, Germany, 2004.
- Pan, J-S., et al. (Editors), Intelligent Watermarking Techniques, World Scientific Publishing Company Singapore, 2004.
- Abraham, A. et al. (Editors), Recent Advances in Intelligent Paradigms and Applications, Springer-Verlag, Germany, 2003.

- Tonfoni, G. and Jain, L.C. (Editors), Innovations in Decision Support Systems, AKI, 2003.
- Shapiro, A. et al., Intelligent and Other Computational Techniques in Insurance, World Scientific Publishing Company Singapore, 2003.
- Tonfoni, G. and Jain, L.C., The Art and Science of Documentation Management, Intellect, UK, 2003.
- Howlett, R., Ichalkaranje, N., Jain, L.C. and Tonfoni, G. (Editors), Internet-Based Intelligent Information Processing, World Scientific Publishing Company Singapore, 2002.
- Seiffert, U. and Jain, L.C. (Editors), Self-Organising neural Networks, Springer-Verlag, Germany, 2002.
- Jain, L.C. and Kacprzyk, J. (Editors), New Learning Paradigms in Soft Computing, Springer-Verlag, Germany, 2002.
- Jain, L.C., et al. (Editors), Intelligent Agents and Their Applications, Springer-Verlag, Germany, 2002.
- Jain, L.C., Ichalkaranje, N. and Tonfoni, G. (Editors), Advances in Intelligent Systems for Defence, World Scientific Publishing Company Singapore, 2002.
- Jain, L.C., Howlett, R.J., Ichalkaranje, N., and Tonfoni, G.(Editors), Virtual Environments for Teaching and Learning, World Scientific Publishing Company Singapore, 2002.
- Schmitt, M. et al. Computational Intelligence Processing in Medical Diagnosis, Springer, 2002.
- Jain, L.C. and De Wilde, P. (Editors), Practical Applications of Computational Intelligence Techniques, Kluwer Academic Publishers, USA, 2001.
- Howlett, R.J. and Jain, L.C. (Editors), Radial Basis Function Networks 1, Springer-Verlag, Germany, 2001.
- Howlett, R.J. and Jain, L.C. (Editors), Radial Basis Function Networks 2, Springer-Verlag, Germany, 2001.
- Teodorescu, H.N., et al. (Editors), Hardware Implementation of Intelligent Systems, Springer-Verlag, Germany, 2001.
- Baba, N. and Jain, L.C., Computational Intelligence in Games, Springer, 2001.
- Jain, L.C., et al. (Editors), Innovations in ART Neural Networks, Springer-Verlag, Germany, 2000.
- Jain, A., et al. (Editors), Artificial Intelligence Techniques in Breast Cancer Diagnosis and Prognosis, World Scientific Publishing Company, Singapore, 2000.
- Jain, L.C. and Fanelli, A.M. (Editors), Recent Advances in Artificial Neural Networks: Design and Applications, CRC Press, USA, 2000.
- Medsker, L., and Jain, L.C. (Editors) Recurrent Neural Networks: Design and Applications, CRC Press, USA, 2000.
- Jain, L.C., Halici, U., Hayashi, I., Lee, S.B. and Tsutsui, S. (Editors) Intelligent Biometric Techniques in Fingerprint and Face Recognition, CRC Press, USA, 2000.

- Jain, L.C.(Editor), Evolution of Engineering and Information Systems, CRC Press USA, 2000.
- Lazzerini, B., et al, Evolutionary Computing and Applications, CRC Press USA, 2000.
- Lazzerini, B., et al., Fuzzy Sets and their Applications to Clustering and Training, CRC Press USA, 2000.
- Jain, L.C. and De Silva, C.W. (Editors), Intelligent Adaptive Control, CRC Press USA, 1999.
- Jain, L.C. and Martin, N.M. (Editors), Fusion of Neural Networks, Fuzzy Logic and Evolutionary Computing and their Applications, CRC Press USA, 1999.
- Jain, L.C. and Lazzerini, B., (Editors), Knowledge-Based Intelligent Techniques in Character Recognition, CRC Press USA, 1999.
- Teodorescu, H.N. et al. (Editors), Soft Computing Techniques in Human Related Science, CRC Press USA, 1999.
- Jain, L.C. and Vemuri, R. (Editors), Industrial Applications of Neural Networks, CRC Press USA, 1998.
- Jain, L.C., Johnson, R.P., Takefuji, Y. and Zadeh, L.A. (Editors), Knowledge-Based Intelligent Techniques in Industry, CRC Press USA, 1998.
- Teodorescu, H.N., et al. (Editors), Fuzzy and Neuro-fuzzy Systems in Medicine, CRC Press USA, 1998.
- Jain, L.C. and Fukuda, T. (Editors), Soft Computing for Intelligent Robotic Systems, Springer-Verlag, Germany, 1998.
- Jain, L.C. (Editor), Soft Computing Techniques in Knowledge-Based Intelligent Engineering Systems, Springer-Verlag, Germany, 1997.
- Sato, M. et al., Fuzzy Clustering Models and Applications, Springer-Verlag, Germany, 1997.
- Jain, L.C. and Jain, R.K. (Editors), Hybrid Intelligent Engineering Systems, World Scientific Publishing Company, Singapore, 1997.
- Vonk, E., et al. Automatic Generation of Neural Networks Architecture Using Evolutionary Computing, World Scientific Publishing Company, Singapore, 1997.
- Van Rooij, A. et al., Neural Network Training Using Genetic Algorithms, World Scientific Publishing Company, Singapore, December 1996.
- Pierre, S. (Editor), E-Learning Networked Environments and Architectures, Springer-Verlag, London, 2007
- Harris, C., Hong, X. and Gan, Q., Adaptive Modelling, Estimation and Fusion from Data, Springer-Verlag, Germany, 2002.
- Chen, S.H. and Wang, P.P.(Editors), Computational Intelligence in Economics and Finance, Springer-Verlag, Germany, 2004.
- Ohsawa, Y. and McBurney, P.(Editors), Chance Discovery, Springer-Verlag, Germany, 2003.
- Deen, S.M.(Editor), Agent-Based Manufacturing, Springer-Verlag, Germany, 2003.
- Ishibuchi, H., Nakashima, T. and Nii, M., Classification and Modeling with Linguistic Information Granules, Springer-Verlag, Germany, 2005.

- Liu, J. and Daneshmend, L., Spatial Reasoning and Planning, Springer-Verlag, Germany, 2004.
- Gasós J. and Thoben, K.D. (Editors), e-Business Applications, Springer-Verlag, Germany, 2003.
- Mentzas, G., et al., Knowledge Asset Management, Springer-Verlag, London, 2003.
- Vazirgiannis, M., et al., Uncertainty Handling and Quality Assessment in Data Mining, Springer-Verlag, London, 2003.
- Gomez-Perez, et al., Ontological Engineering, Springer-Verlag, London, 2004.
- Nguyen, N.T., Advanced Methods for Inconsistent Knowledge Management, Springer-Verlag, London, 2008.
- Meisels, A., Distributed Search by Constrained Agents, Springer-Verlag, London, 2008.
- Camastra, F. and Vinciarelli, A., Machine Learning for audio, Image, and Video Analysis, Springer-Verlag, London, 2008.
- Kornai, A., Mathematical Linguistics, Springer-Verlag, London, 2008.
- Prokopenko, M. (Editor), Advances in Applied Self-Organising Systems, Springer-Verlag, London, 2008.
- Scharl, A., Environmental Online Communication, Springer-Verlag, London, 2007.
- Zhang, S., et. al., Knowledge Discovery in Multiple Databases, Springer-Verlag, London, 2004.
- Wang, J.T.L., et al. (Editors), Data Mining in Bioinformatics, Springer-Verlag, London, 2005.
- Ko, C.C., Creating Web-based Laboratories, Springer-Verlag, London, 2004.
- Grana, M., et al.(Editors), Information Processing with Evolutionary Algorithms, Springer-Verlag, London, 2005.
- Fyfe, C., Hebbian Learning and Negative Feedback Networks,, Springer-Verlag, London, 2005.
- Chen-Burger, Y. and Robertson, D., Automatic Business Modelling,, Springer-Verlag, London, 2005.
- Husmeier, D., et.al. (Editors), Probabilistic Modelling in Bioinformatics and Medical Informatics, Springer-Verlag, London, 2005.
- Tan, K.C., et al., Multiobjective Evolutionary Algorithms and Applications, Springer-Verlag, London, 2005.
- Bandyopadhyay, S., et. al. (Editors), Advanced Methods for Knowledge Discovery from Complex Data, Springer-Verlag, London, 2005.
- Karny, M.(Editor), Optimized Bayesian Dynamic Advising, Springer-Verlag, London, 2006.
- Liu, S. and Lin, Y., Grey Information: Theory and Practical Applications, Springer-Verlag, London, 2006.
- Maloof, M.A. (Editor), Machine Learning and Data Mining for Computer Security, Springer-Verlag, London, 2006.
- Stuckenschmidt, H. and Harmelen, F.V., Information Sharing on the Semantic Web, Springer-Verlag, London, 2005.

- Wang, L. and Fu, X., Data Mining with Computational Intelligence, Springer-Verlag, London, 2005.
- Abraham, A., Koppen, M. and Franke, K. (Editors), Design and Applications of Hybrid Intelligent Systems, IOS Press, The Netherlands.
- Turchetti, C., Stochastic Models of Neural Networks, IOS Press, The Netherlands.
- Loia, V. (Editor), Soft Computing Agents, IOS Press, The Netherlands.
- Abraham, A., et al. (Editors), Soft Computing Systems, IOS Press, The Netherlands.
- Motoda, H., Active Mining, IOS Press, The Netherlands.
- Namatame, A., et al. (Editors), Agent-Based Approaches in Economic and Social Complex Systems, IOS Press, The Netherlands.

1.4.7 Book Chapters

- Pedrycz, W., Ichalkaranje, N., Phillips-Wren, G., and Jain, L.C., Introduction to Computational Intelligence for Decision Making, Springer-Verlag, 2008, pp. 75-93, Chapter 3.
- Tweedale, J., Ichalkaranje, N., Sioutis, C., Urlings, P. and Jain, L.C., Future Directions: Building a Decision Making Framework using Agent Teams, Springer-Verlag, 2008, pp. 381-402, Chapter 14.
- Virvou, M. and Jain, L.C., Intelligent Interactive Systems in Knowledge-Based Environments: An Introduction, Springer-Verlag, 2008, pp. 1-8, Chapter 1.
- Tsihrintzis, G. and Jain, L.C., An Introduction to Multimedia Services in Intelligent Environments, Springer-Verlag, pp. 1-8, 2008, Chapter 1.
- Zharkova, V.V. and Jain, L.C., Introduction to Recognition and Classification in Medical and Astrophysical Images, Springer-Verlag, 2007, pp. 1-18, Chapter 1, ISBN: 10-3-540-47511-7.
- Yoshida, H., Vaidya, S. and Jain, L.C., Introduction to Computational Intelligence in Healthcare, Springer-Verlag, 2007, pp. 1-4, Chapter 1.
- Huang, H.C., Pan, J.S., Fang, W.C. and Jain, L.C., An Introduction to Intelligent Multimedia Data Hiding, Springer-Verlag, 2007, pp. 1-10, Chapter 1.
- Jain, L.C., et al., Intelligent Machines :An Introduction, Springer-Verlag, 2007, pp. 1-9, Chapter 1.
- Jain, L.C., et al., Introduction to Evolutionary Computing in System Design, Springer-Verlag, 2007, pp. 1-9, Chapter 1.
- Jain, L.C., et al., Evolutionary Neuro-Fuzzy Systems and Applications, Springer-Verlag, 2007, pp. 11-45, Chapter 1.
- Do, Q.V, Lozo, P. and Jain, L.C., Vision-Based Autonomous Robot Navigation, in Innovations in Robot Mobility and Control, Springer-Verlag, 2006, pp. 65-103, Chapter 2.
- Tran, C., Abraham, A. and Jain, L., Soft Computing Paradigms and Regression Trees in Decision Support Systems, in Advances in Applied Artificial Intelligence, Idea Group Publishing, 2006, pp. 1-28, Chapter 1.

- Jarvis, B., Jarvis, D. and Jain, L., Teams in Multi-Agent Systems, in IFIP International Federation for Information Processing, Vol. **228**, Intelligent Information Processing III, Springer, 2006, pp. 1-10, Chapter 1.
- Abraham, A. and Jain, L.C., Industry, Evolutionary Multiobjective Optimization, in Springer-Verlag's book, ISBN 1-85233-787-7, 2005, pp. 1-6, Chapter 1.
- Jain, L.C. and Chen, Z., Industry, Artificial Intelligence In, in Encyclopedia of Information Systems, Elsevier Science, USA, 2003, pp. 583-597.
- Jain, L.C. and Konar, A., An Introduction to Computational Intelligence Paradigms, in Practical Applications of Computational Intelligence Techniques, Springer, 2001, pp. 1-38.
- Tedman, D. and Jain, L.C., An Introduction to Innovative Teaching and Learning, in Teaching and Learning, Springer, 2000, pp. 1-30, Chapter 1.
- Filippidis, A., Russo, M. and Jain, L.C., Novel Extension of ART2 in Surface Landmine Detection, Springer-Verlag, 2000, pp.1-25, Chapter 1.
- Jain, L.C. and Lazzerini, B., An Introduction to Handwritten Character and Word Recognition, in Knowledge-Based Intelligent Techniques in Character Recognition, CRC Press, 1999, 3-16.
- Filippidis, A., Jain, L.C. and Martin, N.N., "Computational Intelligence Techniques in Landmine Detection," in Computing with Words in Information/Intelligent Systems 2, Edited by Zadeh, L. and Kacprzyk, J., Springer-Verlag, Germany, 1999, pp. 586-609.
- Halici, U., Jain, L.C. and Erol, A., Introduction to Fingerprint Recognition, in Intelligent Biometric Techniques in Fingerprint and Face Recognition, CRC Press, 1999, pp.3-34.
- Teodorescu, H.N., Kandel, A. and Jain, L., Fuzzy Logic and Neuro-Fuzzy Systems in Medicine: A historical Perspective, in Fuzzy and Neuro-Fuzzy Systems in Medicine, CRC Press, 1999, pp. 3-16.
- Jain, L.C. and Vemuri, R., An Introduction to Intelligent Systems, in Hybrid Intelligent Engineering Systems, World Scientific, 1997, pp. 1-10, Chapter 1.
- Karr, C. and Jain, L.C., Genetic Learning in Fuzzy Control, in Hybrid Intelligent Engineering Systems, World Scientific, 1997, pp. 69-101, Chapter 4.
- Karr, C. and Jain, L.C., Cases in Geno- Fuzzy Control, in Hybrid Intelligent Engineering Systems, World Scientific, 1997, pp. 103-132, Chapter 5.
- Katayama, R., Kuwata, K. and Jain, L.C., Fusion Technology of Neuro, Fuzzy, GA and Chaos Theory and Applications, in Hybrid Intelligent Engineering Systems, World Scientific, 1997, pp. 167-186, Chapter 7.
- Jain, L.C., Medsker, L.R. and Carr, C., Knowledge-Based Intelligent Systems, in Soft Computing Techniques in Knowledge-Based Intelligent Systems, Springer-Verlag, 1997, pp. 3-14, Chapter 1.
- Babri, H., Chen, L., Saratchandran, P., Mital, D.P., Jain, R.K., Johnson, R.P. and Jain, L.C., Neural Networks Paradigms, in Soft Computing Techniques in Knowledge-Based Intelligent Systems, Springer-Verlag, 1997, pp. 15-43, Chapter 2.

- Jain, L.C., Tikk, D. and Koczy, L.T., Fuzzy Logic in Engineering, in Soft Computing Techniques in Knowledge-Based Intelligent Systems, Springer-Verlag, 1997, pp. 44-70, Chapter 3.
- Tanaka, T. and Jain, L.C., Analogue/Digital Circuit Representation for Design and Trouble Shooting in Intelligent Environment, in Soft Computing Techniques in Knowledge-Based Intelligent Systems, Springer-Verlag, 1997, pp. 227-258, Chapter 7.
- Jain, L.C., Hybrid Intelligent System Design Using Neural Network, Fuzzy Logic and Genetic Algorithms - Part I, Cognizant Communication Corporation USA, 1996, pp. 200-220, Chapter 9.
- Jain, L.C., Hybrid Intelligent System Applications in Engineering using Neural Network and Fuzzy Logic - Part II, Cognizant communication Corporation USA,1996, pp. 221-245, Chapter 10.

Innovations in Web Applications by Using the Artificial Intelligence Paradigm

R. Nayak[1] and L.C. Jain[2]

[1] Faculty of Information Technology
Queensland University of Technology
Brisbane, Australia
r.nayak@qut.edu.au
[2] School of Electrical and Information Engineering
University of South Australia
Adelaide, Australia
Lakhmi.Jain@unisa.edu.au

Abstract. Artificial Intelligence techniques are increasingly being used when devising smart Web applications for efficiently presenting the information on the Web to the user. The motivation for this trend is the growth of the internet and the increasing difficulty for users to navigate the web to find useful information and applications. This chapter shows how artificial intelligence techniques can be utilized to improve the Web and the benefits that can be obtained by integrating the Web and Artificial Intelligence.

2.1 Introduction

Simplicity and ubiquity are factors behind the success of the World Wide Web (WWW). Once solely a repository that provides access to information suitable for human interpretation and use, the Web has evolved to become a repository of software components and applications [4, 35, 36]. It has become a common medium to find information and to conduct e-commerce and business applications. This has only been possible due to the successful application of artificial intelligence techniques to the Web data and Web applications [17, 18, 36]. The result is that the volume of data generated on the Web is increasing exponentially. Most data on the Web is unstructured and is huge. This means that it can only be processed efficiently by using machines [6, 48]. Consequently, artificial intelligence techniques have been developed to satisfy this need.

The Web currently contains somewhere between 15 and 30 billion pages of information spread over more than 8 billion websites. There are over 1 billion regular users from various backgrounds with varying degrees of computer literacy and having a wide range of interests and preferences[1]. It is important to be able to present users with the information corresponding to their interests and preferences. This is done by tracking and monitoring the user activities and by analyzing the data using data mining in conjunction with user profiles [16]. The use of artificial intelligence techniques has resulted in many improvements in the Web technologies. Semantic Web

---

[1] http://www.pandia.com/sew/383-web-size.html

R. Nayak et al. (Eds.): Evolution of the Web in Artificial Intel. Environ., SCI 130, pp. 17–40, 2008.
springerlink.com   © Springer-Verlag Berlin Heidelberg 2008

mining is utilised to improve the user browsing. This is accomplished by exploiting the semantic structure of the Web and its usage data [49]. These improvements enhance the use of the Web and generate new business opportunities [34].

Given the competitiveness of the market place, the need to obtain data for business intelligence is now even great. The ability of a business to obtain knowledge for itself and its customers is a valuable asset. Business needs, expanding user interests and the availability of huge amounts of data are all relevant factors. In particular, possible improvements in efficient operation of the Web drive the evolution of the Web in an artificial intelligence environment. The goal of this chapter is to show how artificial intelligence and related techniques can be utilized to improve the Web and to show the benefits which can be gained by integrating the Web and Artificial Intelligence.

This chapter begins by discussing why the artificial intelligence techniques are needed to improve the functionality of the Web. The following section lists the major stages in the development of the Web. In section 2.3 various forms of data contained in the Web are discussed. Some artificial intelligence techniques used to process and deal with the various forms of data are considered. The next section discusses some Web developments obtained by using artificial intelligence techniques. These include Web Personalisation, Semantic Web and Web Services.

2.2   The Web from Its Inception to Now

The Web has an interesting history and some significant historical events leading to its development are of interest [22]. The first of these was in 1858. At this time, a cable was laid under the Atlantic Ocean between Europe and North America for communications. It failed due to technical problems. Cables laid eight years later in 1866 were successful and provided efficient communication. This paved the way for communication over long distances. The next major event was the establishment of ARPA (Advanced Research Projects Agency) in 1958. ARPA was successful in developing America's first earth satellite soon after. A few years later ARPA focused its attention on communication using computers and networks. This was originally done to improve the use of technology by the military. In 1969 ARPA decided to make this technology public in order to provide great benefits to scientists and researchers. Consequently ARPA linked up numerous American University computers in a project called ARPANet. The network was used for better and collaborative research by providing the means of sharing information and research.

The TCP\IP technology was developed in 1977 by ARPA. In 1983 the ARPANet switched to the new TCP/IP. Between 1984 and 1988 CERN began the installation and operation of TCP/IP to interconnect its major computer systems. In 1989, whilst working at CERN, Tim Berners-Lee invented a network-based implementation of the hypertext concept and released it for public use. This was a major breakthrough in that it made the Web more accessible. This pioneer work can be attributed to the success of the Web of today. The growth in the amount of data and the number of users has led to many challenges. One of these is particularly in the area of data comprehension and retrieval. The original Web has evolved into the semantic Web of today.

The Semantic Web provides a 'framework that allows data to be shared and reused across applications, enterprises, and community boundaries'[2].

Today's Web can be considered as a massive information system of interconnected databases with remote applications which provide numerous services. Some of these are related while others are not. These services are becoming more and more user oriented but the concept of smart applications on the Web is considered still in its infancy. Some Web applications adapt techniques evolved from artificial intelligence in order to process the massive amount of both semi-structured or unstructured data [38, 39]. This data is sometimes incomplete or it lacks credentials. The use of these techniques can be utilized to gain previously unknown knowledge or patterns. It can also be used to obtain knowledge structures for business or for user gratification [48].

2.3   Various Forms of Web Data

The web is a stockpile of vast amounts of data stored in different forms that is difficult to comprehend and to access without the use of efficient means of organizing the data logically. After accessing the data it must then be presented in an intelligible form. Over the years Artificial Intelligence and related techniques have evolved so that they can handle the processing of various forms of Web data. Examples of such data include structured and semi-structured texts, hypertext links, multimedia, databases and usage patterns. Techniques have emerged that allow us to successfully interpret the various forms of Web data. The following shows some of the techniques that can be applied to the Web data.

2.3.1   Structured Tables / Databases

Several techniques have been developed and used for organising the semi-structured Web data into a more structured form. These techniques commonly use standard database querying mechanisms and then use artificial intelligence techniques for further analysis [11, 23].

Techniques based on multilevel databases work to create a database where each of the higher-levels is derived from the preceding level [54]. In this way the information is sorted into a higher-level of organisation. These higher-levels provide an overview of the information in the lower-levels. At the highest-level of the database, meta-data or generalisations may be inferred from the lower-levels of information. The data is then organised into structured collections resembling a relational or object-orientated database. Standard database mining or pattern recognition methods can then be applied to group to extract information. Multilevel databases hide their operation from the user and improve the performance of the Web search engines. The categorisation of data into higher-level categories permits easy identification of information. The way in which the topics are merged is at the user discretion. Problems arise when words have multiple meanings. It is necessary for the categorisation algorithm to distinguish between the meanings and to make correct groupings. An example to illustrate this need is the large amount of irrelevant material returned when performing unrelated searches on the internet.

---

[2] http://www.w3.org/2001/sw/

Many applications based on query systems have been developed to mimic standard database query languages such as SQL (Sequential Querying Language). These languages also contain structural information regarding which type of Web documents they are able to relate to. This enables a search to find all the pages in a site that contain a certain string. These types of queries can also be applied to obtain real-time results. An example is the W3QS Query Language[3].. This is an SQL-like language. It provides a mechanism for querying the knowledge using a form of association rules and sequential patterns. The following example provides the code to lookup on data from the 'W3C' home page [8]. This will find all documents with 'XML' in the title and that are linked by paths having a length of 2 or less, and are limited to links local to 'wrc.org'.

> **SELECT** d.url, d.title
> **FROM** Document d **SUCH THAT**
> "http://w3c.org"=->|->.->d
> **WHERE** d.title **CONTAINS** "XML"

The W3QS query language executes searches on the Web in its current state. Other search engines use indexes or databases containing data previously extracted from the source. The current index may not represent the actual content of the page containing dynamically changing contents. If a search is done on a site with frequently changing content, the standard search engine results may be incomplete. W3QS overcomes this problem by directly querying the data from the live Web. It achieves a higher level of precision for returning correct results than that obtained with standard search engines. Search engines usually search based on keywords. In W3QS, the search can be defined in a more precise way by specifying how to navigate the information of interest.

2.3.2 XML (eXtensible Markup Language)

XML has gained popularity for information representation, exchange and retrieval on the Web [53]. XML tags describe the structural and semantic meaning of the information in text documents. They make XML documents semi-structured and self-describing. Many Web data sources have already or are beginning to structure their external view as a repository of XML documents. This is regardless of their internal storage mechanism. As XML material becomes more abundant, its heterogeneity and structural irregularity limit the knowledge that can be gained [38]. The utilization of artificial intelligence and data mining techniques becomes essential for improvement in XML document handling [40]. Due to the fact that the data stored within XML in an unstructured way, the application of normal database or artificial techniques cannot be used. Specialized methods based on the basic concepts of these techniques are needed. For example, a number of clustering techniques have been developed to group similar documents. Several frequent tree mining techniques have been developed to discover commonly occurring patterns within XML data [37].

We concentrate on the method referred to as XTM (XML Topic Maps). This provides a proficient technique for extracting usable data [43]. A XML document alone provides little to no meaning to the reader. A topic map consists of a collection of

---

[3] http://www.cs.technion.ac.il/~W3QS/

topics. Each topic represents a concept. Topics are related to each other by associations. A topic may also be related to any number of resources by noting their occurrences. Another factor of key importance is that topic maps are useful in merging of many XML data files. This is in order to extract useful information from more than one file. Topic maps are used to organise information in such a way that can be optimised for navigation and searching. The first step to obtain useful information from XML is to parse the file and to extract objects and properties. This is done by using the XML document as it is made of elements limited by tags and is hierarchically structured. The second step is to create referencing between the data obtained in the first step. Each object within the topic map is assigned a weight (number of occurrences) which is determined by the importance it holds in the topic map. It can then be used as a basis to matching relevance [45]. A XML lookup table (XTL) is then used to hide the topic map mechanism from the user. The user can then concentrate on the domain model. XTL is not dependant on the ontology and there are no restrictions placed on the domains to which the topic maps can be applied.

Automatic topic mapping is a key feature of topic maps. It allows the feature of collating related data into higher meta-data categories. This helps to group all the lower-level data into meta-data that efficiently represents what is being stored. The mapping of words in a language into concepts is a complex task. The complexities of categorising the words into concepts increase greatly due to synonym and polysemy problems. To overcome this, a standard rule-set is applied to the data to get more meaningful meta-data. Artificial intelligence techniques can be employed to make efficient rule-sets. The use of meaningful variable names in the XML files facilitates the retrieval of useful information.

2.3.3 Log Files

An important form of Web data, although not directly related to the actual content stored within a Web site, are log files that the server creates. The Web log data and the Web site contents are used in many applications such as usage pattern analysis, personalization, system improvement and many others [11]. A user based Web log processing technique is usually comprised of four data pre-processing steps: data cleaning, user identification, session identification and data filtering. The data cleaning is a standard process that removes the unnecessary fields and removes any errors. The user identification process assumes that the users are unique at a certain time when browsing a site based on their IP address. Users can stay inactive for a set period of time. After the elapse of this time the user is assumed to have left the website [27]. Users are considered active if they do not exceed a set inactive time. The active user process identifies unique visitors and an ontology-based session identification process which enables the user's browsing patterns and behaviour on a site to be found. The goal is to find common Web traversing patterns.

The amount of information created by log files is very large, a scalable technique that runs optimally is very important. The output from the user identification process is very useful and it can be used as the input for many different Web related techniques and applications. A fundamental flaw in this technique is that it relies on the assumption that a user is uniquely identified by their IP address. This may be successful if only one person uses an IP address. Generally there are many people using a network that use a single IP address to access the internet. A more accurate, but still flawed, way of gathering user information would be by the use of cookies. This allows more user-centric data to be gathered. The cookies, however, are on the client-side so they can be deleted or disabled by the users [16].

2.3.4 Multimedia

Unlike simple forms of data such as text and log files, the images and video data are much harder to represent in a conceptual form. To give meaning to text is much simpler than to give meaning to pixels. Current technology doesn't have the ability to interpret these images in the same way as text data. To interpret the context and meaning within an image, the information associated with it or contained within it must be analysed. There are numerous techniques that enable the processing and understanding of the multimedia content on the internet. However some of these methods are inefficient [23]. These techniques utilise contextual information as well as multimedia descriptors. Techniques utilising contextual information focus on the information contained in a Web page i.e., HTML within a page. These techniques allow relevant information to be extracted and give meaning and context. By utilising tags such as IMG, ALT, META, significant keywords are used to classify a multimedia element. The text elements can be processed by using standard techniques such as stemming, stop-word removal and natural language heuristics to eliminate irrelevant information [23]. A concept hierarchy can be built based on the textual information. Due to the high dependence on the extracted terms, the concept hierarchy may be inaccurate if the source on which its decisions are based is not interpreted correctly.

The second kind of technique used to make interpretations is based on visual indicators in which the image appears. There are two main descriptors used: feature and layout descriptors. The feature descriptor is a set of vectors for each of the visual characteristic such as colour, the most frequent colour, etc. Layout descriptors contain vectors that encompass edge and colour layout. After this information has been gathered it is compared to a database of stored image segments in an attempt to give meaning to the image. To enable this technique to apply to video, the video is separated into segments. Once a rapid change in scene is detected, a certain number of frames are selected to represent each segment as an image. These techniques are good at identifying the context and objects within an image. They also improve the database being queried with regard to information extracted from the images. This permits a more accurate but still not an infallible way of image interpretation [23].

2.3.5 Link Topology

The most common forms of link topology data are Hypertext and Hypermedia. Generally with link topology analysis, the data is extracted for using in the following applications of digital libraries, product catalogs, reviews, newsgroups, medical reports, customer service reports and homepages. These can benefit as a result of its analytic output. The linkage, whatever form it takes, be it hypertext, hypermedia or any form of coercion can benefit. Artificial Intelligence techniques used for analysing linkages are relatively primitive and benefits arise from its use [10]. A vast amount of data on the Web provides little to no semantic information through HTML documents. This is why an alternative would be to convert many of the documents into XML files that provide information on the content through related tag sets. XML will help the information to conform and to improve inferring and utilising link topology in the applications. XLink[4] standards provide a new conceptual alternative into the process of analysis of linkages. It may have disadvantages, but, it provides a gateway into new possibilities to mine the Web. XLink is similar to that of HTML but it allows for bidirectional links which provide more data and resources to gather extensive related data and key information.

Hyperlinked data mining is multi-relational and it must be encompassed by at least two inter-related types of pages which are structured on the Web as hubs and authorities [21]. The *authorities* are forms of authorising bodies over the Web and present the most highly regarded pages on a topic and *hubs* are able to sort out the authorities and the way in which they are correspond to each other. When a query is posed using a search engine, the authorities return instant occurrences. They are then linked to more data sets through the hubs, which return up to the 10 most specific pages on the topic. Each page p has a hub weight $h(p)$ and an authority weight $a(p)$. These are all initialized to 1. Let $p \rightarrow q$ denote "page p has a hyperlink to page q". The h's and a's for the pages are updated iteratively as follows: $a(p) = \sum q \rightarrow p\ h(q)$, $h(p) = \sum p \rightarrow q\ a(q)$ [20].

Another link topology process is link-based cluster analysis [21]. This segregates the web into subgroups and establishes that the pages have links with in-links and out-links. The research done on this facet uses two approaches: (1) those based on the frequently occurring patterns using apriori-like algorithms [3] and (2) those based on graph mining algorithms such as ANF [41]. The ANF hyperlink mining system condenses a specific graph by identifying the neighbourhood function for every individual node. This enables the information based on its content to be separated into appropriate groups.

2.3.6 Semi-structured Data

In terms of semi-structured data, the more the internet expands the more irregular data is added to the Web. It must be noted that the capabilities of the internet have surpassed many expectations, but the ability to gather meaningful interpretable information from the Web can be questioned. A way to solve this problem is to form closed networks of complete data collaborative databases where only the full details of data are formatted to gather useful information. An issue is to identify and select the useful data for this purpose. Another approach is to utilise constraint-based artificial intelligence techniques. Specific data related algorithms are constructed using user-defined constraints to eliminate unnecessary data and to reduce the number of trivial results from the output. This approach may be useful as it allows for exact analysis of useful information using effective algorithms.

The object exchange model (OEM) is a method used to extract key information. It is presented in a graphical form where each object in the formation is linked through vertices and labels [55]. Each object serves as an identifier with clear connections to the corresponding sub-objects. This establishes the meaning for the specific relationship between the two objects. OEM can also be used in a different way where the

---

[4] http://www.w3.org/TR/xlink/

nodes are represented as objects and the labels are on the edges of the objects. If the graph at hand is acyclic (i.e. a tree) so is the corresponding OEM. This model separates information into the subclasses. It uses the separation capabilities of databases on the Web into many semi-structured data. Through the use of artificial intelligence techniques this model can present useful information. The model endeavors to find a typical structure that could correspond to the targeted objects within a database. This can be used to determine the irregularities of semi-structured data by establishing links within a database which have a useful meaning [1].

2.4   Applications of Artificial Intelligence and Related Techniques to Improve the Web Functionality

The benefits of the artificial intelligence and related techniques can be seen as falling into two broad categories. Those that provide a business value and those that have technical value. Applications having a business value can be used by businesses to assist in making decisions and in maximising effectiveness and minimising costs. Applications with technical value can also be used in devising new Web applications and providing services more effectively and efficiently.

Artificial intelligence and related techniques are usually utilised on the Web for online businesses. They are also used by financial analysts and have an increasing use in Science and Engineering. These techniques provide a way to discover hidden facts and trends that are not obvious in a general perspective or by using simple analysis [34]. Many businesses are finding that as a result of the enormous amount of customer data, these techniques help to use the information collected from visitors and to allow trends in behavior to become obvious. Some examples are (1) discover trends in behavior before customers cancel purchases; and (2) determine what sells well and what doesn't sell [16, 48]. The list of uses of these techniques is extensive and growing. These methods mine from the huge amount of data and learn algorithms that help in finding trends and patterns. They then report and offer conclusions and advice on choices for the business.

Data collection is an essential part of these techniques. The data fall into three categories: Demographics (the name, sex, location, age, job, income of a person); Psychographics, (related to a person's personality, interests or opinions); and Technographics (focuses on the hardware or software a customer is using). The artificial intelligence techniques are then used to decide who to target, what the targets might be interested in, what are the future requirements of people, what is any are the similarities between the target groups. Some examples are:

- Targeting – A method of selecting people for an advertisement which is based on the demographics for brand recognition and sales in an area.
- Personalisation – A method of targeting specific people by optimising an advertisement based on the person's actions. It is mainly applied to businesses sites.
- Association – A method of using user's actions on a site to determine what is most likely to be purchased in a shopping session. Amazon uses this technique particularly well by showing the user related items.

- Estimation and Prediction – A method to predict future trends by using past and present information and to examine the patterns.

Artificial intelligence and related techniques are used to enhance the semantics of the information presented on the Web. Many semantic matching and meta-data management methods have been developed based on artificial intelligence techniques. Some of these methods are for the integration and combination of the data drawn from diverse sources. It is then stored and applied to real world objects. There are many areas in the Web that can benefit by the use of Artificial Intelligence and related techniques. Some examples are listed below.

2.4.1 Web Personalisation

There is an exponential growth in the use of the Web as a major communication channel and information source. There are many Web information providers and much of the information available is large and complicated. Users may receive irrelevant information or fail to find useful information when searching the internet. The means to effectively sift through the information and deliver only relevant content to individual users is imperative [50]. There is also a need to provide the ability to predict the users' requirements to improve the efficiency of information retrieval from the Web. To effectively acquire information from the Web, the personalization techniques are being developed to tailor the output to the end users special requirements. Personalization is defined as the tailoring of a consumer product, services and information to the users' personal needs to improve the quality of service. The aim is to provide improved usability, data consolidation, user retention and targeted marketing [33]. Web personalisation is now becoming a necessity rather than a luxury. This is very true in the areas of e-business and Customer Relationship Management (CRM). Customers are becoming more demanding and the service provided must meet this need. If their needs are not met customers will go elsewhere and the organisation will lose business.

Web personalization is achieved by exploiting explicitly and implicitly captured knowledge obtained by the analysis of the user's search behaviour and interests. The content and structure of the website are also used to provide one-on-one guidance [16]. The personalised recommendations are made by using information previously obtained or collected in real-time. Customization of this information is achieved by means of the manual change of the appearance of the interface. This provides both convenience and lower cost for users [47]. Explicit personalisation is a rudimentary mechanism that makes use of forms. It is reliable but weak on end user acceptance. This is because the user must know beforehand the content of interest. Implicit personalisation is an inference method based on observable behavior patterns and user preferences from indirect sources. It strongly depends upon the means used to collect the user behavior.

Web personalisation mainly exists in three distinct categories: (1) Personalised Web Searches; (2) Personalised Content; and (3) Personalised Advertising Delivery. These categories of personalisation are presented by adapting user profiles. There are several recognized and widely applied techniques to assess user profile personalization. Among these are Collaborative Filtering (CF), Content Based Filtering, Rule-Based Filtering, Adaptive Filtering, Batch and Routing Filtering. A profile is defined by a set of attributes which may be organised into abstract entities. These values can be either user defined or dynamically derived from the user behaviour. A profile should characterize a user's domain of interest and all specific features that help the information system to deliver the relevant data in the correct form at the correct place and time. These attributes are generally ranked and organised in a preference model which drives the query compilation, query execution and data delivery [46]. In conjunction with the information collected from the user from the input, other data is collected by other methods. These methods are automated and do not require input from the user. They are generally done whilst the user is logged into their account. These construct the profile of tastes and preferences.

In general in each of the categories, the data to be mined includes Web clicks, user provided data, URL history, purchase and transaction records. Various methods are used to collect the required information. The Predictive and Association methods are greatly used by online merchants such as Amazon.com to provide their users with personalised content and to promote cross-selling of products. Clustering is also commonly used by marketing firms to segment the target demographics. No one method can be guaranteed to independently work effectively. Therefore, a combination of various methods is often used. Detail of each of the categories is given below.

2.4.1.1 Personalised Web Searches

Present day search engines generally handle search queries without consideration of the contexts in which the users submit the queries. Consequently, it becomes difficult to obtain the desired results due to the ambiguity of user's needs. Previous personalised researches have utilised a hierarchical category list to help users to specify the desired context of search. Due to the necessity for user input, the output of these researches was of limited success. Current research focuses on the use of user profile orientated methods of personalisation to improve the retrieval effectiveness [46]. By using user profiles for personalisation, the issues of ambiguity and neglect of changes in user tastes are eliminated [31]. Automatic learning of the user search preference allows the search engines to adapt to changes in user tastes.

A prime example of this is Google, which previously did not have sign in functions of any sort. Over the years as the technology has progressed they offered user accounts for their extended services such as Gmail. Google begins with a basic search functionality by redirecting the user to localised versions of Google. This displays relevant links according to the geographic area. No personalization occurs at this stage. Once a user has signed in, they receive their own private customised page displaying the existing user specifications given in the 'sign-up' process. The personalisation of web searches now becomes possible. In this environment, the first step of personalized search is accomplished by mapping a user query to a set of categories. These reflect the users intent and serve as a context for the query based on the user profile [7]. First, the similarities between a user query and the categories which represent the user's interests are determined. The categories are then ranked in the descending order of similarities. Finally, the top categories are displayed as search results. If the user clicks on one or more of these top results, the user's intention is explicitly shown on the system. If the user's interest is not among the top results, the user may have developed new interests. The most recent search records are used to personalise future searches. Thus, the user profile reflect the subscriber's most recent interests [31].

By using user profiles based on log data of user search history, user search terms, user tastes and preferences, searching the internet and retrieving relevant information becomes easier. User profiles used in conjunction with data mining outputs make it possible to store information about user preferences and match the searches with the desired results and automatically adapt the changing tastes of the users.

2.4.1.2 Personalised Content

Personalised content is the customisation of pages to suit user tastes and preferences. The level of customisation varies from a simple change of colour and layout to a more sophisticated customisation which allows for display of user specific items in a desired style. The simplest example is the use of multi-lingual functionality which allows the website to reach a broader audience in an understandable manner.

Content data personalisation has been used in different technologies such as information retrieval systems (Web search engines), database systems (online merchant) and human-computer interaction systems (specialised expert systems). In information retrieval systems, the user is involved in the query evaluation which is conducted as part of a stepwise refinement process. The user can decide at each step which data they will include or reject. The process is then considered as a machine learning process based on user feedback. In database systems, it is not common to involve the user in the data retrieval process. A database query generally contains all the necessary criteria for the selection of relevant information. Personalisation of the data is considered from the viewpoint of the query language extensions. The query rewriting processes is done by using user profiles. In a human-computer interaction example the user profiles generally confine the user expertise to the application domain. This is done in order to provide appropriate interfaces and dialogues [7]. The content data delivery modules, graphical metaphors and the level of expertise of the dialogue form the main personalisation items in this context.

The content must be customized to suit the tastes and preferences of the user. It is done on an individual user level by using user profile and the user input. There must be more emphasis on the user input since current technology platforms are incapable of knowing details such as the layout or style desired by the user. When more advanced personalisation technologies will emerge and will integrate into everyday use, everything will be personalised to suit the individual.

2.4.1.3 Personalised Advertising

Personalised advertising is an effective form of marketing which matches the advertising material with the correct target demographic. This will produce a more effective means of promotion suited to the correct demographic of the user group [13]. This means that advertising material relevant to the user is given. This improves the chance that the advertisement will achieve its desired result. This increases the possibility of a successful follow-up visit and a transaction. The method of data mining commonly used in this type of personalisation is Association Analysis. The best example of this method of personalised advertising is Amazon.com. Amazon.com is in common use among the e-commerce community as a result of its clever use of customer profiling and data mining methods. This forms the basis of its ability to sell products to customers. At Amazon.com, after doing a book search the user is informed that "people who purchased this book have also bought these other books". The data needed for this analysis consists of "who are searching", "what they are searching for", and "what other people who may share similar interests and have conducted similar past searches". Clustering and association methods are used to analyse customer behaviour and preferences.

Personalised advertisement delivery systems are of more benefit to the website owners as opposed to the users who will generally try and dismiss the advertisements. Some benefits can be reaped by the user such as "finding products that a user was looking for and the user may not have found them by other means". There are many ways to personalise the advertisements. Some examples are keyword targeting (as used by Google) and content targeting. Advertising based on content targeting uses the metadata of a Webpage and of an advertisement. This ensures that the relevant group of advertisements can be shown to the user and is based on what the user is examining at the time. By targeting advertisements based on the content of the current pages and the information from the advertisement page, advertisements can be personalised better in comparison to personalisation which is based on the keyword targeting. The customisation of the marketing component is based on the detailed demographic, lifestyle behavioural information, user-defined preferences in the profile database and the content of the current page. The customisation is done by the use of components in the content of the provider's servers. Through the application of technology the content providers are able to target desirable profiles and tailor messages to obtain the desired actions. This is done by scaling large demographic or behavioural segments to the unique interests of an individual user.

2.4.1.4 Some Issues with Personalisation

Many technologies have been proposed for dynamically customising the content for a user based on past and present browsing patterns. There remains the potential for significant improvement in this area.

Considering practices in the current field of personalisation using artificial intelligence and related techniques, it may be seen that however accurate a system is in delivering data to a user and accurately profiling that user's interests and preferences, they are limited. Such systems will always be restricted by the quality and availability of data for each particular user. The manipulation of the information must always produce a result that remains domain specific. Current technology demands that each new Web service must learn the preferences of a user despite the availability of data in other services. At present the Amazon's online web service is regarded as a leader in personalisation using data mining. Such an application is limited to a single domain. Despite the accuracy of the Amazon's users' profile, these profiles are restricted to Amazon's services. This suggests that a system of global identification is desirable. A centralised user profile may be used to control access to a multitude of websites and to act as a central repository containing the user's browsing history, interests and preferences. Such a profile could be stored in a centralised database or saved on the user's computer as a 'cookie'. The later introduces problems when using multiple computers, while the former has security complications.

An abstract personalisation profile could be used by various websites to tailor content more accurately and specifically. It would enable a Web service such as Amazon to continue to deliver personalisation and to build the particular user's overall profile that is accessible and compatible with other web services. If the concept of a global identifier and profiling tool were to be developed, it would enable the massive delivery of accurate personalised data to a user for use in all aspects of online work. It could simultaneously rapidly build a cohesive and definitive user profile. Such a centralised personal profile repository will provide a higher quality personalised services and an easier access to user accounts. It would also pose a serious security risk. People must be willing to adapt to the changing technologies and to accept a higher level of intrusion into their lives for progress to be made. This can be considered as the price for accurate, rapid and user-centric browsing. It is necessary to trust the security of the central profile repository. Other issues arising with personalisation include scalability and accuracy. In order to deal with large Websites that have a high degree of activity, the personalisation systems need to be efficient in analyzing the massive amount of data in real-time. The personalisation systems must also be able to deal with false information provided by the users that can lead to inaccurate personalisation.

2.4.2 Semantic Web

The Web doubles in size every eight months and now includes approximately 20 billion content areas[5]. According to [29], 85% of users use search engines to find information. Consumers use search engines to locate and buy goods and to do research for many decisions. These include items such as choosing a vacation destination and the choice of medical treatment and appliances. The search engines currently lack comprehension and dynamicity. They do not index sites regularly or completely. The current state of search engines may be compared to a phone book which is updated at irregular intervals and is biased toward listing popular information, and is missing many pages. It is difficult to collect and to make sense of such information on the web because it is mostly disorganized. The challenge in the next few years is to translate disorganised human-readable data into effective and efficient machine comprehensible data that can easily be retrieved [6].

The semantic web is considered to be the next generation of the Web. It relies on ontologies and meta-data [6]. Semantically enhanced information that computers can process will help the user in the performance of a wide range of tasks involving data, information and knowledge exchange. This will allow intelligent access to heterogeneous and distributed information. It will enable software products and agents to mediate between user needs and the available information sources [15]. The semantic Web describes how, using an analogous example, a mother's physical therapy session can be arranged by a sibling using software agents. The agents can locate an appropriate clinic and checke the calendar schedules. Most importantly, the semantic Web will help users to find the correct information rapidly and efficiently. Semantic based information retrieval techniques such as Latent Semantic Indexing [28] has already improved query precision and recall. This gives improves the results in keyword selection.

---

[5] http://www.pandia.com/sew/383-web-size.html

The idea that machines can communicate with other machines in order to facilitate the everyday tasks of humans indicates an important role for Artificial intelligence and related techniques in conceptualizing a semantic Web [12]. These techniques can pre-process and classify large amounts of data. This produces more effective ontologies and meaningful relationships between machines. The Semantic Web seeks machine-to-machine communication through the use of semantic data. Artificial Intelligence techniques attempt to give a meaning to such data. The combination of these two areas will enable the integration of diverse data and help to discover new meaning about objects in the evolving Web.

2.4.2.1 Artificial Intelligence and Related Techniques Applied to Semantic Web

The aim of the semantic Web is to add semantic annotation to Web documents to access knowledge instead of using unstructured material. This permits knowledge to be managed automatically. Artificial intelligence techniques can help to learn the definitions of structures for knowledge organization (ontologies) and to populate these knowledge structures [49]. All of the approaches discussed here are semi-automatic. They assist humans in extracting the semantics but are unable to completely replace them. This is due to the large amount of tacit knowledge involved in the modeling process.

Ontology Learning. It is a challenging task to extract the ontology from the Web. There is a need for advanced techniques and tools that can help the learning of complex ontologies and make them easier for human understanding and manipulation. One way is to engineer the ontology by hand but this is an expensive exercise. In Ontology learning for the semantic Web, the expression *Ontology Learning* was coined for the semi-automatic extraction of semantics from the Web in order to create ontology [49]. Machine learning techniques are used to improve the ontology engineering process [12]. Ontology learning exploits many existing resources such as texts, thesauri, dictionaries, and databases. In order to discover the 'semantics' in the data and to make them explicit, it builds on the techniques of Web content mining and it combines the machine learning techniques with methods from fields such as information retrieval and agents [14]. Using these techniques, the results obtained must be incorporated in a machine interpretable format such as ontology.

Instance Learning. For new documents, it is expected that the users will manually annotate them to a certain degree. However for old documents which contain unstructured material this does not solve the problem. It would be impossible to mark every document manually. Moreover, additional information may be needed by individual users. It is essential for the semantic Web to be able to extract results from Web documents automatically or semi-automatically. Information Extraction (IE) is one of the most promising areas of human language technologies. IE is a set of automatic methods for locating important facts in electronic documents for subsequent use. This is for annotating documents or for information storing for further use [49]. An example is populating an ontology with examples. There are two commonly used systems for IE, FASTUS [5] and OntoMat Annotizer [25]. These systems provide support for knowledge identification and extraction from Web documents. They can, for example, provide support for document analysis either automatically that is unsupervised extraction of information or semi-automatic extraction as a support for human annotators for locating relevant facts in documents using information highlighting.

Mapping and Merging Ontologies. Domain-specific ontologies by multiple authors are modeled in multiple settings. This lays the foundation for building new domain-specific ontologies by assembling and extending multiple ontologies. The process of ontology merging uses as input two or more source ontologies and returns a merged ontology based on the given source ontologies. The process of merging becomes essential due to the increasing usage of ontologies which causes the overlapping of knowledge in a common domain to occur more often. This method is difficult, labor intensive and error prone when using conventional editing tools without support to do the task of manual onotology construction. Several systems and frameworks are introduced to support the ontology merging task [49]. This depends on the syntactic and semantic matching heuristics derived during the task of merging ontologies. The mapping is done by allocating the concepts and instances of one ontology to the concepts of another. This technique is useful when the ontology chosen is the correct one for the task. Instances of the correct ontology may be classified as part of the target ontology.

Latent Semantic Indexing (LSI). The semantic Web is dependent on the widespread availability of large collections of semantically rich, trustworthy and meaningful resources. Since semantic classification is dependent upon complex ontologies, a difficulty is the steep learning curve is presented to human classifiers when they attempt to utilize these ontologies. Even with access to ontological knowledge, there is still a high level of dispersion in the final tag set. To solve this problem, a tool is needed to ease the manual semantic tagging burden of content creators when using the standardized ontological resources. This tool should help users to explore ontologies and should also automatically suggest a restricted set of tagging terms. These terms are drawn from the ontology to guide the user in the classification tasks. LSI has improved both the query precision and recall [28]. This results in more meaningful selection of keywords from the ontology. This would result in more effective semantically tagged resources.

Using existing conceptualizations as ontologies and for automatic annotation. For most of the sites in the Web, an explicit domain model for the generation of Web pages can be found in a database or in a Content Management System (CMS). The existing formalizations can be used for semantic markup and mining. These annotations derive mappings between the information structures. The mapping are used for querying the semantic information stored in the database underlying the Website [49].. A good example of this is the use of artificial intelligence techniques such as neural networks and clustering to find relationships between multilingual text information in the web. The semantic Web relates information and becomes more integrated as data in different languages can be understood in a common context.

2.4.2.2 Some Issues with Semantic Web

There are a number of issues related to the use of artificial intelligence techniques in conceptualizing the semantic Web. The first issue is the extraction of ontology from the Web. The Web contains a huge number of existing resources which include texts, thesauri, dictionaries and databases. This poses a challenge to extract and engineer the ontology manually. Additionally, many old Web resources contain unstructured material. It is impossible to manually markup every document.

Another factor is the growing use of ontologies which lead to overlaps between knowledge in a common domain. Domain-specific ontologies by different authors are modeled in multiple settings. These ontologies lay the foundation for building new domain-specific ontologies in similar domains. This may be done by assembling and extending multiple ontologies from repositories. If the user is to merge the ontologies, conventional editing tools are required. Without support of advanced technologies, this task will be difficult, labor intensive and error prone. We next consider the issue of the tagging burden in the semantic Web. A human will find it hard to understand the complex ontologies when trying to use them [19]. The more complex the ontologies are, the more it will confuse potential users trying to use them. This illustrates the clash between the ontology conceptualization and the user's different predetermined conceptualization when considering the classification task.

Lastly, there are many different languages used in the Web today. Semantics which are fully multilingual cannot be created. Multilingual text mining can be used to cluster conceptually related terms and texts in various languages and to create ontology. This will take the time and resources [30]. The predominant use of English in the Web content is an issue for non-native English speaking Web users. English is currently the main language globally used on the internet content. Only 44 percent of internet users are native English speakers [52]. There is a need to create tools that can explain the content in other languages. Semantic Web information should enhance the performance of natural language processing tools to produce a good understanding of text content [9]. The issue pertains to establishing appropriate inter-ontology translators that can plan ontologies and content in other languages. The developed ontologies should also acknowledge the cultural requirements of other countries and areas [52]. Multilingual and multicultural phenomenon will play an important role in establishing and maintaining a common understanding of human knowledge.

2.4.3 Web Services

The concept of a "Web Service" refers to an application or a program provided on the Internet by a supplier. It can be accessed by the customers through the standard Internet protocols. It works as an Application Programming Interface (API) that sends users' requests to the supplier's database and returns the information required to provide the service. The Web service describes a standardized way of integrating Web-based applications using the XML, SOAP (Simple Object Access Protocol), WSDL (Web Services Description Language) and UDDI (Universal Description Discovery Integration). These are open standards over an Internet protocol backbone [35].

To give readers a sense of Web Service discovery, a real-world example is given. Consider the service on a ticket selling business's web site. Here the Web Service retrieves the information requested by the user. This could include information such as the number of tickets, time, location, and destination. It then returns the desired result. In this example, the Web Service will probably send a request to the business's database first to receive the data about the available ticket. It then calls another web service that provides the "check outs" service to complete the purchasing process.

Web services adhere to a concept known as loose coupling. This means services are platform-independent, discoverable and interoperable. Web services enhance current web functionality by altering the nature of documents to service-oriented. As the number of web services increases, it becomes increasingly important to provide a scalable infrastructure of registries that allows developers and end-users to perform service discovery [36]. Web services technology represents the next generation of distributed computing. It builds on and extends the current traditional model of artificial intelligence applications. Since the Web service is a relatively new technology, there are many areas where improvements can be made to improve the current state of the art. Not only do these improvements enhance the technical workings of Web services but they can also generate new business opportunities. Researchers have worked on the problem of addressing the shortcoming of UDDI by finding relationships between search terms and service descriptions. Other research areas include WSDL which is used to represent the semantics while describing the service [35, 36].

2.4.3.1 Issues and Applications in Web Services

Artificial intelligence techniques can be used to facilitate and improve the use of Web services such as to assist management in making strategic decisions; to optimizing staffing levels and minimizing costs; or by assisting technical staff in devising new services or in improving existing services.

Web Services Cost and Savings Prediction. It is difficult for businesses to gauge the costs and savings of a Web service deployment with little or no experience in their deployment. By using the data collected by research firms such as Nemertes [26], businesses can learn from the experience of similar organizations to obtain a good indication of these values. Value Prediction is suitable to model the investment versus the return functions to obtain a prediction of costs and savings. Regression techniques may be used to derive the predicted continuous values obtained from best fit functions [24]. For predicting the overall costs of a deployment, the input data required consists of the number of staff involved, the time taken, and the complexity of the deployment. The complexity of the deployment can be quantified in terms of the lines of code used in the programs and the annual revenue obtained from the operations. The costs of the proposed deployment can be predicted using these parameters. Once the costs are known, the prospective savings can be predicted. Using inputs such as the cost of the deployment, and the original and the new cost of the operation, the savings can be determined. Having determined the costs and the savings that can be gained, the return of investment for Web services deployments can then be calculated. Businesses can then determine the size best suited to them. These discovered insights can be used to decide future actions.

Performance Monitoring. Strategic placement of human resource plays a crucial role in the effective monitoring of the performance and mangement of events. This leads to the need to prioritise tasks. A service being used by many clients when a problem occurs should have a higher priority than a service which is only being used by a few clients. By knowing the usage pattern of services, training programs on groups of services with similar usage patterns can be developed. This allows staff monitoring the services at certain times to have a more in depth knowledge of the requirements of particular services. To identify services with similar usage patterns, similar time sequence analysis can be used [2]. The input for such an operation is in the form of a time-series data recording the number of clients using a particular service. Although such data is not normally collected explicitly, it is implicitly recorded in the web server access logs. Specialized processes are required to infer this information from the collected data. Algorithms [2, 24] for approximate subsequence matching in time-series can now be applied to find Web services with similar usage patterns. These patterns can then be used to help to design of roster schedules that optimize staffing levels and skill requirements while minimizing the number of employees that need to be present.

Service Innovation. It is important for service providers to establish themselves in the market by offering a range of quality services. The set of queries used by potential clients to find suitable Web services is a rich source of clues about the clients' requirements. When an unusual search term is used with other more common search terms in the queries and that the search terms are all related, it is then a good indication that there is a demand for a new service. The unusual search term may represent a new concept, or a specialization of a general service currently being offered. As an example, SMS (Short Message Service) sends text messages to mobile phones while a more recent technology MMS (Multimedia Message Service) sends multimedia messages. SMS is a frequently used search term but MMS is not. As the technology becomes more prevalent, the demand for MMS Web services will grow and the appearance of MMS in query data will provide evidence of this. The simplest approach to discover uncommon search terms is by deviation analysis [24].

Service Recommendation. Web services providers can recommend services to clients based on the services that other similar clients have used in the past when using artificial intelligence techniques. This is because similar clients are likely to have similar service needs. Service providers have information such as the line of business, size of the business and the services their clients use. They can use this information as inputs for predictive modelling operations and to make recommendations to new clients. Inputs such as the interfaces, functionality, the security offered, the cost and other resources required by the service should also be considered in the analysis. Decision trees [44] can be used to build rules on service requirements and subscriptions. The only information service providers have about their clients are those used for billing purposes, but the number of attributes available is small. Consequently, the structure of the resulting decision tree will be relatively simple and easily comprehensible to a human analyst. To further enhance the success rate of the recommendations, service providers can find dissociations among the services they offer. Dissociations [51] capture negative relationships between services with rules such as the use of services X and Y implies that it is unlikely service Z will also be used. This is even though services X and Z are often used. That is, $X \Rightarrow Z; X \wedge Y \Rightarrow \neg Z$. Incorporation of these dissociations in the recommendation process enables more specific recommendations to be made.

Service discovery. The discovery of services requires seven steps these are namely description, presentation, publication, request, discovery, brokering and execution. Description is the process of defining meta-data for a service, such as the interfaces it offers, the operations and arguments for an interface and the binding of network protocols and endpoints. Presentation is concerned with mechanisms that allow the retrieval of the service descriptions. This requires a means of uniquely identifying a service on a global scale. Publication makes a service known to interested clients via the use of a registration system. Request and discovery involves the user formulating a request for a service while discovery involves searching the registry for candidate services that implement the requested functions. Brokering and execution are concerned with the scheduling and allocation of resources for the execution of the services. Artificial intelligence techniques can help in each of these steps.

WSDL and the UDDI are two standards designed for the description, presentation, publication, request, and discovery of Web services. However, they have limitations when addressing the problem. On this account, the UDDI especially has drawn much criticism. The WSDL facilitates the description of the service parameters, messages to be exchanged between applications and how to connect to the services. However it provides limited information to the service consumer as to exactly what the service does. The only parts of the description that may give hints on the functionalities of a service are the name attributes of the parameters, messages and operations. This lack of semantics led to the development of DAML (DARPA Agent Markup Language)-based languages for service description. This is where the service capability matching is based on the inputs, outputs, preconditions and effects. Ontology are used to encode relationships between the concepts [32, 42]. The current state of Web services does not provide automatic service matching. At present the manual discovery of services avaialble must suffice. Effort is needed to improve its efficiency.

Similar to WSDL descriptions, the UDDI does not make use of semantic information to describe and discover the appropriate services based on their capabilities [42]. Apart from the text descriptions, there is no provision for specifying the capabilities of the service. The categorisation of businesses and services is of little help in the discovery of suitable services. The classification schemes used – NAICS and UNSPSC, were designed for the broad-based classification of industries, products and services. There is little or no differentiation between products or services available in the area. These schemes do not provide the specificity needed for service discovery. Searching in UDDI is restricted to keyword matching using names, locations, business, bindings and tModels (a unique identifier for reusable concepts). There is no provision for inference or flexible matching of the keywords. This means service providers and requesters must choose the names and description of services very precisely when using the UDDI as a repository for the Web service discovery.

Normally, developers have to search the service from tens or hundreds of service entries in UDDI. This process is very time consuming and is a loss of profits in terms of business. As the single most pervasive source for the discovery of Web services, there is a great need to address the shortcomings of the UDDI by the use of artificial intelligence techniques. An intuitive way to improve the situation is by predicting what the user of a service may be looking for and then suggesting the appropriate services. This can be done by extracting common behaviours of users searching the UDDI by mining the user query logs. Thus artificial intelligence can help Web search engines in finding high-quality Web pages and it enhances the Web click-stream analysis.

The potential to achieve dynamic, scalable and cost-effective infrastructure for electronic transactions in business and public administration has caused recent research efforts to consider Semantic Web services. That is obtained by enriching the Web services with machine-processable semantics. Semantic information aims to enhance the integration and the web service discovery by utilizing the machine readable constructs of the representation. Semantic Web services aim at an integrated technology in the next generation of the Web by combining Semantic Web technologies and Web services. This will turn the Internet from an information repository for human consumption into a world wide system for distributed Web computing. A number of ontologies such as OWL, OWL-S, WSDL-S, WSMO have been proposed to address the semantic heterogeneity among Web resources and services. Each one of these has its own strength and can be used in a specific case. The use of ontology has enhanced the discovery process but the problem of accurately finding a match persists. This is because the current Web searching techniques don't consider the semantics and consequently are unable to provide accurate results.

2.5 Conclusion

The Web is widely used by the public and forms a part of our daily life. People search the Web using engines such as Google and Wikipedia. There is an increasing trend to online shopping. As the Web and its usage grow, the opportunity to develop smart Web applications and to analyse Web data to extract useful knowledge from it increases. The past few years have seen the emergence of Web intelligence as a rapidly growing area due to the efforts of the research community and the various commercial organizations that are using it.

This chapter provides a comprehensive analysis of Web and its development from birth to the use of artificial intelligence techniques. The Web is a massive information system with interconnected databases and remote applications providing numerous services. These services are becoming more user-friendly but the concept of smart Web applications is still in its infancy. This chapter details three applications which are Web Personalisation, Semantic Web and Web Services The basic concepts, main challenges and other issues with these applications are identified.

Web personalization is a way to meet the Web users' requirements in order to provide precise information that best suits the user's needs and to provide useful recommendations. Commercial organizations are investing heavily on personalization tools to provide a better service to the public and to obtain a competitive advantage. Better methods and techniques are needed to increase the accuracy of personalization. We assume that in the future these techniques built into current Web applications will be 'off-the-shelf' package software.

The amount of data already available on the Web has made it difficult to retrieve the existing required information. Most of the data on the Web is so unstructured that it can only be understood by humans. It is so huge that is can only be processed efficiently by machines. These factors have highlighted the importance of utilising artificial intelligence techniques in the semantic Web. The semantic Web's success is largely dependent on the availability of semantically rich resources that both provide meaning and trust between machines. There is much more research, development and work needed in semantic Web before it can effectively achieve its ultimate goal of producing a better usable Web.

Service oriented components, or Web Services, are emerging as the key to enabling technology for today's e-commerce and business applications. These are transforming the Web into a distributed computation and application framework. The implication of this transformation is that the volume of data generated on the Web is increasing exponentially. The future appears to need artificial intelligence techniques for accessing thesemantic Web services and ontology to automate Web service discovery processes.

Research shows that artificial intelligence techniques have influenced the current state of the Web by improving the way different computers applications and services communicate and relevant information is identified.

References and Further Reading

1. Abiteboul, S., Buneman, P., Suciu, D.: Data on the Web: From Relations to Semistructured Data and XML. Morgan Kaumann, California (2000)
2. Agrawal, R., Srikant, R.: Fast Algorithms for Mining Association Rules, IBM Research Report RJ9839, IBM Almaden Research Center (1994)
3. Agrawal, R., Srikant, R.: Mining Sequential Patterns: Generalizations and Performance Improvements. In: Apers, P.M.G., Bouzeghoub, M., Gardarin, G. (eds.) EDBT 1996. LNCS, vol. 1057. Springer, Heidelberg (1996)
4. Alesso, P., Smith, C.: Developing the Next Generation Web Services - Semantic Web Services. A K Peters Ltd. (2004)
5. Appelt, D.E., Hobbs, J.R., Bear, J., Israel, D., Tyson, M.: FASTUS: A Finite-state Processor for Information Extraction from Real-world Text. In: IJCAI (2003)
6. Berners-Lee, T., Hendler, J., Lassila, O.: The Semantic Web. Scientific American Magazine (2001)
7. Bradley, K., Rafter, R., Smyth, B.: Case-based user profiling for content personalisation. In: Proceedings of the International Conference on Adaptive Hypermedia and Adaptive Web-based Systems, Trento, Italy (2000)
8. Cafarella, M.J., Etzioni, O., Suciu, D.: Structured Queries OverWeb Text. In: Bulletin of the IEEE Computer Society Technical Committee on Data Engineering (ICDE) (2006)
9. Calzolari, N.: Language resources in the Semantic Web vision. In: International Conference on Natural Language Processing and Knowledge Engineering (2003)

10. Chakrabarti, S.: Data Mining for Hypertext: A Tutorial Survey. SIGkDD Explorations 1(2), 1–11 (2000)
11. Cooley, R., Mobasher, B., Srivastava, J.: Web Mining: Information and Pattern Discovery on the World Wide Web. In: Proceedings of the IEEE International Conference on Tools with Artificial Intelligence (ICTAI 1997) (1997)
12. Cristani, M., Cuel, R.: A Survey on Ontology Creation Methodologies. Journal of Semantic Web and Information Systems (IJSWIS) 1(2), 49–69 (2005)
13. Cristo, M., Golgher, P., Moura, E., Ribeiro-Neto, B.: Web search 3: Impedance coupling in content-targeted advertising. In: Proceedings of the 28th annual international ACM SIGIR conference on Research and development in information retrieval SIGIR 2005 (2005)
14. David, J.F.G., Briand, H.: Association Rule Ontology Matching Approach. International Journal on Semantic Web and Information Systems 3(2), 27–49 (2007)
15. Ding, Y., Fensel, D., Klein, M., Omelayenko, B.: The Semantic Web: Yet Another Hip? Data & Knowledge Engineering 41(2-3), 205–227 (2002)
16. Eirinaki, M., Vazirgiannis, M.: Web Mining for Web Personalization. ACM Transactions on Internet Technology 3(1), 1–27 (2003)
17. Finin, T., Norvig, P.: Special Track on Artificial Intelligence and the Web. In: Proceedings of the Twenty-Second AAAI Conference on Artificial Intelligence. The AAAI Press, Vancouver, British Columbia (2007)
18. Finin, T., Radev, D.: Special Track on Artificial Intelligence and the Web. In: Proceedings of the Twenty-first AAAI Conference on Artificial Intelligence. The AAAI Press, Boston, Massachusetts (2006)
19. Forno, F., Farinetti, L., Mehan, S.: Can Data Mining Techniques Ease The Semantic Tagging Burden? In: VLDB 2003. First International Workshop on Semantic Web and Databases, Berlin, Germany (2003)
20. Garofalakis, M.N., Rastogi, R., Seshadri, S., Shim, K.: Data Mining and the Web: Past, Present and Future. In: WIDM (1999)
21. Getoor, L.: Link Mining: A New Data Mining Challenge 4(2), 1–6 (2004)
22. Gromov, G.R.: The Roads and Crossroads of Internet History (1995)
23. Han, J., Hou, J., Li, Z., Zaiane, O.R.: Mining multimedia data. In: Proceedings of the 1998 conference of the Centre for Advanced Studies on Collaborative research (1998)
24. Han, J., Kamber, M.: Data Mining: Concepts and Techniques. Morgan Kaufmann, San Francisco (2006)
25. Handschuh, S., Staab, S., Volz, R.: On Deep Annotation. In: WWW 2003, Budapest, Hungary (2003)
26. Johnson, J.T.: State of the Web services world (2003)
27. Khasawneh, N., Chan, C.-C.: Active User-Based and Ontology-Based Web Log Data Preprocessing for Web Usage Mining. In: Proceedings of the 2006 IEEE/WIC/ACM International Conference on Web Intelligence (2006)
28. Landauer, T.K., Foltz, P.W., Laham, D.: An introduction to latent semantic analysis. Discourse Processes 25, 259–284 (1998)
29. Lawrence, S., Giles, L.: Accessibility of Information on the Web. Nature 400, 107–109 (1999)
30. Lee, C.H., Yang, H.C.: Acquisition of Web Semantics based on a Multilingual Text Mining Technique. In: Proceedings of 2nd Semantic Web Mining Workshop of the ECML/PKDD 2002 (2002)
31. Liu, F., Yu, C., Meng, W.: Personalized Web Search For Improving Retrieval Effectiveness. IEEE Transactions on Knowledge and Data Engineering 16(1), 28–40 (2004)

32. McIlraith, S.A., Son, T.C., Zeng, H.: Semantic Web services. IEEE Intelligent Systems 16(2), 46–53 (2001)

33. Mobasher, B.: Data Mining for Personalization. In: Brusilovsky, P., Kobsa, A., Nejdl, W. (eds.) The Adaptive Web: Methods and Strategies of Web Personalization, pp. 90–135. Springer, Heidelberg (2007)

34. Nayak, R.: Data Mining for Web-Enabled Electronic Business Applications. In: Nanshi, S. (ed.) Architectural Issues of Web-Enabled Electronic Business, pp. 128–139. Idea Group Publishers (2002)

35. Nayak, R.: Facilitating and Improving the Use of Web Services with Data Mining. In: Taniar, D. (ed.) Research and Trends in Data Mining Technologies and Applications, pp. 309–327. Idea Group Publishers (2007)

36. Nayak, R.: Using data mining in web services planning, development and maintenance. International Journal of Web Service Research 5(1), 62–80 (2008)

37. Nayak, R.: XML Data Mining: Process and Applications. In: Song, M., Wu, Y.-F. (eds.) Handbook of Research on Text and Web Mining Technologies. Idea Group Inc., USA (2008)

38. Nayak, R., Iryadi, W.: XML schema clustering with semantic and hierarchical similarity measures. Knowledge Based System 20(4), 336–349 (2007)

39. Nayak, R., Tran, T.: A Progressive Clustering Algorithm to Group the XML Data by Structural and Semantic Similarity. International Journal of Pattern Recognition and Artificial Intelligence 21(4), 1–23 (2007)

40. Nayak, R., Zaki, M.J. (eds.): KDXD 2006. LNCS, vol. 3915. Springer, Heidelberg (2006)

41. Palmer, C.R., Gibbons, P.B., Faloutsos, C.: ANF: A Fast and Scalable Tool for Data Mining in Massive Graphs. In: Proceedings of 2002 Internation conference in Knowledge Discovery and Data mining, Edmonton, Alberta, Canada (2002)

42. Paolucci, M., Kawamura, T., Payne, T.R., Sycara, K.: Semantic Matching of Web Services Capabilities. In: Horrocks, I., Hendler, J. (eds.) ISWC 2002. LNCS, vol. 2342. Springer, Heidelberg (2002)

43. Park, J., Hunting, S.: XML Topic Maps: Creating and Using Topic Maps for the Web. Addison Wesley, Reading (2006)

44. Quinlan, R.: Simplifying decision trees. Int. J. Hum.-Comput. Stud. 5(12), 497–510 (1999)

45. Rath, H.H.: The topics Map Handbook (2003)

46. Sieg, A., Mobasher, B., Burke, R.: Web Search Personalization with Ontological User Profiles. In: Proceedings of the 16th ACM Conference on Information and Knowledge Management (CIKM 2007), Lisboa, Portugal (2007)

47. Sollund, A.: Personalisation Through The Use of Personal Profile, ePerSpace, Fornebu, Norway (2007)

48. Srivastava, J., Desikan, P., Kumar, V.: Web Mining: Accomplishments and Future Directions. In: Proc. US Nat'l. Science Foundation Workshop on Next-Generation Data Mining (2002)

49. Stumme, G., Hotho, A., Berendt, B.: Semantic Web Mining: State of the art and future directions. Web Semantics: Science, Services and Agents on the World Wide Web 4(2), 124–143 (2006)

50. Tam, K.Y., Ho, S.Y.: Web Personalization: Is It Effective? IEEE IT Professional 5(5), 53–57 (2003)

51. Teng, C.M.: Learning from Dissociations. In: The 4th International Conference on Data Warehousing and Knowledge Discovery DaWak, Aix-en-Provence, France (2002)

52. Tjoa, A.M., Andjomshoaa, A., Shayeganfar, F., Wagner, R.: Semantic Web Challenges and new requirements. In: The Sixteenth International Workshop of Database and Expert Systems Applications (2005)
53. Yergeau, F., Bray, T., Paoli, J., Sperberg-McQueen, C.M., et al.: Extensible Markup Language (XML) 1.0 (Third Edition) W3C Recommendation (2004)
54. Zaiane, O.R., Han, J.: Resources and knowledge discovery in global information systems. In: Proceedings of 1995 Internation conferences in Knwoledge Doscovery and Data mining (1995)
55. Zhau, A., Zhou, J., Tian, Z.: Incremental Minign of Schema for Semi-structured data. In: The Third Pacific-Asia Conference on Knowledge Discovery and Data Mining (1999)

3

Modeling Interests of Web Users for Recommendation: A User Profiling Approach and Trends

Daniela Godoy and Analía Amandi

[1] ISISTAN Research Institute
Universidad Nacional del Centro de la Prov. de Bs. As.
CP 7000, Tandil, Bs. As., Argentina
[2] CONICET, Consejo Nacional de Investigaciones Científicas y Técnicas
CP 1033, Capital Federal, Bs. As., Argentina
{dgodoy,amandi}@exa.unicen.edu.ar

In order to personalize Web-based tasks, personal agents rely on representations of user interests and preferences contained in user profiles. In consequence, a critical component for these agents is their capacity to acquire and model user interest categories as well as adapt them to changes in user interests over time. In this chapter, we address the problem of modeling the information preferences of Web users and its distinctive characteristics. We discuss the limitations of current profiling approaches and present a novel user profiling technique, named *WebProfiler*, developed to support incremental learning and adaptation of user profiles in agents assisting users with Web-based tasks. This technique aims at acquiring comprehensible user profiles that accurately capture user interests starting from observation of user behavior on the Web.

3.1 Introduction

To provide effective personalized assistance, information agents depend on the knowledge about individual users contained in user profiles, i.e. models of user interests, preferences and habits gathered mainly through observation. User profiles represent long-term information preferences users might have regarding their work (e.g. *computer software*), their hobbies (e.g. *sports*) or current issues. The accurate representation of user interests is crucial to the performance of personal agents in satisfying user information needs and, consequently, also are the learning approach used to acquire user profiles and the adaptation strategy used to cope with changes in user interests over time.

In spite of the agreement on the importance of user profiling as an integral component of personal agents, existing approaches have only partially addressed the characteristics that distinguish user profiling from tasks such as text classification or information retrieval. These characteristics include modeling specific interest topics which are likely to change over time as well as modeling

R. Nayak et al. (Eds.): Evolution of the Web in Artificial Intel. Environ., SCI 130, pp. 41–68, 2008.
springerlink.com

contextual information about these interests. Moreover, little attention has been paid to the assessment of readable descriptions of user interests to be interpreted by either users or other agents.

In this chapter we present a novel user profiling technique, named *WebProfiler*, for incremental learning and adaptation of user profiles in personal agents assisting users with Web-based tasks. This technique aims at acquiring comprehensible user profiles that accurately capture user interests starting from observation of user behavior on the Web. It addresses representational issues of user profiles, such as the description of interests at different abstraction levels, the adaptation to interest changes and the incorporation of contextual information to enable the development of context-aware agents.

The rest of the chapter is organized as follows. The problem of determining the content of user profiles is first addressed in Section 3.2. In Section 3.3, we review the state of the art in user profiling in the context of intelligent Web agents. Section 3.4 presents an overview of *WebProfiler* technique for acquiring user interests and learning profiles. Section 3.5 compares results of *WebProfiler* with those of more conventional algorithms applied to user profiling. Finally, a summary and conclusions of this chapter are given in Section 3.6.

3.2   What Profiles Should Model about Web Users?

To progress in the specification of a user profiling technique for information agents, the knowledge about users a profile should represent needs to be first discussed. The more personal agents know about users, the better they are expected to fulfill user information needs. Beyond this simply fact, there appears to be little consensus on three key issues of user profiling: (1) what information constitutes a profile; (2) how profiles are acquired; and (3) how the profile content is used to assist users.

Users usually have diverse information interests regarding, for example, their work (e.g. *finances*), their hobbies (e.g. *sports*) or current issues (e.g. *presidential debates*). Naturally, generalization across all of them is not effective from neither the predictive nor the representational point of view. Instead, interest topics have to be represented in separate categories and generalizations obtained from these categories.

In addition to modeling multiple interests in several domains, profiles have to model the different abstraction levels of such interests. For example, a user interest in *skiing* may respond to a more general interest in *winter sports*. This approximation to the real constitution of user interests suggests a hierarchical organization of topics according to different degrees of generality from the most general ones, those representing broad interests, to the most specific ones, those referring to particular aspects. This hierarchical organization enhances the semantic of profiles and allows having a temporal view of user interests since those at the top levels can be seen as long-term interests, while more specific ones can be seen as short-term interests.

Interest categories in the profiles are abstracted from the observation of individual experiences belonging to these categories such as Web pages users browsed through, news they read in on-line newspapers, articles they obtained from newsgroups and others. These experiences define categories in an extensional sense and at the same time provide both contextualization and high level of granularity to the description of user interests.

Even though all categories in a profile are interesting for the user to some extent, their relevance may vary according to the evolution of user interests. The changes in the relevance of categories during their lifespan can be gradual or abrupt and depend on both time and external circumstances. For example, after some time a user may start to gradually lose interest in articles related to some recent events (e.g. a presidential election) because of the natural process of forgetting. In parallel, some other interest categories may start receiving more attention.

The task of personal agents is not just choosing the right information for users, but also presenting this information at the right time. In consequence, in addition to knowing user information preferences and interests, personal agents are required to know user habits, routines and patterns of behavior regarding such interests. For example, the information an agent scanning news wires, retrieving breaking news and compiling personal newspapers can provide will be valuable for a user only early in the morning. If the agent misses the opportunity of presenting the personalized newspaper at that time, the user will reach the same information by other means.

Modeling of contextual information into user profiles enables pro-active and adaptable behavior of personal agents which will be able to predict and anticipate user information needs. In a context-aware setting, agents discover contexts and present information to the user accordingly. Indeed, pro-activeness is usually considered context triggered, i.e. an agent receives some context data and decides to inform the user or perform some action. Multiple contextual elements can be included in a user profile and effectively combined to enhance assistance, including location, time and activities.

To summarize, the aspects taken into consideration in the definition of *WebProfiler* for the content and structure of user profiles are the following:

- multiple and dynamic interest categories which can change over time, each referring to an individual and distinguishable user interest
- experiences of user interests populating the categories and exemplifying the user interests
- a hierarchical organization of categories denoting general to specific relationships among them
- a model of relevance over experiences and categories in the user profile representing the popularity of user interests and individual experiences
- the representation of one or more contexts related to categories allowing the identification of future similar contexts for recommendation

Figure 3.1 depicts the proposed knowledge and structure of user profiles which can be divided in two layers. In a first layer the content of the profile is placed,

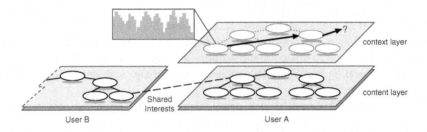

Fig. 3.1. The two-layers view of user profiles

consisting of several hierarchically organized categories. In a second layer the profile is augmented to model the knowledge about multiple contexts, such as temporal or activity contexts. By overlapping both layers, personal agents can provide both precise and timely information to users.

Even though different users are expected to have different world views and, consequently, different interest hierarchies, agents are still expected to reconcile such heterogeneous views in order to cooperate in satisfying information needs. If common interests exist among users, links between these interests can be potentially established at the content level of profiles, fostering information exchange between agents.

3.3   User Profiling in Web Agents

The user profiling approaches personal information agents are based on are typically drawn on work from the Information Retrieval (IR) and Machine Learning (ML) communities, since both communities have explored the potential of established algorithms for user modeling purposes [3, 34]. However, the use current user profiling approaches have made of techniques in these paradigms has a number of shortcomings [11].

In user profiling approaches based on IR techniques, both documents and user interests are represented as vectors of weighted terms according to the Vector Space Model (VSM) [29]. Interests are represented by either a single vector embracing all preferences or, more frequently, multiple vectors representing user interests in several domains (e.g. *sports*, *movies*, etc.). Even though multiple vectors allow agents to closely track user interests, the effectiveness of profiles depends on the degree of generalization to represent such interests. Fine-grained profiles can be achieved at expenses of increasing the number of vectors, whereas coarse-grained profiles require fewer vectors to represent the same interests. In the first case, the effectiveness of profiles can be high, but also its complexity, while in the second case the effectiveness of profiles is limited since several interests co-exist in individual vectors.

In addition to the level of generalization to represent user interests, a problem related to the application of statistical keyword analysis in user profiling is that words belonging to interesting documents are analyzed in isolation from any contextual information. This affects the completeness of user profiles, preventing agents from providing users with context-relevant information. In other words, documents are considered as independent entities without taking into account the situation the user is actually experiencing, such as the topics of recently consulted documents.

User profiling approaches based on ML techniques generate a predictive model using a learning algorithm. In this case, agents are engaged in a supervised learning task which requires positive and negative examples of user interests in order to identify their salient features and classify new examples, i.e. the acquisition of profiles is seen as a text classification problem in which a classifier is trained to recognize documents belonging to one or more categories. Machine learning algorithms which have been used for user profiling include naïve Bayes, k-NN, genetic algorithms and neural networks.

Even though user profiling appears to be a good candidate for straightforward application of standard ML techniques, it states a number of challenges to traditional algorithms in this area [34, 39, 24]. The main difficulty algorithms have to face is the computational complexity arising from the amount of features, words or phrases, to be considered in conjunction with the necessity of efficient, on-line user profiling. This rules out many computationally expensive learning algorithms which are only suitable for off-line scenarios.

The interaction of agents and users over time also imposes some constrains over the learning method to be applied, since not only the learning process should be incremental, but also algorithms need to be capable of adjusting models to changes in user interests and, consequently, in the target concepts. From a ML perspective, this is a difficult issue known as concept drift [35]. To cope with this problem, dual user models consisting of both a short-term and a long-term model of user interests [4] as well as adaptation as result of genetic evolution [22] have been considered in the literature. However, in most personal agents adaptation has been limited to the incorporation of new information, preserving past interests untouched. Mitchell [20] proposed to learn the user interests from only the latest observations using a time window, whereas a gradual forgetting function was introduced by [34]. The forgetting mechanisms proposed in current approaches simply forget old examples according to their age without taking into account their utility (e.g., if they provided good recommendations).

Finally, little attention has been paid to the assessment of explicit, comprehensible models of user interests feasible to be interpreted by users and other agents. It has been argued that the lack of transparency in recommender systems may result in a loss of trustworthiness, because humans need to feel they are in charge and this requires the system to be understandable by them [15]. However, most personal agents generate predictive models which act as black boxes to users. Even though ML algorithms provides a clear and unambiguous semantics of their output formats, the implicit representations of user interests can be difficult to understand by non-experts users. In order to be truly useful, a user profile needs to be readable by users so that they can explore their profiles and verify their correctness. Moreover, since user profiles are a starting point for the creation of user communities based on shared interests, meaningful user profiles can be easily compared in the search for common interests in groups of users.

In recent years, ontology-based user profiling approaches have been proposed to take advantage of the knowledge contained in both general-purpose and domain ontologies instead of acquiring and modeling user profiles from scratch [9, 18, 31]. In these profiling approaches, user interests are mapped to concepts in a reference ontology by classifying the examples of user interests into the ontological concepts. Thus, profiles are represented in terms of which concepts from the ontology a user is interested in, irrespective of the specific examples of such interests.

The use of ontologies for the representation of user interests promises to close the semantic gap between former approaches, which make use of low-level features extracted from documents (e.g., words), and the more abstract, conceptual view users may have of their interests. In spite of this advantage, committing to a general ontology has a number of pitfalls. First, dealing with the high number of concepts most general-purpose ontologies comprise may become rather expensive for modeling a single user profile (e.g. classifiers are needed for every concept in the reference ontology). Second, ontologies often mirrors the ontology of common-sense of a user community, but they fail to capture highly specific concepts that may be interesting for individual users and the kind of documents users like to read in these concepts.

3.4 *WebProfiler* Technique

WebProfiler is a novel technique for modeling and learning the knowledge and structure of user profiles. The subsequent sections describe the different components of profiles, the algorithms for extracting and using each of these components and their empirical evaluation. Section 3.4.1 describes how examples of user interests are gathered by observation. Section 3.4.2 specifies how interest categories are identified and organized into a hierarchy. Section 3.4.3 presents the adaptation strategy used to deal with interest changes and Section 3.4.4 shows the integration of contextual elements in user profiles.

3.4.1 Collecting Interesting Experiences

Ideally, an agent should be able of acquiring user profiles without user intervention through non-intrusive observation of user behavior on the Web. In the observation process, agents can capture experiences regarding user interests such as Web pages a user read. From the analysis of these experiences emerges the knowledge to be modeled in user profiles.

Each experience encapsulates both the specific and contextual knowledge that describes a particular situation denoting a user interest in a certain piece of information. Experiences can be divided into three main parts: the description of the Web page content, the description of the associated contextual information, and the outcome of applying the experience in personalization. The first part enables agents to carry out content-based experience comparison for discovering and retrieving topical related information, whereas the second and third parts allow agents to act based on both the current user contexts and its own confidence in recommendations.

The content of Web pages is represented using a bag-of-words approach, in which each page is described by a feature vector in a space in which each dimension corresponds to a distinct term associated with a weight indicating its importance. The resulting representation of a Web page is, therefore, equivalent to a t-dimensional vector $d_i = \langle (t_1, w_1), ..., (t_t, w_t) \rangle$, where w_j represent the weight of the term t_j in the document d_i.

Experiences also describe the contextual information of the situation in which a Web page was deemed interesting, including the page address, date and time it was registered and the level of interest the user showed in the page according to some agent criteria. This level serves as an initial degree of confidence in the experience and is estimated by agents based on either explicit or implicit feedback mechanisms. In the first case, the interest in the page is given by the user according to some quantitative scale. In the second case, the user interest is deduced by the agent based on a number of implicit interest indicators such as the time spent in a page, scrolling, etc.

Finally, experiences also keep track of the patterns of received feedback regarding actions undertaken based on the knowledge they provide. Basically, each experience has an associated relevance which is a function of the initial interest of the experience, the number of successful and failed decisions an agent takes based on this experience and the time that passes from the moment in which the experience was captured or used for the last time.

3.4.2 Inferring User Interest Hierarchies

WebProfiler builds a user interest hierarchy starting from experiences representing user interests. In order to describe user interests, the proposed user profiling technique is built upon a clustering algorithm, named *WebDCC* (Web Document Conceptual Clustering) [12], with structures and procedures specifically designed for user profiling. The unsupervised learning process provided by this algorithm offers to the profiling task the advantage that categories can be incrementally discovered from scratch.

WebDCC is an algorithm belonging to the conceptual clustering paradigm that carries out incremental, unsupervised concept learning over Web documents. First introduced by [17], conceptual clustering includes not only clustering, but also characterization. This form of clustering refers to the task of, given a sequential presentation of examples and their associated descriptions, finding clusters that group these examples into concepts or categories, a summary description of each concept and a hierarchical organization of them [33].

Experiences representing user interests that agents can capture through observation are presented to *WebDCC* algorithm, which is concerned with forming hierarchies of concepts starting from them. Hierarchies of concepts produced by this algorithm are classification trees in which internal nodes represent concepts and leaf nodes represent clusters of experiences. The root of the hierarchy corresponds to the most general concept, which comprises all the experiences the algorithm has seen, whereas inner concepts become increasingly specific as they are placed lower in the hierarchy, covering only subsets of experiences by themselves. Finally, terminal concepts are those with no child concepts but clusters.

Hierarchies are composed of a set of concepts, $C = \{c_1, c_2, ..., c_n\}$, which are gradually discovered by the algorithm as new experiences become available. In order to automatically assign experiences to concepts, a text-like description given by a set of weighted terms denoted $c_i = \langle (t_1, w_1), ..., (t_m, w_m) \rangle$, is associated to concepts during the process of hierarchical concept formation. Leaves in the hierarchy correspond to clusters of experiences belonging to all the ancestor concepts. In general terms, a set of j experiences or documents belonging to a concept c_i and denoted $D_i = \{d_1, d_2, \ldots, d_j\}$, is organized into a collection of k clusters below c_i, $S_i = \{s_{1i}, s_{2i}, \ldots, s_{ki}\}$, containing elements of D_i such that $s_{li} \cap s_{pi} = \emptyset$, $\forall l \neq p$.

Hierarchies describing user interests are incrementally built by characterizing gradually discovered categories. In this characterization process, *WebDCC* generates text-like descriptions of categories by observing the common features of experiences in these categories and those a novel experience should have in order to belong to them. This kind of descriptions facilitate both interpretation since they are easy to understand for users, and classification as standard text classification methods can be applied to classify novel experiences in the existing categories.

WebDCC builds a hierarchical set of classifiers, each based on its own set of relevant features, as a combined result of a feature selection algorithm for deciding on the appropriate set of terms at each node in the tree and a supervised learning algorithm for constructing a classifier for such node. In particular, an instantiation of Rocchio [28] algorithm is used to train classifiers, the same used in [7], with parameter fixed to $\alpha = 1$ and $\beta = 0$, yielding:

$$p_{c_i} = \frac{1}{n_{c_i}} \sum_{d \in D_{c_i}}$$

as a prototype for each concept $c_i \in C$. Hence, a classifier for a concept c_i is the plain average or centroid of all experiences belonging to this category.

WebDCC integrates classification and learning by sorting each experience through the concept hierarchy and simultaneously updating it. Upon encountering a new experience, the algorithm incorporates it below the root of the existing hierarchy and then recursively compares the experience to each child concept as it descends the tree. Below a given concept, the algorithm considers incorporating the experience into a cluster belonging to this concept as well as creating a new singleton cluster.

In other words, at the bottom levels of the hierarchy the algorithm performs a centroid-based clustering. Given the cluster s_{ji} belonging to the category c_i, which is composed of a set of documents and their corresponding vector representations, the centroid vector $p_{s_{ji}}$ is the average point in the multidimensional space defined by the cluster dimensions. When an experience has reached a given concept in the hierarchy, either because it is a terminal node or because it cannot be further classified down, it is placed in the most similar cluster below this concept. In order to predict which this cluster is, the closest centroid is determined by comparing the new experience with all centroids of the existing clusters using the cosine similarity.

The incorporation of experiences to the hierarchy is followed by an evaluation of the current hierarchical structure in order to determine whether novel concepts can be created or some restructuring is needed. In this step, meaningful concepts can be extracted to refine the hierarchy and previously discovered concepts may be reorganized. Thus, every time a novel concept is defined in the hierarchy, its structure is revised in order to take into account the recently acquired knowledge.

The formation of concepts in *WebDCC* is driven by the notion of conceptual cohesiveness. Highly cohesive clusters are assumed to contain similar experiences belonging to a same category, whereas clusters exhibiting low cohesiveness are assumed to contain experiences concerning to distinctive aspects of more general categories. In this situation, a concept summarizing the cluster is extracted, enabling a new partitioning of experiences and the identification of sub-categories.

Intra-class similarity or cohesiveness is defined in terms of how well individual experiences match the prototypical description given by the centroid of the cluster they are assigned to. It is assumed that the ability to classify experiences and make inductive inferences increases with the similarity of the experiences to the prototypes. The method used to compute the cohesiveness of a cluster s_r is the average pairwise similarity of experiences in this cluster, i.e.

$$\frac{1}{|s_r|^2} \sum_{d_i, d_j \in s_r} sim(d_i, d_j) = \|p_{s_r}\|^2 \qquad (3.1)$$

Every time a novel concept is defined *WebDCC* applies three local operators in order to reduce the effect of example ordering during learning and produce better structured hierarchies for clustering and classification. These operators are merging, splitting, and promotion of concepts in the hierarchy.

Merging and promotion operations are performed according to a criterion of similarity between concepts which is assessed through the overlapping descriptions of these concepts. Merging takes place when the novel concept incorporated to the hierarchy overlaps its description with one of more concepts in the same hierarchical level, whereas a concept is promoted if it is more similar to its grandparent than to its parent. Finally, splitting takes place when a concept is no longer useful to describe experiences in an interest category and then it is removed.

WebDCC aims at finding meaningful categories or clusters to summarize experiences. Like other inductive tasks, clustering solutions serve two different purposes, namely, prediction and description. In the context of user profiling, description focus on finding comprehensible user profiles, whereas prediction on guiding agent actions. This section details experiments we carried out to evaluate the *WebDCC* clustering solutions, whereas the recommendation approach is evaluated in Section 3.4.3.

Evaluation of personal agents and personalization systems in general is a difficult issue since it involves purely subjective assessments. Most datasets are assembled to evaluate learning algorithms and, consequently, do not provide relevance judgment of users about documents. A especially suited dataset for evaluating learning from relevance feedback is the *Syskill&Webert* Web Page Ratings dataset[1], since it contains the rating of a single user about the interestingness of individual pages.

Syskill&Webert Web pages belong to four different categories: *Bands* (61 pages), *Goats* (136 pages), *Sheep* (70 pages) and *BioMedical* (65 pages). In addition, a single user manually rated each page in a three point scale: *hot* or very interesting (93 pages), *medium* or quite interesting (11 pages) and *cold* or not interesting at all (223 pages). From the original collection we removed empty pages, *Not Found* pages and those which have not assigned rating, leading to a total of 327 pages.

In order to evaluate *WebDCC*, we compared its performance in clustering Web pages with the performance of other clustering algorithms, including agglomerative and divisive hierarchical clustering approaches. For this purpose, Web pages in the *Syskill&Webert* collection were transformed into frequency vectors after removing stop-words and applying Porter stemming algorithm. For agglomerative clustering, three variations were included in the comparison corresponding to the single-link (*slink*), complete-link (*clink*) and group-average (*upgma*) cluster similarity measures; whereas six variations of divisive clustering were included in the comparison corresponding to two internal ($\mathcal{I}_1$ and $\mathcal{I}_2$), one external ($\mathcal{E}_1$), two hybrid ($\mathcal{H}_1$ and $\mathcal{H}_2$) and one graph-based ($\mathcal{G}_1$) criterion functions as defined in [38].

Table 3.1 summarises the performance of *WebDCC* by the minimum, maximum and average entropy achieved in ten algorithm runs. Entropy is a external validity measure that evaluates the homogeneity of clusters in regards to the classes examples belong to. *WebDCC* hierarchies are evaluated at two different level of abstraction considering those clusters at terminal concepts as forming a single cluster and considering all leaf clusters. These two entropy values as well as the average number of concepts extracted in the ten algorithm runs are summarized in the table.

WebDCC outperforms most agglomerative and divisive solutions when approximately the same number of clusters is considered, whereas the minimum entropy is smaller than any result obtained with the hierarchical approaches. The hierarchy corresponding to the minimum entropy value is depicted in Figure 3.2(a). This experiment shows that *WebDCC* is able to obtain clustering results that are comparable to the results of more computational expensive, non-incremental algorithms.

WebDCC, on the other hand, generates readable descriptions of the hierarchical solutions. Figures 3.2(a) and (b) depict two hierarchies obtained when running the algorithm. The different orders of experiences caused the algorithm to generate different hierarchies. In some cases, more concepts than the ones

---

[1] http://kdd.ics.uci.edu/databases/SyskillWebert/SyskillWebert.html

Table 3.1. Comparison of *WebDCC* with hierarchical approaches

k	$\mathcal{I}_1$	$\mathcal{I}_2$	$\mathcal{E}_1$	$\mathcal{H}_1$	$\mathcal{H}_2$	$\mathcal{G}_1$	slink	clink	upgma
	Agglomerative and Divisive Algorithms								
4	0.229	0.210	0.336	<u>0.171</u>	0.222	0.235	0.946	0.660	0.946
50	0.108	0.077	0.115	<u>0.076</u>	0.105	0.088	0.803	0.081	0.085
60	0.098	0.069	0.112	0.076	0.098	<u>0.055</u>	0.763	0.076	0.068
65	0.090	0.068	0.109	0.072	0.096	<u>0.055</u>	0.751	0.071	0.064
70	0.088	0.067	0.108	0.072	0.093	<u>0.051</u>	0.732	0.070	0.057

		WebDCC Algorithm				
# clusters	65.4±7.3	Entropy	0.075±0.025	Min	0.027	Max	0.106
# leaves	208.2±15.3	Entropy	0.016±0.005	Min	0.009	Max	0.027
# concepts	4.1±1.3					

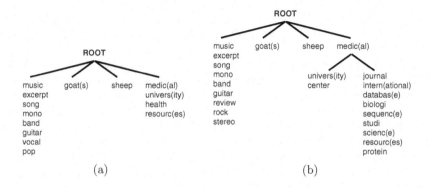

(a)   (b)

Fig. 3.2. Hierarchies for the *Syskill&Webert* collection

existing in the target hierarchies were obtained. Nonetheless, this does not mean than the additional concepts are incorrectly extracted, but they usually belong to more specific categories existing in certain categories of the collections. For example, the hierarchy in Figure 3.2(b) has two sub-categories inside *BioMedical*, one grouping Web pages related to universities and one grouping Web pages about international journals.

3.4.3 Handling Changes in User Interests

WebProfiler tries to capture user interests and represent them into a conceptual hierarchy using the previously described clustering algorithm. However, the interaction of users with their agents may extend over long periods, during which the user interests cannot be assumed to be invariable. By learning from relevance feedback, agents are able to capture novel user interests as well as track changes in already known interests.

In the presence of an interest change, agents have to incrementally gather evidence before they can adapt user profiles. Until detecting a change, agent predictions refer to the user past interests and their quality may not be optimal. Interest changes can be characterized as gradual or abrupt, sometimes referred to as interest drifts and interest shifts, respectively. Interest changes in information domains usually take place gradually, then the relative error caused by the learning delay is small. The error caused by an abrupt change in the same period of time is higher and is usually induced by events.

In order to cope with drifting interests, *WebProfiler* introduces the notion of relevance for experiences, which is directly connected to their aptitude to generate successful recommendations [14]. The relevance of experiences changes over time according to the evolution of user interests. It is controlled using relevance feedback and a further important dimension in user profiling such as the time. In those cases in which both positive and negative feedback is available, the adaptation of profiles can be performed using only this information. However, time-forgetting becomes extremely valuable in situation where only positive feedback can be obtained as it becomes the unique means to forget irrelevant experiences.

WebProfiler deals with the interestingness of individual experiences by maintaining an indication of their relevance in representing a given information preference. In this aspect, the proposed technique diverges from most user profiling approaches based on machine learning algorithms. Personal agents such as *Syskill&Webert* [26] or *NewsDude* [4] treat training experiences equivalently. Likewise, forgetting mechanisms for recommender systems proposed in the literature forget old experiences based on either adaptive or fixed time windows or time-based forgetting functions, but consider remaining experiences equally important during learning of user profiles.

In *WebProfiler* the relevance of experiences is used as a mechanism for both gaining confidence in experiences for recommendation as well as forgetting experiences. In a profile, each experience e_i is attached with a relevance value denoted rel_i, i.e. profiles consist of pairs $\langle e_i, rel_i \rangle$ where e_i is the user experience encoding the Web page the user found interesting, and rel_i represents the evidence about the interest in that experience, confined to the $[0, 1]$ interval. Initially, rel_i is calculated as a function of the relevance an agent was able to perceive about the experience based on the mechanism implemented to obtain feedback from users.

The lifespan of experiences in the user profile is controlled by their relevance, which combines the validity of the experience according to its age and to the feedback received. If the relevance of an experience drops below a certain value, the agent loses confidence in the ability of that experience to both represent an interest category and predict the interest of new information items. Therefore, the experience is removed since it is likely to degrade the effectiveness of the overall profile.

Relevance-Based Recommendation. In order to evaluate whether a new page should be recommended, agents search the user profile for similar experiences assuming that the user interest in a new item will be close to the interest in similar items read in the past. The comparison between experiences and items to be recommended is performed across a number of dimensions that describes them. A similarity function sim_i needs to be defined for each of these dimensions, being the most important the one that measures the similarity between the item contents, which is estimated using the cosine similarity.

To obtain the global similarity between an item to be recommended r_j and a retrieved experience e_i a numerical evaluation function combines the matching of each dimension with the importance value assigned to that dimension as follows [16]:

$$S(e_i, r_j) = \frac{\sum_{k=1}^{n} w_k * sim_k(f_k^{e_i}, f_k^{r_j})}{\sum_{k=1}^{n} w_k} \qquad (3.2)$$

where n is the number of dimensions, w_k is the importance of the dimension k, sim_k is the similarity function for this dimension, and $f_k^{e_i}$, $f_k^{r_j}$ are the values for the feature f_k in the experience e_i and the item to be recommended r_j respectively.

To recommend a Web page, agents retrieve similar experiences in the user profile to obtain a set of the best matching past experiences. Hierarchical categories in the profile bias the search toward the most topical relevant experiences. To assess the confidence in recommending r_i given the experience e_k, a weighted sum of the confidence value of each similar retrieved experience is then calculated as follows:

$$conf(r_i) = \frac{\sum_{k=1}^{n} w_k * rel_k}{\sum_{k=1}^{n} w_k} \qquad (3.3)$$

where n is the number of similar experiences retrieved, rel_k is the relevance in the profile of the experience e_k and w_k is the contribution of each experience according to its similarity. This method to estimate the confidence in a recommendation is based on the well-known distance-weighted nearest neighbor algorithm [19].

Each experience has a weight w_k according to the inverse square of its distance from r_i, $\frac{1}{(1-S(e_k, r_i))^2}$, where $S(e_k, r_i)$ is the similarity between the item to be recommended r_i and the experience e_k. Thus, the more similar and relevant items are the more important for assessing the confidence in a recommendation. If the confidence value of recommending r_j is greater than a certain confidence threshold, which can be customized by users during their interaction with agents, the item is recommended.

Relevance-Based Forgetting. From the moment that users provide either explicit or implicit feedback about recommendations, agents start learning from their actions. If the result of a recommendation is successful, then agents learn from the success by increasing the relevance of the corresponding experiences in the profile and, possibly, incorporating new experiences. If the result of a recommendation is a failure, agents learn from the mistake by decreasing the relevance of the experiences that have led to the unsuccessful recommendation and eventually removing them.

In *WebProfiler* the changes in the relevance of experiences are controlled by the bipolar sigmoid function in Equation 3.4:

$$f(x) = \frac{2}{1 + \exp^{-\beta x}} - 1 \qquad (3.4)$$

The sigmoid function constraints the relevance values in the range of $[-1, 1]$ and captures the reluctance of user interests to change in the long run. It has been used for updating the weights of long-term descriptors of user interests in [36]. For experiments in this section we consider $\beta = 2$ so that an experience with the maximum level of perceive interest ($=1$) is added to the profile with a reasonable high relevance ($\simeq 0.76$), taking into account the uncertainty of making inferences about the user interest by observation.

The initial relevance value for experiences is calculated using the sigmoid function according to the interest a user has shown in the item, i.e.

$$rel_i^{new} = f \text{ (perceived interest in } e_i) \qquad (3.5)$$

where the perceived interest in e_i is in the range $[0, 1]$ since those experiences which perceived interest is less than zero are not included in the profile. Given a feedback f_k in the range $[-1, 1]$ that corresponds to a recommendation r_k made based on a set of experiences $E = \{e_1, e_2, \ldots, e_m\}$, the relevance of each of these experiences is updated as summarizes Equation 3.6:

$$rel_i^{new} = f\left(f^{-1}\left(rel_i^{old}\right) + f_k * w_i\right) \ \forall i : 1 \le i \le m \qquad (3.6)$$

where $f(x)$ is the bipolar sigmoid function in Equation 3.4 and w_i the weight of experience e_i.

Experiences are rewarded if their use results in successful recommendations and penalized otherwise. In both actions, the reward or penalization depends on the contribution of each experience in the recommendation, which was in turn a consequence of the similarity between the experience and the recommended item. Thus, positive and negative feedback is magnified so that less similar experiences to the recommended item are less affected by feedback than more similar ones.

In order to emulate the natural forgetting process of users losing interest in some topics, the relevance of experiences also varies according to time. Particularly, an experience relevance decreases by a constant value each time a new experience becomes available. In this case, the relevance of every experience in the profile is decreased as follows:

$$rel_i^{new} = f\left(f^{-1}\left(rel_i^{old}\right) - \lambda\right) \qquad (3.7)$$

Experiences which are no longer relevant either because they went out of use or because recommendations starting from them received too much negative feedback, are removed from the user profile. In other words, when an experience e_i reaches a negative relevance value, i.e. $rel_i \le 0$, it is completely forgotten since agents can not longer be confident in the experience for generating recommendations.

Each time an experience is forgotten from the profile, the conceptual hierarchy describing the user interests also requires adaptation. First, the experience is subtracted from the centroid vector describing the cluster it belongs to. Second, the description of all concepts defined in the path from the leaf category the experience belongs to, if exists, to the hierarchy root are also updated to guarantee that each category summarizes all experiences below it. As classifiers for categories are extracted from their centroid vectors after local feature selection, the updating of classifiers to forget an experience is limited to updating the corresponding classifier weights to subtract the experience. Third, if the experience being removed is the last one in a given category, the concept describing the category is also removed from the profile.

Interest Drift Scenarios. In order to evaluate the adaptation strategy of *WebProfiler* technique, we simulated a user interacting with an agent and obtaining recommendations. For each simulation the *Syskill&Webert* collection was divided into a training set used to learn the profiles and a testing set used to evaluate the results of recommendation. From the 104 pages rated as *hot* or *medium* in this collection, approximately 70% (72 pages) were used to incrementally build a user profile. The remaining 30% rated as *hot* or *medium* not used in profile construction as well as all pages rated as *cold*, were used to test the performance of the learned profile in predicting the relevance of Web pages. For experimentation, ratings of Web pages in the collection were mapped to numerical values in order to update the relevance of experiences. The feedback f_i of an experience e_i is defined according to its rating as follows: $f_i = 1$ if the page is *hot*, $f_i = 0.5$ if the page is *medium* and $f_i = -1$ if the page is *cold*.

For testing the adaptation of user profiles to drifting interests, we simulated interest drift scenarios using the Web pages and categories of the *Syskill&Webert* collection. These scenarios reflect the behavior of users interacting with agents to obtain information regarding their interests. In each scenario, Web pages in the training set were randomly split into 10 batches of similar size maintaining the proportion of examples in each category. Hence, several simulated user feedback behaviors were specified over each batch in order to configure the different scenarios. Table 3.2 shows the relevance of the four categories in each batch for two simulated scenarios. Initial relevance of experiences and subsequent feedback is based on these values.

In the first scenario, a gradual change of user interests is simulated by inverting the relevance of *Bands* and *BioMedical* categories. Initially, pages about *Bands* were considered relevant whereas all other pages were considered irrelevant. This situation changes slowly from batch 4 to 6, in which *BioMedical* starts to be relevant and all other pages irrelevant. Figure 3.3(a) shows the variation in accuracy of recommendations as batches were processed in this scenario when no feedback is considered, when both positive and negative feedback are obtained and when different values for λ parameter in Equation 3.7 are considered.

Table 3.2. Relevance of categories in the different scenarios

		Batches									
		1	2	3	4	5	6	7	8	9	10
		Scenario I									
Bands		1	1	1	0.5	-0.5	-0.5	-1	-1	-1	-1
Goats		-1	-1	-1	-1	-1	-1	-1	-1	-1	-1
Sheep		-1	-1	-1	-1	-1	-1	-1	-1	-1	-1
BioMedical		-1	-1	-1	0.5	0.5	0.5	1	1	1	1
		Scenario II									
Bands		1	1	1	1	1	-1	-1	-1	-1	-1
Goats		-1	-1	-1	-1	-1	1	1	1	1	1
Sheep		-1	-1	-1	-1	-1	1	1	1	1	1
BioMedical		1	1	1	1	1	-1	-1	-1	-1	-1

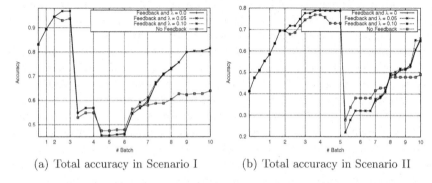

(a) Total accuracy in Scenario I   (b) Total accuracy in Scenario II

Fig. 3.3. Results of adaptation based on feedback and time forgetting

Forgetting according to time is a slow but constant process that enables agents to forget those experiences which are inactive for long periods. A value of $\lambda = 0$ means that no change in the relevance of experiences takes place since no movement is given over the abscissa of the sigmoid function. Instead, a value of $\lambda > 0$ leads to some decrease in the relevance of the whole set of experiences in a profile. Thus, the higher the value of λ, the faster experiences are forgotten.

The global accuracy falls after batch 3 in scenario I because pages about *BioMedical* start to be interesting, but they are not yet recommended. As experiences of batch 4 are included in the profile, accuracy starts to recover slowly. However, a second change in user interests occurs in batch 5 causing a second fall in accuracy. In this case, pages about *Bands* start to receive negative feedback, but the category is still relevant in the profile and then pages about *Bands* continue being recommended.

Figures 3.4(a) and (b) compares the evolution of accuracy for *Bands* and *BioMedical* categories individually. In the *Bands* category, accuracy recovers from the interest drift very quickly, reaching the perfect accuracy even before the

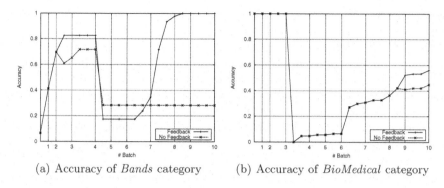

(a) Accuracy of *Bands* category   (b) Accuracy of *BioMedical* category

Fig. 3.4. Accuracy of individual categories in scenario I

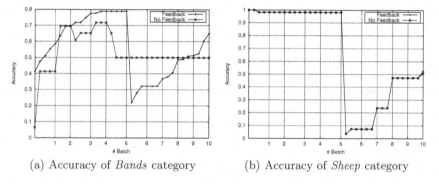

(a) Accuracy of *Bands* category   (b) Accuracy of *Sheep* category

Fig. 3.5. Accuracy of individual categories in scenario II

last batch. Indeed, pages about *Bands* are no longer recommended in batch 10. In contrast, the accuracy of *Bands* category when not feedback is learned remains stable after falling in batch 5. No further experiences belonging to this category are included in the profile, but the accumulated experiences are not forgotten either. Therefore, pages about *Bands* continue being erroneously recommended.

Experiences in the *BioMedical* category start to be learned and produce recommendations in batch 4. Before that batch the accuracy is optimal because no page about this subject is recommended and it is not interesting to the user. Afterward, accuracy of recommendations increases as novel experiences are incorporated in the profile. Learning from positive feedback causes a faster increase in the relevance of this category and then a more significant improvement in accuracy in the last batches.

In the second scenario, user interests in *BioMedical* and *Bands* shift abruptly in batch 5 to *Goats* and *Sheep*. This scenario simulates user interests that, possibly because of external circumstances, suffer a sudden change. Figure 3.3(b) shows the behavior of the global accuracy in this scenario. From its break-down after the interest shifts in batch 5, the predictive accuracy of recommendations shows a steady recovery. Nevertheless, more batches are still needed to reach the levels of accuracy prior to the interest shifts.

Figure 3.5(a) and (b) depict the accuracy of *Bands* and *Sheep* categories. For *Bands*, the influence of feedback is essential to recover accuracy. If no feedback is provided, accuracy remains unchanged over time, causing a detriment in performance of the user profiles in the long run. For *Sheep*, on the other hand, the influence of feedback is only noticeable in the last batch.

In both scenarios gradual forgetting results in an improvement of the average predictive accuracy. Particularly, the faster recovery is noticeable in the intervals in which a given previously relevant category is no further interesting to the user in the scenario. In these intervals, prediction is higher until the same effect is reached through negative feedback. For example, between batches 6 and 8 in Figure 3.3(a) the profile is being adapted to forget the user interest in *Bands*. The time forgetting mechanisms accelerate this process, but in batch 8 the accuracy is restored as consequence of the strong impact of negative feedback. Figure 3.3(b) depicts the same process for the *BioMedical* category in scenario II.

The forgetting of experiences can be the only mechanism some agents will be able to use in adapting profiles, since many agents do not consider negative feedback, but only positive one. Indeed, it has been stated that inferring negative feedback is difficult and is prone to misinterpretations [30]. Likewise, in some situations agents received only a few relevance judgments as users tend to be unwilling to provide them [37]. In this case, user interests which do not receive further positive feedback remain in the user profile until being slowly forgotten because of time.

3.4.4 Modeling Context for User Interests

The main goal of personal information agents is to present relevant information to users based on the knowledge of their interests. However, agents have to be also aware of the user context in order to provide this information in the time and in the place it is more relevant for users. In *WebProfiler*, the representation of profiles is augmented to consider the temporal and activity context of users. Thus, data about time and activities in regards to individual user interests is explicitly modeled in user profiles enabling agents to tailor their assistance according to the user situation.

Temporal Context Modeling. Making recommendations demands predicting what is of interest to a user at a specific point in the time. In general, users have certain repetitive behaviors over time such as reading the newspaper in the morning or looking for travel information before holidays. This repetitive and, therefore, predictable behavior allows agents to anticipate future information needs matching previously observed regularities.

In addition to the continuous interpretation of time and the time-stamp itself, the time dimension can have multiple aggregations that are overlapping such as time of the day, day of the week, time of the week, months, and seasons. Each of these attributes entails a number of conditions that are related to the context the interests of users are attached to. For instance, most jobs are on weekdays but not on weekends, then the day of the week may determine when certain activities take place.

In *WebProfiler* experiences are augmented with the information about the temporal context in which they were captured. This context is specified by a number of attributes for which the current value can be determined at the moment the user-agent interaction takes place, including: *time of the day* (morning, afternoon or night), *day of the week* (the seven days of the week are valid values), *time of the week* (weekdays or weekend), *month* (the twelve months of the year) and *season* (winter, spring, etc.). The choice of suitable attributes and their granularity depends on the type of assistance which is expected from agents and the application domain.

By annotating user interest experiences with the temporal context in which they were collected or used, simple statistic data can be used for reasoning about time. For example, it is feasible to calculate the frequent time-contexts associated to categories in the profile in order to recognize them in the future and determine whether a recommendation is contextually relevant based on past experiences. Thus, agents can discover access routines regarding each user interest, which in turn can be used to provide timely recommendations in a pro-active fashion.

In order to extract frequent time-contexts for user interests, concepts and clusters in the user profile have an associated summary of the time-contexts their experiences were located in. For each attribute, this summary specifies a probability distribution over the attribute values given the category. By taking advantage of the hierarchical organization of user interests, a category in the hierarchy possesses the probabilistic summarization of all its descendants and context patterns can be extracted at different levels of specificity.

For an attribute $t_i \in T$ and a given value $v_{ij} \in V_i$, the larger the probability $P(t_i = v_{ij} \mid S)$ the greater the proportion of members of S sharing the same value for this attribute and, therefore, the more predictable this value is of other time-relevant information. The predictive power of a given attribute t_i in the set of temporal attributes T is then defined as follows:

$$\sum_{v_{ij} \in V_i} P(t_i = v_{ij} \mid S)^2 \tag{3.8}$$

which can be seen as the intra-class similarity of the experiences in the cluster S regarding the attribute $t_i \in T$. The higher this probability, the more chances experiences belonging to the cluster have to be interesting to the user at the time the value of the attribute specifies. Thus, routines involving single attributes can be extracted from a cluster by imposing a threshold over the intra-class similarity an attribute should exceed in order to be considered as representative of a user behavioral pattern.

Rules are obtained such as the antecedent of a rule is a pair attribute-value and its consequent is the cluster or category the pattern is defined for, i.e.

$$(t_i = v_{ij}) \Rightarrow^{p_i} c_k$$

This rule can be interpreted as, if the temporal attribute t_i takes the value v_{ij}, the category c_k is contextually relevant to the user with a probability of p_i. To extract these rules the algorithm iterates over the attributes in T and, for those with predictive power exceeding a threshold ε, its most probable value is used as antecedent of a rule.

Finally, to extract more complex patterns regarding all attributes in T or combinations of them, the expected number of attribute values that can be correctly guessed for a member of S is given by

$$\frac{1}{|T|} \sum_{t_i \in T} \sum_{v_{ij} \in V_i} P\left(t_i = v_{ij} \mid S\right)^2 \tag{3.9}$$

This utility measure [10] is used by COBWEB clustering algorithm to guide classification and tree formation to predict the number of attributes that can be correctly guessed for a given partition of examples.

The combinations of two or more attributes exceeding the intra-class similarity evaluated in Equation 3.9 can be considered a relevant time-context for recommendation. Thus, the most significant single attributes are extracted using Equation 3.8 and combinations of them can be systematically tested using Equation 3.9 to extract more specific patterns.

Each time a new experience is incorporated or removed (forgotten) from the hierarchy, the validity of the rules involving the cluster the experience was added into as well as all their ancestor categories is verified. Thus, no longer supported rules are removed and new rules are added if new patterns are detected. If the contexts in which a user access information change (e.g. because the user changes its work shift), the addition and forgetting of experiences in the profile will cause the assumptions about previous patterns to be also updated.

To pro-actively recommend items based on context, the values of temporal attributes need to be checked at constant time intervals. If the current time matches the antecedent of some extracted pattern or rule, a set of candidate recommendations is analyzed in order to select those recommendations belonging to the category specified by the consequent of the rule.

The set of candidate recommendations can be constructed using diverse methods. The agent can either search the Web for pages matching the user profile during computer idle time and saved them for being recommended in the appropriate context or trigger a Web search when the current time matches a rule antecedent using the category in the consequent for obtaining similar documents.

In order to evaluate the extraction of frequent time-contexts we used the *Syskill&Webert* collection. In this collection, each Web page has not only the rate it received from the user (*hot*, *cold* or *medium*), but also the date and time in which the rated was given. From the 332 Web pages in the collection, 212 pages have the information about the moment they were rated by the user, including complete date and time. The earliest rate given to a page was on October 4 1995, whereas the latest was on June 24 1996, so that rates are distributed in a period of nine months.

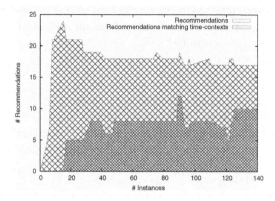

Fig. 3.6. Percentage of predictable recommendations

The mentioned 212 pages were divided into a training set of 148 pages (approx. 70%) and a testing set of 64 pages (approx. 30%). The training set was used to incrementally build a user profile using the 140 pages rated as *hot* or *medium*, while the test set was used to generate recommendations. The most frequent time contexts for the pages in the profiles were extracted using the information about the date and time in which each page was rated, i.e. the moment in which the page received the rate was used as the time-context of the experience in the profile. Hence, as pages were incorporated to the profile the frequent time-contexts for clusters and categories were found.

The aim of this experiment was to evaluate whether the algorithm is able of accurately infer the contextual relevance of experiences. To accomplish this goal we measured the percentage of recommended Web pages in the testing set that matched one or more of the extracted time-contexts using the date of their rate. If recommended pages are seen as the candidate set of recommendations, which elements have been previously identified as relevant to the user interests, the experiment measures the number of these recommendations whose relevance could have been anticipated by an agent because of their matching with the extracted time-contexts.

Figure 3.6 depicts the relationship between the number of recommendations and the number of these recommendations that matched some of the frequent time-contexts in the profile. In the figure, the area representing the recommended Web pages is overlapped by the area of the recommended Web pages that matched the time-contexts. The no overlapping area corresponds to Web pages that either do not belong to any of the clusters or categories the patterns were extracted for (i.e. there is no rule with the category as consequent), probably because the user has not a defined habit related to the category; or the recommendations do not matched the extracted patterns for the category they belong to.

It can be observed that after some training experiences the number of predictable recommendations in terms of their time-context starts to increase. For this experiment we extracted patterns for clusters having at least five experiences since no conclusive results can be extracted from an inferior size and we set the threshold for pattern extraction to $\varepsilon = 0.5$. Only the attributes time of the week and time of the day were considered for this experiment since there is only six month data about month and season.

Activity Context Modeling. In contrast to other contexts, the semantic of the activity context –information about the current ongoing task– can be rather complex. In *WebProfiler*, user profiles are gathered through observation of user browsing behavior and, therefore, browsing is considered the main activity of users. By activity it is understood a page visit which takes place during the course of browsing, while groups of these activities can be referred to as sessions.

Information agents can take advantage of the knowledge gained from observing user browsing in conjunction with long-term user interests to retrieve context-relevant information. If an agent detects the user is browsing through certain interest categories, it can anticipate the categories in which the user is likely to be interested in during the session. In user profiles, browsing habits are represented by association of the form $A \Rightarrow B$, where A and B are groups of categories and the association indicates that, if the user current activities include visiting pages about the categories in A, the next activities are likely to include visiting pages about B.

A browsing session is a set of page references that takes place during one logical period, e.g. from a log in to a log out of the browser. By identifying the session boundaries, it is ensured that the information collected from one session is within the same context, which provides a good foundation for inferring and applying context in recommendation. A session S_j is a list of pages a user accessed to ordered by time-stamp as follows:

$$S_j = \{(p_1, time_1), (p_2, time_2), \ldots, (p_n, time_n)\}$$

where $time_i$ is the time the user accessed the page p_i such that $time_i \leq time_j, \forall i \leq j$. Then, the user browsing activities are partitioned into a set of sessions $S = \{S_1, S_2, \ldots, S_k\}$ containing individual page references.

The process of segmenting the activity of a user into sessions is performed using a time-oriented heuristic in which a time-out establishes a period of inactivity that is interpreted as a signal that the session has ended. If the user did not request any page for a period longer than max_time (30 min. is used as default time-out) subsequent requests are considered to be in another session. In addition, the active session is finished when the browser is closed and a new session is started when the browser is re-opened.

The notion of session is further abstracted by selecting a subset of pages that are significant or relevant for analysis. Each semantically meaningful subset of pages belonging to a user session is referred to as a transaction. To identify semantically meaningful transactions, content pages are considered as those belonging to one or more categories in the profile, unlike content pages in other approaches which are identified simply based on the time spent on a page or on backtracking during the user navigation [6]. Web pages not belonging to any category are considered irrelevant for usage mining since they do no entail information about the user habits regarding interests. Then, a content-only transaction is formed by all the content pages in a session.

The resulting transactions are further divided using a time window approach, which divides each transaction into time intervals no longer than a specified threshold. This approach assumes that meaningful transactions have an overall average length associated with them. If W is the length of the time window, then two pages p_i and p_j are in the same session if $p_i.time - p_j.time \leq W$. In this way, the set of pages $P = \{p_1, p_2, \ldots, p_n\}$, each with its associated $time_i$, appearing in the set of sessions S are partitioned into a set of m transactions $T = \{t_1, t_2, \ldots, t_m\}$ where each $t_i \in T$ is a subset of P. The problem of mining association rules is defined over this collection of subsets from the item space where an item refers to an individual page reference.

To incorporate the knowledge of the user interests in pattern extraction, further processing of user activities is needed to map individual Web page references to one or more user interest categories. The enriched version of transactions leads to set of rules referring to categories instead of individual Web pages. To integrate content and usage data, each page p_i in a transaction t_j is considered to have an associated set of categories it belongs to, denoted $C_i = \{c_1, c_2, \ldots, c_p\}$, where C_i is extensionally defined by all the categories c_j in the path from the root of the hierarchy to the leaf cluster in which the page p_i was classified into.

If only the cluster a page belongs to is used to describe sessions, the discovered association rules will relate clusters but not categories. Instead, the inclusion of the ancestors in the path from the cluster the page was classified into until the root, makes it possible to find rules at different hierarchical levels. The result of replacing the elements of the transactions in T by categories in the user profile is a set of transactions $T' = \{t'_1, t'_2, \ldots, t'_m\}$ where each $t'_i \in T'$ is a subset of C.

The problem of mining generalized association rules in a database of transactions, potentially across different levels of the taxonomy, consists in finding all rules $X \Rightarrow Y$ that have support greater than a user-specified minimum support and confidence [32]. For each rule, the support threshold describes the minimum percentage of transactions containing all items that appear in the rule, whereas the confidence threshold specifies the minimum probability for the consequent to be true if the antecedent is true. *Apriori* [1] algorithm over the set of extended transactions obtained as pages are classified in the concept hierarchy allows to obtain such associations.

In the recommendation phase, the active session is compared with the discovered rules. If the active session matches the antecedent of an association rule, recommendations are finding by retrieving Web pages belonging to the categories in the rule consequent. A fixed-size sliding window is used over the active session to capture the current user activity. For a sliding window of size n, the active session ensures that only the last n visited pages influence recommendation. Experimentation about recommendation using the activity context of users can be found in [13].

3.5 Comparison with Other Profiling Approaches

In this section, *WebProfiler* is compared to several algorithms from machine learning and information retrieval fields which have been used for user profiling in different personal agents. Experiments regarding user profiling approaches basically involve filtering a set of documents using a learned profile and evaluating the filtering performance using standard measures. In order to calculate these measures, we employed the user ratings associated to Web pages in the *Syskill&Webert* collection.

For each individual trial, a randomly selected group of 229 pages (approx. 70%) were used as a training set and the remaining 98 pages (approx. 30%) were reserved for testing. Web pages were transformed into frequency vectors after stop-word remotion and stemming. The different learning algorithms were run to create a representation of user preferences and the resultant profiles were used to determine whether pages in the testing set can be considered as *cold, hot* or *medium* for recommendation. In order to make results of learning algorithms and *WebProfiler* comparable, the same sequence of examples was presented to all algorithms. Each value in the results averages ten trials across the different number of examples presented to the algorithms.

Figure 3.7(a) shows the performance of several algorithms over the frequency vectors. These algorithms include a decision tree algorithm such as *J48* (a *C4.5* implementation), explored in a Web agent such as *Syskill&Webert*, the instance based learner *IB1* [2], variations were explored in *Law* [25], naïve Bayes, *NewsDude* [4], *iFile* [27] and *Personal WebWatcher* [21] are agents that use this algorithm, *Rocchio* algorithm, also used in several agents like *Butterfly* [8] and *WebMate*, a radial basis function (RBF) network, neural networks were also explored in *Syskill&Webert*, and *JRip* which implements the propositional rule learner RIPPER used in *OyS-TER* [23] to induce user models describing interests in Web pages.

The superior results obtained with *WebProfiler* are due to several reasons. First, all algorithms used for comparison tried to learn a model using both positive and negative evidence of user interests in the same way they were applied to the aforementioned agents, i.e. the models involved learning the three classes *cold, hot* and *medium*, whereas *WebProfiler* learns only interesting categories. Second, *WebProfiler* uses user feedback for updating the relevance of the experiences in the model. Third, for prediction *WebProfiler* considers the relevance of experiences instead of treating them as equally important to determine the relevance of a new example.

Further disadvantages of the experimented algorithms are that some of them are non-incremental so that the model needs to be completely re-built after adding each new experience (e.g. decision trees). Then algorithms become computationally expensive for being used in user profiling. In addition, most algorithms require a controlled vocabulary in order to lean a model which needs to be either established beforehand or extracted as a previous step to learning. Finally, the most important disadvantage of these algorithms is their lack of adaptation to changes in the user interests. If a change in the interests takes

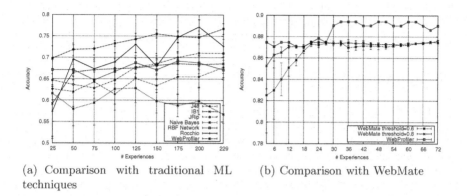

(a) Comparison with traditional ML   (b) Comparison with WebMate
techniques

Fig. 3.7. Performance of *WebProfiler* and different algorithms in user profiling

place the examples regarding the old and new interests can reduce the model ability to predict the relevance of new examples.

In a qualitative analysis, user profiles generated with machine learning algorithms depend on the formats of learning results which are specific to each algorithm (decision trees, probabilities, etc.). Even though some of these algorithms provide a clear and unambiguous semantics of their output formats, these representations can be difficult to understand to non-expert users. For example, when using the *J48* and *JRip* algorithms user profiles are decision trees or a set of rules. Moreover, the interpretation of results from *IB1* and naïve Bayes algorithms is even harder since a profile in *IB1* is simply the set of training experiences whereas naïve Bayes only reports the statistics it collects during training. *WebProfiler*, on the other hand, builds conceptual hierarchies as the ones shown in Figures 3.2(a) and (b) to represent the same examples.

The proposed user profiling technique was also compared with the profiling mechanism of *WebMate*[2] [5], an agent that monitors user browsing behavior and learns user interests. *WebMate* categorizes document vectors, using a *tf-idf* weighting scheme, into a fix number of categories which are learned automatically by the agent, assuming that a user has at the most N interest domains. For each positive example (i.e. a Web page that the user has marked as *I liked it*), the agent calculates the similarity with the current interests, combines the page to the most similar vector in the profile or creates a new vector if the maximum value N has not been reached.

Like *WebProfiler*, *WebMate* learns exclusively from positive examples. Then, we used the training and testing sets of the *Syskill&Webert* collection for experimentation. To overcome the ordering effects, results average 10 trials with different experience ordering. In Figure 3.7(b) it is possible to see that *WebProfiler* outperforms *WebMate* in accuracy of recommendations. Other aspect to observe in this figure is the high deviation in accuracy values caused by the order in which the N

[2] http://www-2.cs.cmu.edu/~softagents/webmate/

first examples are presented. Moreover, it is possible to infer that this agent will perform poorly if more than N classes have to be modeled since more than one class or user interest will be represented by the same vector in the profile.

3.6 Summary

User profiles are the basis of personalized assistance and, consequently, the capacity to acquire, model and maintain representations of user interests is a crucial component of personal agents. The knowledge about users described in their profiles enables agents to predict the relevance of new, unseen pieces of information as well as to discover new information matching the user interests proactively. Thus, agents become a valuable tool to relieve part of the burden of finding and exploring new information on the Web.

In this context, we have focused our attention on building user profiles starting from observation of user readings and behavior on the Web. A novel user profiling technique, named *WebProfiler*, which enables agents to build user profiles by incrementally specifying a hierarchy of user interests has been presented. The characteristics that distinguish user profiling from related tasks, such as text learning in a broad sense, have been thoroughly analyzed and addressed in the definition of this technique.

WebProfiler provides a step towards the assessment of more comprehensible, semantically enhanced user profiles, which application can lead to more powerful agents capable of accurately identifying user interests, adapting their behavior according to interest changes and providing assistance based on the context of users. In addition, this technique opens new possibilities regarding the interaction of users with their profiles as well as the collaboration with other agents at a conceptual level.

Acknowledgments

This research has been partially supported by ANPCyT PICT No. 34917.

References

1. Agrawal, R., Srikant, R.: Fast algorithms for mining association rules in large databases. In: Bocca, J.B., Jarke, M., Zaniolo, C. (eds.) Proceedings of the 20th International Conference on Very Large Data Bases, Santiago de Chile, Chile, pp. 487–499. Morgan Kaufmann, San Francisco (1994)
2. Aha, D.W., Kibler, D., Albert, M.K.: Instance-based learning algorithms. Machine Learning 6(1), 37–66 (1991)
3. Belkin, N., Kay, J., Tasso, C.: Special issue on user modeling and information filtering. In: User Modeling and User Adapted Interaction, vol. 7. Kluwer Academic Publishers, Dordrecht (1997)
4. Billsus, D., Pazzani, M.J.: A hybrid user model for news story classification. In: Kay, J. (ed.) Proceedings of the 7th International Conference on User Modeling (UM 1999), Banff, Canada, pp. 99–108. Springer, Heidelberg (1999)

5. Chen, L., Sycara, K.: WebMate: A personal agent for browsing and searching. In: Sycara, K.P., Wooldridge, M. (eds.) Proceedings of the 2nd International Conference on Autonomous agents, St. Paul, USA, pp. 132–139. ACM Press, New York (1998)
6. Chen, M.-S., Park, J.S., Yu, P.: Efficient data mining for path traversal patterns. IEEE Transactions on Knowledge and Data Engineering 10(2), 209–221 (1998)
7. Dumais, S., Platt, J., Heckerman, D., Sahami, M.: Inductive learning algorithms and representations for text categorization. In: Proceedings of the 7th International Conference on Information and Knowledge Management (CIKM 1998), Bethesda, USA, pp. 148–155. ACM Press, New York (1998)
8. Dyke, N.W.V., Lieberman, H., Maes, P.: Butterfly: A conversation-finding agent for internet relay chat. In: Proceedings of the 1999 International Conference on Intelligent User Interfaces, Los Angeles, USA, pp. 39–41. ACM Press, New York (1999)
9. Gauch, S., Chaffee, J., Pretschner, A.: Ontology-based personalized search and browsing. Journal of Web Intelligence and Agent Systems 1(3-4), 219–234 (2003)
10. Gluck, M.A., Corter, J.E.: Information, uncertainty, and the utility of categories. In: Proceedings of the Seventh Annual Conference of the Cognitive Science Society, Irvine, USA, pp. 283–287 (1985)
11. Godoy, D., Amandi, A.: User profiling in personal information agents: A survey. The Knowledge Engineering Review 20(4), 329–361 (2005)
12. Godoy, D., Amandi, A.: A conceptual clustering approach for user profiling in personal information agents. AI Communications 19(3), 207–227 (2006)
13. Godoy, D., Amandi, A.: Learning browsing patterns for context-aware recommendation. In: Bramer, M. (ed.) IFIP International Federation for Information Processing. Artificial Intelligence in Theory and Practice, vol. 217, pp. 61–70. Springer, Heidelberg (2006)
14. Godoy, D., Amandi, A.: Interest drifts in user profiling: A relevance-based approach and analysis of scenarios. The Computer Journal 2008; doi: 10.1093/comjnl/bxm107
15. Herlocker, J.L., Konstan, J.A., Riedl, J.: Explaining collaborative filtering recommendations. In: Proceedings of ACM 2000 Conference on Computer Supported Cooperative Work, Philadelphia, USA, pp. 241–250. ACM Press, New York (2000)
16. Kolodner, J.: Case-Based Reasoning. Morgan Kaufmann, San Francisco (1993)
17. Michalski, R.S., Stepp, R.E.: Learning from observation: Conceptual clustering. In: Michalski, R.S., Carbonell, J.G., Mitchell, T.M. (eds.) Machine Learning: An Artificial Intelligence Approach, pp. 331–363. Morgan Kaufmann, San Francisco (1984)
18. Middleton, S.E., Shadbolt, N.R., Roure, D.C.D.: Ontological user profiling in recommender systems. ACM Transactions on Information Systems (TOIS) 22(1), 54–88 (2004)
19. Mitchell, T.M.: Machine Learning. McGraw-Hill, New York (1997)
20. Mitchell, T.M., Caruana, R., Freitag, D., McDermott, J., Zabowski, D.: Experience with a learning personal assistant. Communications of the ACM 37(7), 80–91 (1994)
21. Mladenic, D.: Using text learning to help Web browsing. In: Smith, M., Salvendy, G., Harris, D., Koubek, R.J. (eds.) Proceedings of the 9th International Conference on Human-Computer Interaction, HCI International 2001, Usability Evaluation and Interface Design, New Orleans, USA, vol. 1, pp. 893–897. Lawrence Erlbaum Associates, Mahwah (2001)
22. Moukas, A., Maes, P.: Amalthaea: An evolving multi-agent information filtering and discovery system for the WWW. Autonomous Agents and Multi-Agent Systems 1(1), 59–88 (1998)

23. Müller, M.E.: Inducing content based user models with inductive logic programming techniques. In: Proceedings of the UM 2001 Workshop on Machine Learning for User Modeling, Sonthofen, Germany, pp. 67–76 (2001)
24. Müller, M.E.: Can user models be learned at all? inherent problems in machine learning for user modelling. The Knowledge Engineering Review 19(1), 61–88 (2004)
25. Payne, T., Edwards, P., Green, C.: Experience with rule induction and k-nearest neighbour methods for interface agents that learn. IEEE Transactions on Knowledge and Data Engineering 9(2), 329–335 (1997)
26. Pazzani, M., Muramatsu, J., Billsus, D.: Syskill&Webert: Identifying interesting Web sites. In: Proceedings of the 13th National Conference on Artificial Intelligence and 8th Innovative Applications of Artificial Intelligence Conference (AAAI/IAAI 1996), vol. 1, pp. 54–61. AAAI/MIT Press, Portland, USA (1996)
27. Rennie, J.D.M.: iFile: An application of machine learning to E-mail filtering. In: Proceedings of the 6th ACM SIGKDD International Conference on Knowledge Discovery and Data Mining, Workshop on Text Mining (KDD 2000), Boston, USA (2000)
28. Rocchio, J.: Relevance feedback in information retrieval. In: Salton, G. (ed.) The SMART Retrieval System, pp. 313–323. Prentice Hall, Englewood Cliffs (1971)
29. Salton, G., Wong, A., Yang, C.S.: A vector space model for automatic indexing. Communications of the ACM 18, 613–620 (1975)
30. Schwab, I., Pohl, W., Koychev, I.: Learning to recommend from positive evidence. In: Proceedings of the 5th International Conference on Intelligent User Interfaces, New Orleans, USA, pp. 241–247. ACM Press, New York (2000)
31. Sieg, A., Mobasher, B., Burke, R.: Ontological user profiles for personalized Web search. In: Proceedings of the 5th Workshop on Intelligent Techniques for Web Personalization, Vancouver, Canada, pp. 84–91. AAAI Press, Menlo Park (2007)
32. Srikant, R., Agrawal, R.: Mining generalized association rules. In: Proceedings of the 21th International Conference on Very Large Data Bases, pp. 407–419. Morgan Kaufmann, San Francisco (1995)
33. Thompson, K., Langley, P.: Concept formation in structured domains. In: Fisher, D., Pazzani, M., Langley, P. (eds.) Concept Formation: Knowledge and Experience in Unsupervised Learning, Morgan Kaufmann, San Francisco (1991)
34. Webb, G.I., Pazzani, M.J., Billsus, D.: Machine learning for user modeling. User Modeling and User-Adapted Interaction 11(1-2), 19–29 (2001)
35. Widmer, G., Kubat, M.: Learning in the presence of concept drift and hidden contexts. Machine Learning 23(1), 69–101 (1996)
36. Widyantoro, D.H., Ioerger, T.R., Yen, J.: Learning user interest dynamics with a three-descriptor representation. Journal of the American Society for Information Science and Technology 52(3), 212–225 (2001)
37. Widyantoro, D.H., Ioerger, T.R., Yen, J.: Tracking changes in user interests with a few relevance judgments. In: ACM CIKM International Conference on Information and Knowledge Management (CIKM 2003), New Orleans, USA, pp. 548–551. ACM Press, New York (2003)
38. Zhao, Y., Karypis, G.: Evaluation of hierarchical clustering algorithms for document datasets. In: Proceedings of the 11th International Conference on Information and Knowledge Management, pp. 515–524. ACM Press, New York (2002)
39. Zukerman, I., Albrecht, D.W.: Predictive statistical models for user modeling. User Modeling and User-Adapted Interaction 11(1-2), 5–18 (2001)

4

Context-Aware Web Content Adaptation for Mobile User Agents

Timo Laakko

VTT Technical Research Centre of Finland, Finland

There is a growing need for mobile users to access Web content. In this chapter, adaptation approaches and techniques are described that lets mobile users access Web content that's not directly targeted to user agents of mobile devices. Typical Web applications are designed to be used on a wide range of target devices. In practice, these target devices may include anything from a mobile phone to a Web browser in PC environment. However, the user agent capabilities vary and are generally modest in mobile environments. Content adaptation is necessary, for instance, when a mobile client requests a large document from a Web server. Due to target devices' limitations, the adaptation process must often break input Web documents into suitable units to improve users' experiences on mobile devices. User agents render perceivable units and enable interaction. Content adaptation can be based on the information of delivery context, which can include static or dynamic information including device capabilities, network characteristics, user preferences, and other application-specific parameters. Content selection may allow selection between versions of the material. Web content can be adapted as is or on the basis of author-provided or any additional metadata related to the input. Content authors may choose among several authoring methods related to adaptation. This chapter also outlines typical use cases of Web content adaptation.

4.1 Introduction

Adaptation is needed, for instance, while a mobile client requests a large HTML document from a Web server. Web documents are commonly designed for desktop computers with large display and fast network connection, so considerable processing and reorganization of document is often necessary.

In general, Web applications are to be designed so that they are usable in a wide range of possible target devices. In practice, these target devices may include anything from a mobile phone to a Web browser in PC environment. The design of applications for mobile browsers is quite different from that for the conventional browsers. The HTML application developer may rely that the user agent has a large color display, a pointer and an alphanumeric keyboard. In

R. Nayak et al. (Eds.): Evolution of the Web in Artificial Intel. Environ., SCI 130, pp. 69–99, 2008.
springerlink.com

mobile world, however, the user agent capabilities vary and are generally very modest. Because of this heterogeneity, the application may need to be tailored to the particular device to guarantee good quality.

Mobile data transfer typically costs money, so the mobile user would rather like to retrieve relevant information only. Further, mobile devices often support limited types of content. So, the user may accidentally click a Web link that causes content retrieved that cannot be displayed on his/her mobile phone or rendered correctly to provide reasonable user experience.

Mobile device use can be said to be usage of the services the mobile device can provide. Mobile users intentions are more likely immediate. When the user takes the device out of a pocket, the user has some intention, what kind of a service (s)he is about to use. From a user point of view it means that relevant services have to be found quickly. When the percentage of anticipated positive experience increases the likelihood of using a service also increases, while anticipated negative experiences decrease the likelihood. Thus, it is essential to provide a user quick access to content that is most relevant to current context the user is in. In particular, the retrieved content should be suitable and adaptable for his/her mobile device.

Mobile browsing and usage include several unique properties not included in desktop browsing, such as context-awareness (including location-awareness), one-handed operation and always-on feature. All these bring several challenges for content adaptation and delivery.

4.1.1   Mobile User Agents

A (mobile) *user agent* can be defined as a client within a (mobile) device that performs rendering [1]. Browsers are the most typical instances of user agents used to access Web content. A browser may also embedded with a mobile application.

Mobile browsers are not only used for accessing web-based services but also for 3rd party software applications directly targeted for the mobile devices.

4.1.2   Mobile Web Content

Mobile user agents may access different kinds of Web content such as documents (e.g., HTML or SMIL documents) and other media types (images, audio, video).

Current browsers of mobile devices support increasingly XML based content, such as XHTML MP [2], WML [3], SVG (Tiny and Basic) [4], SMIL [5] and VoiceXML [6]. For example, XHTML MP is a compact core module that is based on the modularization of XHTML.

Mobile user agents will support new media formats increasingly. For instance, the 3rd Generation Partnership Project (3GPP)[1] has defined an architecture of coding technologies and formats based on open standards. It include, for instance, 3GP container file format and H.263, MPEG-4, AMR-NB, AMR-WB

---

[1] http://www.3gpp.org/

and AAC formats. For instance, a video clip in a 3GP media file can be composed of H.263 video and AMR-NB audio tracks. The recommended media formats and codecs to be used within MMS messages are specified in the technical specification [7].

4.1.3   Content Delivery

Content delivery to mobile device users may be *pull* or *push* type [8]. In the pull model, the server responds to client requests by returning content to client. In the push model, server automatically delivers content to mobile clients. Typically, the push model is used in mobile applications such as news and alert services and mobile advertising. In the push model, to prevent users from getting irrelevant content selective content delivery is required. In particular, methods of personalization (see Section 4.3) including user profiling are commonly used in selective content delivery.

Web content is delivered to the user terminal via many different access mechanisms that cause several challenges to the delivery and adaptation processes. How to match Web content to the the needs, capabilities and limitation of the delivery environment? For example, the recently formed W3C Ubiquitous Web Applications Working Group[2] aims to simplify the creation of distributed Web applications involving a wide diversity of devices, including desktop computers and mobile devices (phones) amongst others.

A core concept in content delivery and adaptation is *delivery context* that can be defined as "a set of attributes that characterizes the capabilities of the access mechanism, the preferences of the user and other aspects of the context into which a web page is to be delivered." [1]. The delivery context [9] can include information such as device capabilities, network characteristics, user preferences, and other optional application-specific parameters.

4.1.4   Content Adaptation

W3C defines *content adaptation* as "a process of selection, generation or modification that produces one or more perceivable units in response to a requested uniform resource identifier in a given delivery context." [1]

A *perceivable unit* is a "a set of interact material which, when rendered by a user agent, may be perceived by a user and with which interaction may be possible." [1] Often, decomposition of the input Web document into suitable delivery or perceivable units is required due to the limitations of the target device and to achieve improved user experience on a particular device.

Web content can be adapted as is (without any additional metadata related to the input) or on the basis of author-provided metadata or adaptation hints.

Adaptation software adapts Web content on the basis of the delivery context information (Figure 4.1). The information can be static, such as the concerning device and browser capabilities, or dynamic, such as the properties of the currently active network connection or the user's situation [10].

---

[2] http://www.w3.org/uwa/

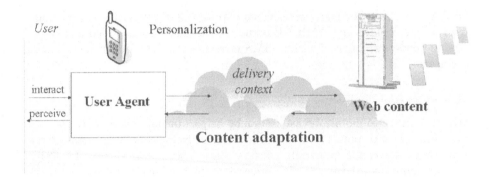

Fig. 4.1. Content adaptation

Mobile user agents deliver relevant information used by the adaptation. For instance, the user agent profile (Section 4.4.4) includes device hardware and software characteristics, information about the network the device is connected to, and other attributes. The adaptation software should also have available other information available in the delivery context, such as user preferences and application-specific parameters.

4.2   Context-Awareness

Context-awareness is a fundamental paradigm for mobile computation. Further, context-awareness is regarded as an enabling technology for ubiquitous computing systems.

Varying contexts of use are typical of mobile services. Utilizing context information provides a basis for context-aware mobile services that are not just reduced duplicates of fixed network services but services that are specifically intended for mobile use. These services can give the user quick access to the information or services that (s)he needs. Mobile services can adapt or even activate themselves based on the context information. [11]

Although the term "context" is sometimes equated with "location" in the scope of mobile applications, it have several other meanings. Schilit et al. decomposed context into three categories[12]: computing context, user context, and physical context. Later, Dey and Abowd provided very general and widely referenced definitions [13]:

– *"Context* is any information that can be used to characterize the situation of an entity. An entity is a person, place, or object that is considered relevant to the interaction between a user and an application, including the user and applications themselves."
– "A system is *context-aware* if it uses context to provide relevant information and/or services to the user, where relevancy depends on the user's task."

Location and time have been the commonly used elements of the context, and these two factors alone make it possible to implement various new kinds of context-aware services [14]. However, to provide more versatile context-aware behavior, more measurement data is needed or, alternatively, the users should be tempted to participate in context definition and activation. The contexts may be very personal and even if the context would be generic, each user's needs regarding the context may differ. Thus, to get the full potential out of context-awareness, the users should be activated to take part in defining, sharing and activating the contexts and the context-aware behavior related to them. Previous experiences with personalization point out that even though the user would appreciate the outcome, there is little or no desire to expend much effort in setting up personalization [15]. In mobile environments this challenge is especially demanding, as user interaction with mobile devices is tedious. However, if the users would feel context-awareness beneficial, and enjoyable, they might want to participate in defining the contexts.

Context-aware systems are concerned with the acquisition of context (e.g. using sensors to perceive a situation), the abstraction and understanding of context (e.g. matching the perceived situation to a context), and application behavior based on the recognized context (e.g. triggering actions based on context).

4.2.1 Acquiring and Processing Context Information

The main challenge with processing of context information is the reliable measurement or identification of the context that may include several physical, technical, social as well as other elements. In addition to time and location, other sources of information can be consulted to complete the picture and to insure a more detailed recognition of contexts. Abowd and Mynatt [16] describe the handling of different sources of context recognition as context fusion. Not only can an individual context source (sensor reading) identify a separate context, it can also add to the accumulated information from other sources to increase the reliability of the recognition of a particular context. It is important that the system can manage several operational contexts without limiting itself to certain aspects of the context.

4.2.2 Context Types and Ontologies

An efficient model for handling, sharing and storing context information is essential for a context-aware system [17]. Context information can be systematically structured by the creation of an *ontology*, i.e., a data model that represents a set of concepts within a domain and the relationships between those concepts. In particular, the use of ontologies can facilitate the reuse, sharing and communication of context information between different information providers of collaboration devices [18]. Further, an ontology-based approach allows the description of contexts semantically in a way which is independent of programming language, underlying operating system or middleware [19].

The Semantic Web[3] aims to provide a common framework that allows data to be shared and reused across application, enterprise, and community boundaries.

---

[3] http://www.w3.org/2001/sw/

Further, it is an effort to provide richer and explicit descriptions of Web resources. In particular, Resource Description Framework (RDF) is a language for representing information about resources and Web Ontology Language (OWL) [20] enables the definition of domain ontologies and sharing of domain vocabularies. OWL builds on RDF and RDF Schema and adds more vocabulary for describing properties and classes. [21] Further, W3C has recently established a new working group[4] to refine and extend OWL.

For example, Toivonen et al. [22] described a frame-based ontology in RDF. The ontology contains concepts for describing time, location, social aspects and device characteristics. Device characteristics are categorized as part of context information, but are not included in the ontology. Instead, UAProf (cf. Section 4.4.4) is utilized for describing device properties. The combining operators (and, or, not) used for adaptation purposes, and reasoning mechanisms, are not included as a part of the ontology.

As an other example of implemented ontologies, Gu et al. [19] proposed a context model based on ontology using OWL to address issues including semantic representation, context reasoning, context classification and dependency. Their generalized ontology defines the basic concepts. The upper layer of the ontology is illustrated in Figure 4.2. Each instance of ContextEntity presents a set of descendant classes of Person, Location, CompEntity and Activity. Then, the details of these basic concepts are defined in domain-specific ontologies.

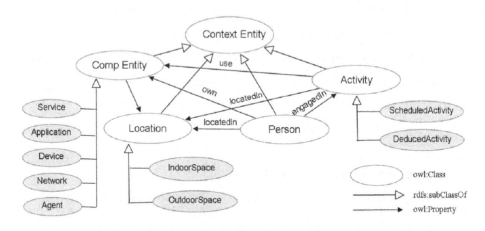

Fig. 4.2. An example context ontology [19]

4.2.3   Activating Contexts

Activating a context means to determine whether a context instance is in an active state. In principle, activating can be automatic or user-initiated. A context can be automatically activated by a system which is able to sense the user's environment and situation and reason the context based on the sensing data.

---

[4] http://www.w3.org/2007/OWL/

For example, the state (active or inactive) of location-aware contexts can be determined by the related positioning method (e.g. GPS). In automatic context activation, the limited accuracy of the related method (e.g. positioning) may sometimes cause ambiguous interpretations of the context [23]. Context validation may be required to check the context hypothesis, and this can be done by asking the user for confirmation or by making additional context measurements. Sometimes it is beneficial that the user can him/herself activate the context. For instance, user-controlled context awareness may better obtain users' acceptance and thus facilitate adoption of the services.

When contexts are based on environmental measurements, contexts are basically defined by the system. Another approach is to allow the users to define the contexts themselves from the very beginning. For instance, in the Kontti platform [24, 11], a *user-defined context* includes the following basic attributes: context name, public name (by default same as the context name), description, public description (by default same as description), and persons ("friends") or groups that have access to the context information. Then, the user can select a recognition method for the new context, for instance, a context could be based on measured context data such as location. A location-based context requires the coordinates and radius of the circular area where the context is in active state. However, it is extremely difficult for the user to give such attributes directly without automatic attribute determination. Thus, auxiliary facilities are required, for instance, a facility can let the user to apply the context type -related positioning method to locate his/her current position that gives direct input to the required attribute values of the context.

4.3 Personalization

Personalization aims to improve the user's experience of a service. Further, it helps the user to get relevant content in the current situation. Personalization involves a process of gathering user information during interaction with the user. The primary goal behind personalization is to "make usage easier and the perception of the communication space richer, and enable personalized filtering of the global communication space to each individual communication space." [25] Consequently, content and services become increasingly tailored to the user's personal preferences.

4.3.1 User Profiling

A user profile contains information about the user for personalization. It can consists of information such as [8]:

- Information interests (e.g., by keywords);
- Browsing history (recently visited Web sites, access times, frequencies);
- Content presentation preferences;
- Quality of Service (QoS) preferences;

- Access privilege options;
- Demographic information.

Profile information can be static and manually entered, but can also be collected automatically, for instance, using automated learning functionality. For instance, collaborative filtering [26] is used to make predictions about the interests of a user by comparing and combining preferences with groups of similar profiles.

4.3.2   Utilizing Contexts

Personalization can be based on the stored and semantically refined context information of the user. Further, a generic user context can contain parts such as [27]:

- Environment context
- Personal context
- Task context
- Social context
- Spatio-temporal context.

Therefore, there is first need to sense the user context as an input, and then react to it. The context information may also be related to content for personalized content selection and delivery, and recommendation systems. The key issues are then how to cather relevant context information and to utilize semantically refined context information of the user.

Different ontologies can be designed to be modular so that one ontology encapsulates one functional element or function. For example, there can be ontologies for: devices, users, context data, and services. When generating profiles, information from several ontologies can be mapped together as one whole. This is depicted in Figure 4.3.2. For example, the user can select a device she wants to use from the device ontology, a set of services she subscribes to from the service ontology and information about time and place from the context ontology. She can then incorporate the information together as her contextual profile. Depending on the implementation-specific details, these ontologies can be grouped together or stored separately. Functionally they are nevertheless separate units.

For example, Korpipää et al. [18] described an application personalization tool, Context Studio, for semiautomated context-based adaptation. With Context Studio, personalization is accomplished with a user interface that allows the user to bind contexts to application actions. Further, a context ontology was defined to offer scalable representation and easy navigation of context and action information in the UI. Furthermore, the ontology supports the use of context scripts as a rule model.

Weißenberg et al. [28] described a prototype integration platform for intelligent personalized Web services, FLAME2008, which is intended to serve as a conceptual basis for a platform to be used during the Olympics 2008 in Beijing. In FLAME2008, user profiles, context- and situation-awareness, and semantic

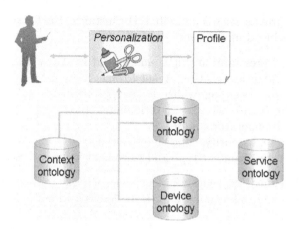

Fig. 4.3. Creating a contextual profile [24]

service descriptions provide the basis for a demand-driven personalized information and service logistics. In FLAME2008 most system logic is implemented in the ontology. Sensors gather context values of users, an inference engine derives the new situation of a user while the context of the user changes meaningfully. As a result, the service matching provides all offers fitting to the user's situation and profile, and the system updates the actual set of recommended services.

Lee [29] described a *multi-agent* methodology that considers different contexts to support the development of personalized services in a mobile environment. In the platform, client representatives, content providers, and service operators are all considered as software agents that communicate and interoperate to request and deliver mobile services. The implemented applications include a shopping-assistance service, which helps mobile users to search products and negotiate with selling agents, and a personalized multimedia instances selection service, which delivers device-dependent and user-dependent multimedia instances to individual users.

4.4 Delivery Context

Adaptation is based on the information about the delivery context. Accordingly, there are several aspects that may influence to the delivery of Web content. In this section, we describe potential characteristics that might be expressed in the delivery context, and formats and protocols to exchange and access delivery context information.

4.4.1 Delivery Context Information

Delivery context includes information that could have effect on content adaptation and, thus, on the user's experience on the delivered content. Therefore, the list of possible characteristics is unlimited. For instance, the following character-istics could be included into delivery context [30]:

- Interaction and presentation (input and output modality characteristics)
- User agent capabilities (cf. User Agent Profile; Section 4.4.4)
- Connection properties (bandwidth, latency, protocols)
- Location (coordinates, accuracy, positioning method)
- Environment (temperature, noise)
- Trust (privacy and security, content restrictions)
- Context properties (context type and parameter values; cf. Section 4.2.2).

Generally, well-defined, application-independent delivery context characteristics enable the delivery of web content effectively across a wide range of target user agents.

4.4.2   Profiles

The delivery context information can be defined in profile structure. The profile in-formation can be communicated between clients, intermediaries and origin servers. Further, the profile information should be easily accessed and handled by applica-tions and services. The W3C has specified the Composite Capabilities/Preferences Profiles (CC/PP) [31], which provides a structure that can be applied to commu-nicating delivery context information between clients, intermediaries and origin servers. CC/PP is vocabulary neutral allowing different vocabularies to be devel-oped and implemented. It is the basis for UAProf (User Agent Profile) [32], which is used to express the capabilities of mobile devices and preference information in practice. UAProf is specified by Open Mobile Alliance (OMA)[5]. CC/PP and UAProf are described in Sections 4.4.3 and 4.4.4, respectively.

Other approaches to define delivery context include Media Queries (W3C Candidate Recommendation[6]) and WURFL (the Wireless Universal Resource File)[7].

4.4.3   Composite Capabilities/Preferences Profile (CC/PP)

A CC/PP profile is a description of device capabilities and user preferences [31]. It is based on RDF[8], which provides a domain-neutral mechanism to exchange and process metadata. RDF is an application of XML and defines a model for describing interrelationships among resources in terms of named properties and values.

A CC/PP profile is structured and contains a number of attribute names and associated values. It allows a client to describe its capabilities by reference to a accessible standard profile. CC/PP vocabulary consists of a set of CC/PP attribute names, permissible values and associated meanings.

[5] http://www.openmobilealliance.org/
[6] http://www.w3.org/TR/css3-mediaqueries/
[7] http://wurfl.sourceforge.net/
[8] http://www.w3.org/RDF/

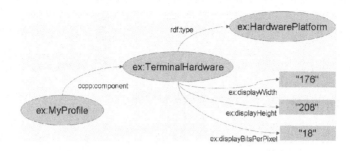

Fig. 4.4. A CC/PP profile including one component with three attributes

A profile contains one or more components, and each component contains one or more attributes. Typical components include the hardware platform, the software platform, or an individual application, such as a browser. In the case of Figure 4.4, the profile consists of one component (ex:TerminalHardware) and three attributes specifying display width, height and bits per pixel value.

The corresponding XML code segment of the above figure is as follow:

```
<?xml version="1.0"?>
<rdf:RDF xmlns:rdf="http://www.w3.org/1999/02/22-rdf-syntax-ns#"
         xmlns:ccpp="http://www.w3.org/2006/09/20-ccpp-schema#"
         xmlns:ex="http://www.example.com/schema#">
  <rdf:Description
       rdf:about="http://www.example.com/profile#MyProfile">
    <rdf:type rdf:resource=
      "http://www.w3.org/2006/09/20-ccpp-schema#Client-profile"/>
    <ccpp:component>
      <rdf:Description rdf:about=
           "http://www.example.com/profile#TerminalHardware">
        <rdf:typerdf:resource=
           "http://www.example.com/schema#HardwarePlatform"/>
        <ex:displayWidth>176</ex:displayWidth>
        <ex:displayHeight>208</ex:displayHeight>
        <ex:displayBitsPerPixel>18</ex:displayBitsPerPixel>
      </rdf:Description>
    </ccpp:component>
  </rdf:Description>
</rdf:RDF>
```

Different applications may use different vocabularies. However, in order to applications to work together a common vocabulary is required, or, alternatively, methods to convert between different vocabularies.

4.4.4 User Agent Profile (UAProf)

UAProf [32] uses the CC/PP model to describe classes of device capabilities and preference information. The profile contains information for content formatting purposes.

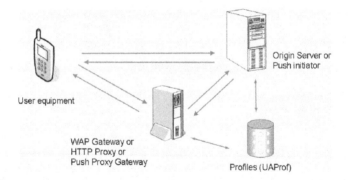

Fig. 4.5. A use case of User Agent Profile

In particular, the User Agent Profile architecture enables the end-to-end flow of delivery context information between the client device, the intermediate network points, and the origin server. Wireless Session Protocol (WSP) clients connect to servers via a WAP gateway and Wireless Profiled HTTP clients may connect to an origin server directly or via proxies (see Figure 4.5).

Capability and Preference Information (CPI) contained in UAProf is transmitted within the protocol headers (cf. Section 4.4.5). The device may not have all this information, but it may indicate the location by a URI. Then, the gateway or proxy forwards the request to the origin server and includes the profile information within HTTP header. A server (or proxy) may extract the profile information and resolves all indirect references (i.e., to utilize the information in content adaptation). The UAProf may be cached in the gateway or proxy, which may also add information to the profile.

The User Agent Profile schema has six key components: HardwarePlatform, SoftwarePlatform, BrowserUA, NetworkCharacteristics, WapCharacteristics and PushCharacteristics. Each of these resources has a collection of properties that describe the component. The HardwarePlatform describes the hardware characteristics of the terminal device such as screen size, model etc. The SoftwarePlatform has properties that describe the operating system software such as OSVersion etc. The BrowserUA has a collection of attributes to describe the browser application. The NetworkCharacteristics has information about the network's capabilities: for instance, current network bearer service and a supported security. The WapCharacteristics has properties to describe WAP capabilities on the device, e.g., WmlDeckSize and WapVersion. PushCharacteristics describes the push connection capabilities of the terminal device.

It is also possible to add definitions outside of the current UAProf descriptions. It is possible to define descriptions by telling only the location where the actual description can be found, and, so, to limit the amount of wireless traffic. Further, additional components can be added to the schema to describe other capabilities and preferences used by applications and services.

In the following a UAProf code segment is shown for the HardwarePlatform component:

```
<rdf:RDF xmlns:rdf="&ns-rdf;" xmlns:prf="&ns-prf;">
 <rdf:Description rdf:ID="MyDeviceProfile">
  <prf:component>
   <rdf:Description rdf:ID="HardwarePlatform">
    <rdf:type rdf:resource="&ns-prf;HardwarePlatform"/>
    <prf:ScreenSizeChar rdf:datatype="&prf-dt;Dimension">15x6
      </prf:ScreenSizeChar>
    <prf:BitsPerPixel rdf:datatype="&prf-dt;Number">2
      </prf:BitsPerPixel>
    <prf:TextInputCapable rdf:datatype="&prf-dt;Boolean">Yes
      </prf:TextInputCapable>
    <prf:ImageCapable rdf:datatype="&prf-dt;Boolean">Yes
      </prf:ImageCapable> ...
   </rdf:Description>
  </prf:component>
  <prf:component>...</prf:component>
 </rdf:Description>
</rdf:RDF>
```

4.4.5   Transferring and Updating Delivery Context Information

UAProf have the following two alternative transport mechanisms [32]: Wireless Profiled HTTP (W-HTTP) or Wireless Session Protocol (WSP)

W-HTTP defines extension headers to transport the UAProf profile over HTTP format: a profile header (*x-wap-profile*), a profile difference header (*x-wap-profile-diff*) and a warnings header (*x-wap-profile-warning*). The profile header is a general header that contains URI referencing to CPI or a reference to a profile difference or a combination of multiple instances of these two types of data. The profile difference header may be generated by the mobile terminal or an intermediate proxy to enchange or alter the capability and preference information (CPI) in UAProf. There may be several profile differences, each of them must have a reference in the *x-wap-profile* header to define the order in which the differences are applied.

4.4.6   Delivery Context Interfaces

W3C has specified an interface for delivery context, *DCCI* (*Delivery Context: Client Interfaces*)[33]. It defines platform and language neutral interfaces that provide Web applications with access to a hierarchy of dynamic properties representing device capabilities, configurations, user preferences and environmental conditions.

DCCI defines an interface for accessing delivery context by the assumption that the overall form of the delivery context is a hierarchy (for instance, UAProf has such a hierarchical arrangement of properties). DCCI is defined as an extension to the Document Object Model (DOM)[9], and the delivery context is represented as an instance of the DOM.

Because property values of delivery context may change at any time, dynamic access to the delivery context is required. DCCI provides interfaces that allow code to:

– query the value of properties within the delivery context
– subscribe/unsubscribe to/form from notifications about changes to properties.

Open Mobile Alliance has specified an interface, *OMA Standard Transcoding Interface* (STI) [34], to provide a standardized interface between application platforms and a transcoding platform. The target is to allow transcoding of media content files based on application specific parameters and device capabilities. Thus, the delivery context information should be first mapped into STI parameters, which are transferred to transcoding platform in order to perform the adaptation based on the parameters. The STI interface uses the SOAP protocol over HTTP or HTTPS transport.

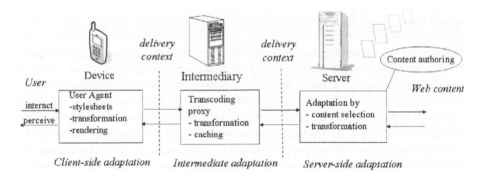

Fig. 4.6. Content adaptation approaches

4.5   Content Adaptation Approaches

The approaches to content adaptation for mobile user agents may be categorized into *server-side*, *intermediate* (or proxy-side) and *client-side content adaptation* that are described in the following subsections 4.5.1- 4.5.3, respectively.

An other broad division of adaptation approaches is based on whether the adaptation is done by selection or not (, i.e, by transformation). The *adaptation by selection* means choosing between different versions of content according to some set of criteria. The content selection methods are considered in more detail in Section 4.6. This section focuses on transformation type-of adaptation that involves creating a new version of some material according to some set of criteria.

[9] http://www.w3.org/DOM/

4.5.1 Server-Side Adaptation

Server-side adaptation provides the possibility for maximum author control to deliver content for mobile devices.

XSL Transformations are commonly used to generate appropriate delivery markup from an XML content representation [35].

A dynamical application may include explicit variants (related to delivery context) within the source code such as within Java ServerPages and PHP.

Multimedia resources may be encoded with specific parameters (i.e. quality, frame rate and resolution) or converted to an other format [36]. Scalable encoding formats need only to be encoded once to suit a variety of characteristics.

There are different content authoring methods related to server-side adaptation. Content authoring is considered in more detail in Section 4.7.

Content selection methods are typically applied in server-side. They are considered in more detail in Section 4.6.

4.5.2 Intermediate Adaptation

Intermediate adaptations are typically applied in proxy-based architectures. Web content can be adapted as such or, for example, on the basis of author-provided adaptation hints. The proxy-based approach naturally address the heterogeneity problem of clients and servers.

Heuristics can be applied to content adaptation in several ways. For example, Hwang et al. [37] introduces heuristics for structure-aware Web transcoding that consider a Web page's structure and the relative importance of Web components. The supported transcoding heuristics include the image reduction and elision transforms, the restricted first sentence elision transform, the indexed segmentation transform, the outlining transform, the generalized outlining transform, and the selective elision transform. Further, the heuristics exploit common layout characteristics of complex Web pages.

Lum and Lau [38] describes a decision engine, which automatically negotiates for the appropriate content adaptation decisions that the transcoder will use to generate the optimal content version. The engine takes into account the computing context, focusing on the user's preferences.

In the server-directed transcoding approach by Knutsson et al. [39], the origin server guides the transcoding system about whether and how to convert between representations.

Gupta et al. [40] describes a framework to DOM-based content extraction. Their HTML content extractor navigates a DOM tree recursively using customizable filters to remove and modify specific nodes and leave only content behind. There are two sets of filters with different levels of granularity. The first set of filters is used to ignore tags or specific attributes within tags, and can be applied to remove elements such as images, links, scripts, styles. The second set of filters consists of the advertisement remover, the link list remover, the empty table remover, and the removed link retainer.

Shilit et al. [41] describes a middleware proxy system (mLinks). They developed a model that supports navigation and action in separate interfaces. mLinks separates links from page content, and, thus, makes navigation a matter of selecting a link from a list. Their link engine processes Web pages into a data structure for link collections. A request to the navigation interface for a Web page involves four steps: page parsing, data detection, link naming and link categorization.

A framework for context-aware content adaptation, FACADE, is described in [42]. The FACADE architecture is designed as a hybrid between server and proxy-side adaptation. In order to improve adaptation results FACADE applies extensions to content representation. Into XHTML documents, the following tags are included to conduct the adaptation: allow adaptation, content relevance, adaptation unit specification, pagination, navigation, inclusion/exclusion tags and selection tags. UAProf was used as the source of the context information. User preferences and network properties were added to UAProf to achieve a more complete context characterization.

4.5.3   Client-Side Adaptation

A client device can utilize style sheets (i.e. CSS) to format content into a browser. Style sheets inherently support the separation of presentation and content. In general, the user agent may also provide facilities to transcode media and documents suitable for rendering. For example, an approach of adaptable mobile browser is described in Section 4.9. However, there may be several efficiency issues depending on properties of the client.

Further, client-side adaptation may be used in conjunction with intermediate and server-side adaptation to provide improved user experience. The user agent may also provide several ways to personalize the user interface and layout [43].

4.6   Content Selection

Content selection (i.e. adaptation by selection) involves "choosing between different versions of materials according to some set of criteria" [50]. The selection may require picking one particular variant of a specific resource. For example, suppose that different variants of a video clips have been prepared for different delivery contexts. Then, the adaptation process may select one particular variant as the most appropriate for a particular mobile device according to device characteristics from the delivery context. Variants need not to be of same media type.

W3C has recently specified a syntax and processing model for general purpose content selection and filtering, *DISelect (Content Selection for Device Independence)*[46]. The DISelect module provides the ability to select between different versions of materials and is designed to be used within other markup languages, and, in particular, within DIAL (described in Section 4.7.1 in relation to single authoring). According to the model, selection is a specific type of transformation applied to an XML information set[10]. It involves conditional processing of

---

[10] http://www.w3.org/TR/xml-infoset/

various parts of an XML information set according to the results of the evaluation of expressions. After, some parts of the information can be selected for further processing and others can be suppressed. The model includes elements, attributes and XPath expressions.

In the following, a code segment is shown where DISelect **when** elements are used in selection:

```
<sel:select>
    <sel:when expr="dcn:cssmq-width('px')> 300">
        <object sel:selid="fig1" src="image1"/>
    </sel:when>
    <sel:when expr="dcn:cssmg-color() > 8">
        <object sel:selid="fig1" src="image2"/>
    </sel:when>
    <sel:otherwise>
        <p sel:selid="fig1">No images!</p>
    </sel:otherwise>
</sel:select>
```

If the display on the device is more than 300 pixels wide, the image1 will be included in the result infoset. If display suppors more than 8 bits per color, the image2 will be included. Otherwise, the text "no images" will be displayed.

4.7 Content Authoring Techniques

Content authors can choose among several authoring methods related to adaptation. However, there are several authoring challenges when supporting a large scale of different target devices and browsers, and delivery context properties. In addition to challenges, new devices offer additional opportunities for authors. For example, a device may support interaction with the user via a range of modalities. An authored application may require textual input on other devices, and voice input (e.g. VoiceXML) on the other devices, or both modalities are supported.

4.7.1 Single, Multiple and Flexible Authoring

The authoring techniques can broadly be classified as: single, multiple and flexible authoring [44] (cf. Figure 4.7).

Multiple Authoring. In multiple authoring, the author creates a different version of the content for each user agent class.

The ability to select from a set of possible versions of content can be supported by several adaptation techniques in server-side, proxy and client-side. Examples of such techniques include URL redirection (server-side adaptation) and server selection (intermediate adaptation). Also, client-side selection-based adaptation can be implemented using techniques supported by the device platform.

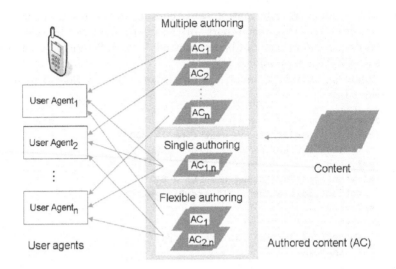

Fig. 4.7. Content authoring techniques

Single Authoring. In single authoring, the author creates only a single version of the content. Then, the applied adaptation solution translates the single authored content into a form appropriate to the user agent. The author typically have to provide additional information to assist the adaptation.

W3C has published a working draft on a *Device Independent Authoring Language (DIAL)* [45]. DIAL is an XML language profile of XHTML[11] version 2, XForms[12] and Content Selection for Device Independence (DISelect)[46]. In particular, DIAL aims to support the creation of web pages delivering a harmonized user experience across multiple delivery contexts. This is achieved by allowing authors to declare "authorial intent". A DIAL conformant adaptation engine must include a DISelect processor, and all elements/attributes from DISelect are available in DIAL. Thus, DIAL strongly relies on DISelect module. In Section 4.6 the DISelect module is described and an example shown where DIAL profile is used with DISelect.

In addition to DIAL, several other languages has developed for single authoring. For example, Harumoto et al [47] developed a language for content authors to specify application-level QoS policies. They describe an adaptive content delivery mechanism where the adaptation is based on the QoS policies that describe transmission time threshold, quality prioritization, format conversion conditions, and transmission order. The description of policies is embedded in an HTML document.

One Web Approach. One Web means "making the same information and services available to users irrespective of the devices they are using" [48]. This

---

[11] http://www.w3.org/MarkUp/
[12] http://www.w3.org/MarkUp/Forms/

approach does not mean that precise the same information is available in precise the same representation for all devices. "Best practices" can be applied in content authoring to provide suitable Web content for mobile and other user agents.

The W3C Mobile Best Practices working Group have created a recommendation [48] on best practices for delivering Web content to mobile devices. The objective of which is to improve the user experience of the Web when accessed from mobile devices. The best practice statements are grouped under the following headings: Overall Behavior, Navigation and Links, Page Layout and Content, Page Definition and User Input. For example, a navigation related practice explained: "Provide basic navigation, which should be placed on the top of the page. Any other secondary navigational element may be placed at the bottom of the page if really needed. It is important the users should be able to see page content once the page has loaded without scrolling ."

Flexible Authoring. In flexible authoring, the techniques of single and multiple authoring are combined. Thus, the author may create single versions of some and multiple versions of other documents. Further, a part of the document may uses single and the rest multiple versions.

4.7.2 Author Awareness of Context

Several technologies allow the author to insert rules into the content related to some delivery context parameters (such as core device characteristics, selection of layouts). However, the more parameters are offered to the author, the more confusing the authoring process can become. Thus, to facilitate the authoring, separate techniques have been defined for different delivery contexts to vary: style, layout, content, structure and navigation [44].

In the One Web approach, a "default delivery context" may be applied to provide a reasonable experience of the Web in different devices [48].

4.7.3 Client-Side Content Authoring

Authoring typically occurs in the server-side. However, it is very probable that content will be authored increasingly in mobile devices. Currently, mobile applications are used to produce content (i.e. capture images and other media content).

For example, Palviainen et al. [49] introduced a new framework for mobile component-based XML editors (called FEdXML). In this framework, mobile browsers present and contained editor components enable users to edit contents (and relate captured media files and sources). Editor components could be adapted for various kinds of devices and contexts. A mobile XML editor should offer efficient editing controls and present the content as illustratively as possible. The FEdXML framework facilitate the construction of various kinds of mobile XML editors (using Java MIDP). The advanged property of FEdXML is that the editor can utilize and adapt to the context information available. Thus, the presentation in browser can be adapted according to the content and contexts.

4.8   Context-Directed Browsing

Browsing with mobile devices usually involves a lot of horizontal and vertical scrolling. Thus, the navigation in small display can be time-consuming and cumbersome. The navigation often cause several usability problems [51][43].

The user is often only interested in a fragment of a Web page, which again may not fit on the limited-size screens of mobile devices, requiring more scrolling in both dimensions. Borodin et al. [52] address the navigation problem during mobile Web access using geometric segmentation of Web pages and the notion of context. They developed a *context-directed browsing* system, called CMo. Their prototype aims to reduce information overflow and, thus, to allow its users to see the most relevant fragment of the page and navigate between other fragments if necessary. It operates as an adaptation proxy and facilitates mobile browsing by logically dividing a Web page into several pages, showing the most relevant page. A geometric segmentation algorithm is used to segment semantically-related information of Web pages. Then, it uses Natural Language Processing and Machine Learning techniques to identify the most relevant segment to be displayed in a mobile Web browser. After the user selects a link to be followed, the context identification algorithm collects the context of the link, i.e., the content around the link that maintain the same topic.

Their experiments show that the use of context can potentially save browsing time and improve the user experience in browsing. They are also planning to use other features for enhancing the performance of context-browsing, for example, by distinguishing between parts of speech and text-formatting, the accuracy of the model could be improved.

Lee et al. [53] described a topic-specific Web content adaptation method with user personalization. In their proxy-based system, the process of Web content adaptation consists of the following phases: filtering blocks, extracting block titles, summarizing block contents, and reordering of the content list through personalization. The block filtering module uses an algorithm to partition the Web page into blocks, and filters out unnecessary blocks using heuristic rules. The content block is processed to extract the title of the block and to summarize the content. Learning is initiated when the user selects the full content menu from

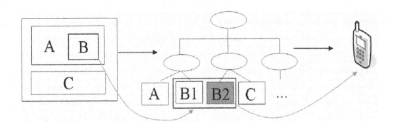

Fig. 4.8. Context-directed browsing

the content summary page. As a result of the personalization by learning, the most relevant block is shown at the top of the content list.

An approach of adaptive content presentation has been implemented in the Mobile Web system [8]. By following the principle of "overview first, then zoom in for details," Mobile Web first parses a Web page and generates a DOM-tree of its content. The DOM-tree provides a hierachical view to the content. In generating a view the system identifies content blocks and their relationships in a Web page. A user can either expand any branch of the tree, or view the dynamically generated summary (heuristic rules applied) of a selected section. Mobile Web stores users' personal interest as keywords in their user profiles. When a Web page is displayed, those specified keywords appearing in the page can be highlighted. Focus+context visualization techniques are applied to select high-detail and low-detail areas. Further, environmental context can be used to adapt the content presentation, for example, if the light level is low the font size of textual content can be automatically enlarged.

4.9 Adaptable Mobile Browser

The mobile environment sets demanding and various requirements for contents, browsers, and visualization. Thus, it may not be sufficient that only the browsing content is adapted. Also, mobile browsers and applications need to be adaptable to the content.

Palviainen et al [49] described a framework, MIMEFrame, for statically and dynamically composed mobile user agents. It provides core interfaces for presenting various types of content in mobile user agents. With *task-based adaptation* technique, a browser can be composed dynamically for different contexts of use. Browser and browsing content can be adapted both client and the server side. The software components performing adaptation utilise delivery context information.

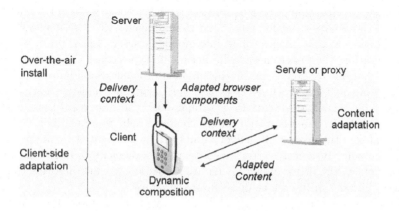

Fig. 4.9. An adaptable mobile browser use case

4.10   Case: Content Adaptation Proxy

As a case example, this section describes an implemented adaptation proxy system [10], which main function is adapting Web documents to mobile devices so that users can more easily view and navigate them. To accomplish this, the adaptation framework at the system's core conducts the process and provides a set of elementary tools for element conversions and base classes for the adaptation modules. Content adaptation is governed by a set of adaptation options, which are derived from the current delivery context.

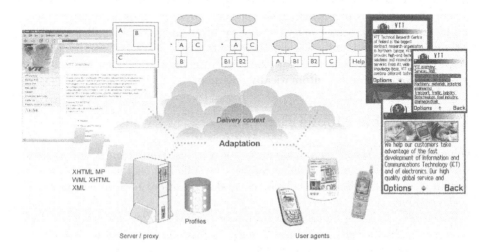

Fig. 4.10. Web content adaptation proxy system

Figure 4.10 outlines the content adaptation process. The proxy adapts Web content to different kinds of terminals, using user agent profiles and other delivery context information obtained either dynamically or from a storage of profiles.

The media adaptation module handles WBMP, GIF, JPEG, and PNG images, providing the ability to scale images and tune image quality parameters during the adaptation, so that images will fit into mobile browsers. In the proxy's current implementation, the proxy checks whether media types other than these default types are suitable for rendering by the target user agent. See Figure 4.11).

Navigating in large documents is a major usability challenge with mobile browsing [43]. Thus, generating navigational aids is an important task in adaptation. By default, the adaptation proxy creates a help document that contains a page index with links to all delivery units that were created by adapting the input document. Instructive error messages were implemented. Each time an error takes place, the proxy aims to generate a document, which informs the user and gives him/her instructions on avoiding the error.

The adaptation system is completely implemented using the Java programming language. The system has been evaluated in several use cases and field trials (Section 4.10.5). The adaptation proxy is described in detail in [10].

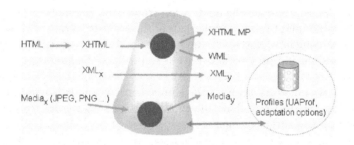

Fig. 4.11. Adapting different types of content

4.10.1 Design Principles

The design principles included following items.

- *Functional user experience.* This involves providing access to all components of a web page and presenting the mobile user with all the navigation links to external resources. No assumptions about the web page structure is made, since optimizing for some particular page type would probably weaken the quality of the other pages.
- *Extendability.* New source and target XML languages can be introduced with relatively little effort. The adaptation software consists of a language-independent framework and one or more adaptation modules. Only the modules need to be modified or added when extending the transcoder's input and/or output language set.
- *Decomposition.* The adapted web page is often longer than accepted by mobile devices, so segmentation into multiple delivery units is required. The navigation and viewing is more comfortable if the page is divided into several semantically coherent perceivable units.
- *Structure preservation.* The parent/child relationships between elements are maintained over the adaptation. The decomposition breaks the structure, but it would be recovered if the links referring to the relocated content were replaced by the content itself.
- *Delivery context.* The adaptation software discovers the capabilities of the requesting mobile device as well as other options (Section 4.10.4). The capabilities should be used efficiently, but never exceeded.
- *Performance.* The adaptation should perform adequately in realtime multiuser environment.

4.10.2 The Algorithm

The adaptation software proxy was primarily designed for mobile devices, and, thus, emphasis is on the reliable delivery of textual content and interactive functionality, rather than on reproducing the original visual appearance.

Adaptation Framework. The adaptation framework provides the basic functionality that the adaptation modules extend in order to carry out the adaptation. It is independent of the source and target XML languages. It controls the adaptation module and processes its output to produce the final target language presentation.

The adaptation module analyzes the source content and extracts some useful information in the process. This information is passed to the framework by adding metadata to target elements.

A breakpoint suggests suitable position for splitting the delivery unit for the framework. A label is a short piece of text that describes the content of an element. Labels are used, for example, as a content of link heads, and as keywords in help document.

The decomposition is necessary in order to ensure that the size of delivery units do not exceed the device capabilities. It also helps the user to access only the parts of the document that (s)he finds interesting. So, the relevant information may be found without transferring all delivery units.

The adaptation algorithm performs decomposition in two phases. First, the adaptation module builds the target content as several perceivable units, according to a semantic interpretation of the content. After, during postprocessing, the algorithm places the perceivable units into delivery units in such a way that the sizes of delivery units remain below the given maximum size. Perceivable units are split when necessary. The adaptation process is illustrated in Figure 4.12.

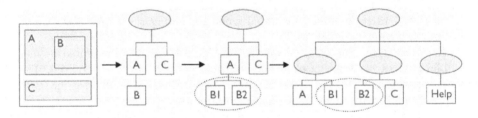

Fig. 4.12. An adaptation process

Adaptation Modules. An adaptation module translates the markup, and decomposes the source web page into several perceivable units. Currently, modules for adaptation from XHTML to XHTML MP and from XHTML to WML 1.x are implemented. The adaptation module consists of a module manager and a number of element conversion rules.

The module manager assigns an element conversion rule for each element type of the source language, and implements a number of content creation functions used by the framework. The functions construct the markup for the hypertext links and help document. The module manager also implements the module API visible to the framework.

Element Conversion Rules. Element conversion rules create the target language markup. Each element conversion rule handles one or several element types of the source language. Applying element-conversion rules results in an input for the segmentation algorithm and produces the main body of the target document and associated metadata, such as preferred split points.

Conversion Sequence. The adaptation framework takes a source document's DOM presentation and creates a new one for the target documents it creates. It traverses the source tree and calls conversion rules to construct a tree of perceivable units (cf. Figure 4.12).

Estimation of Computational Requirements. The number of elements in the target DOM tree is approximately equal to that in the source DOM tree. The source tree is traversed once. The target tree is traversed four times. The number of metadata items is about half of the number of elements in the target. To this is added the processing by HTML Tidy, creation of DOM tree, and serialization. The computational complexity increases linearly with the input size. The memory and CPU requirements of the adaptation allow deployment in realtime server context.

4.10.3 Caching

The proxy provides a user session-based cache for the resulting documents. In the case an original document produces several partial documents (delivery units), each part should be stored with a generated unique URI. For high performance, the most resent entries are kept in system memory (within preserved maximum space) and addressed through a hashtable. This provides very efficient access to the delivery units within the active session.

However, the use of only a single proxy may suffer of a limited scalability. The efficiency of the adaptation and caching can be improved by proxy chaining and systems of cooperative proxy servers [54].

4.10.4 Profile Information and Adaptation Options

The adaptation is governed by a set of adaptation options, which are derived from the current delivery context. The options are further divided into four categories:

- system default options,
- user preferences
- UAProf attributes, and
- device class specific options.

The adaptation options are bound to the user's session. The XML structure for the default options look like the following:

```
<adaptation-defaults>
  <feature name="AutomaticNextLinks" value="true"/>
  <property name="MaxDocumentSize" type="integer" value="1250"/>...
</adaptation-defaults>
```

The allowed option names correspond directly with the system inner option names (cf. above "AutomaticNextLinks" and "MaxDocumentSize"), and any of them can be included in the defaults. There are 49 different options handled by the system. In the same way than the default options, also the user preferences are be defined. The selection of these options is based on the user session.

The system includes methods to map UAProf descriptions (see Section 4.4.4) to adaptation options. However, the UAProf is not supported by all the mobile devices. Thus, the system includes a database of the capabilities and characteristics of the most common user agents.

4.10.5   Use Cases

The developed adaptation software have been evaluated in different use cases. In addition to the basic case of a standalone adaptation proxy it has also been used as a module within a context-aware service platform, and a module within an editor toolkit.

Context-Aware Service Platform. The research project Kontti [24, 11]used the adaptation proxy as a module of a context-aware service platform (Kontti). In this case, the operation of adaptation was evaluated with context-aware applications developed on the top of the platform.

The Kontti platform offers the users a possibility to define, utilize and share contexts. The current implementation of the platform manages the following basic types of contexts: time, location and user-defined type. In the Kontti framework, the user-defined type has a particular importance, because it gives the user the freedom to define the context. The context may then have little or no connection to measurable context identification such as location or time.

Contexts can be associated with other information, called contextual information (see Figure 4.13). Contextual information can be viewed and managed within the system. The contextual information is further divided into objects and messages.

The platform deals with the following kinds of contextual objects:

- Content (link, note, file and directory objects) for the user's personal use and sharing them with others, linked to specific contexts.
- Contextual information as social media; associating and publishing contexts with friends and groups.
- Context-based messages; sharing of contexts and objects.

The platform manages repositories of the user's personal and shared contexts and contextual information. Further, the Kontti framework enables the flexibility in defining contexts as well as in combining of contexts, objects and messages by user.

In particular, all the content was adapted suitable to the user's device and the context of use according to the current delivery context. Thus, the user was able to store one version of the content and view it with different devices.

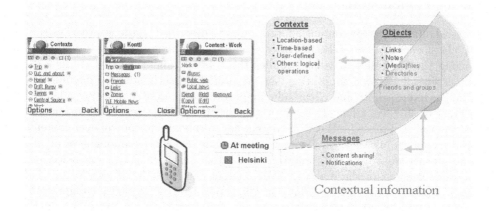

Fig. 4.13. Kontti platform: context, objects and messages

Context-Aware Service Package. The Kontti project also elaborated the scenario where a package (container) binds all relevant contexts, messages and content into a whole. The package should be easily distributed and shared between users. The user could create and modify the packages and distribute them. An early version of such package concept was evaluated with end-users during the Kontti project. The concept appeared to be very promising and widely applicable.

An example application was implemented to evaluate a context-aware festival guide during the Tampere Theatre Festival. The guide was created as a distributable package with contexts and contextual information. For example, the users could click on the programme category to view upcoming events within the active time-based context period. The contents of the guide were links to the festival programme on the Web, notes and media files. The content was adapted suitable to the client device. The package provided also target contexts (locations) where the users could drop messages to other visitors. So, the messages were delivered and adapted context-aware.

The core advantage of the context-aware package is that once the package is created, it could be straightforwardly distributed and shared as a message attachment to each user of the Kontti service. In particular, this approach is very useful for content providers.

Mobile Application Development Toolkit. The adaptation software was also used as a module in the a development toolkit for mobile applications, mPlaton [55]. mPlaton uses the adaptation module to provide input (XHTML MP and WML) for the editor tool on the basis of existing HTML documents, and can simulate different mobile user agents by using UAProf information associated with the user agent. The adaptation module provides a flexible mechanism for finalizing content for selected user agents.

4.11   Conclusions

Several approaches for content adaptation were described. Basic approaches can be classified as: server-side, intermediate, and client-side. The main difference among the three lies in what supplies the content, and what the content is. In server-side content adaptation, servers provide directly adapted documents, whereas intermediate and client-side adaptations use input content provided by servers or proxies. Client-side adaptation occurs in the client device. Server-side adaptation provides the Web content author maximum control over content delivery for mobile devices. Content authors can also choose among several authoring methods related to adaptation. Content selection may allow selection between versions of the material. Web content can be adapted as is or on the basis of author-provided metadata or adaptation hints. Proxy-based adaptation methods are typically based on some heuristics; for example, a set of rules to perform adaptation tasks. Client-side adaptation avoids the problems of server or intermediate adaptation in transmitting client properties to the server or proxy. Depending on the client's properties, however, the adaptation might face several efficiency issues related to processing time and memory usage. Client-side adaptation can also be used in conjunction with intermediate and server-side adaptation to improve the user experience. Additionally, a user agent can provide several ways to employ user preferences and delivery context information to personalize a mobile device's user interface and layout.

References

1. Lewis, R. (ed.): Glossary of Terms for Device Independence. W3C (2005),
   http://www.w3.org/TR/2005/WD-di-gloss-20050118/
2. Open Mobile Alliance: XHTML Mobile Profile (Version October 29, 2001) (2001),
   http://www.openmobilealliance.org/
3. Open Mobile Alliance: Wireless Markup Language Specification, Version 1.3.
   (2000), http://www.openmobilealliance.org/
4. Capin, T. (ed.): Mobile SVG Profiles: SVG Tiny and SVG Basic. W3C (2003),
   http://www.w3.org/TR/2003/REC-SVGMobile-20030114/
5. Bulterman, D., Grassel, G., Jansen, J., Koivisto, A., Layaida, N., Michel, T., Mullender, S., Zucker, D.F. (eds.): Synchronized Multimedia Integration Language (SMIL 2.1). Number Recommendation REC-SMIL2-20051213. W3C (2005), http://www.w3.org/TR/2005/REC-SMIL2-20051213/
6. Oshry, M., Auburn, R.J., Baggia, P., Bodell, M., Burke, D., Burnett, D.C., Candell, E., Kilic, H., Kusnitz, J., McGlashan, S., Lee, A., Porter, B., Rehor, K.G. (eds.): Voice Extensible Markup Language (VoiceXML) 2.1. Number Recommendation REC-voicexml21-20070619. W3C (2007), http://www.w3.org/TR/2007/REC-voicexml21-20070619
7. The 3rd Generation Partnership Project (3GPP): Technical Specification Group Services and System Aspects; Multimedia Messaging Service (MMS); Media formats and codecs (Release 7) (2007), http://www.3gpp.org/ftp/Specs/html-info/26140.htm
8. Zhang, D.: Web content adaptation for mobile handheld devices. Communications of the ACM 50(2), 75–79 (2007)

9. Butler, M., Giannetti, F., Gimson, R., Wiley, T.: Device independence and the web. IEEE Internet Computing 6(5), 81–86 (2002)
10. Laakko, T., Hiltunen, T.: Adapting web content to mobile user agents. IEEE Internet Computing 9(2), 46–53 (2005)
11. Laakko, T., Virtanen, T., Niemelä, M., Kaasinen, E.: User role in defining and utilizing contexts. In: Workshop on Innovative Mobile Applications of Context (IMAC) at MobileHCI 2006, Espoo, Finland (2006)
12. Schilit, B., Adams, N., Want, R.: Context-aware computing applications. In: Workshop on Mobile Computing Systems and Applications, Santa Cruz, CA, U.S. (1994)
13. Dey, A., Abowld, G.: Towards a better understanding of context and context awareness. In: The CHI 2000 Workshop on the What, Who, Where, When, Why and How of Context-Awareness (2000)
14. Kaasinen, E.: User acceptance of location-aware mobile guides based on seven field studies. Behaviour and Information Technology 24(1), 37–49 (2005)
15. Kaasinen, E.: User Acceptance of Mobile Services: Value, Ease of Use, Trust and Ease of Adoption. PhD thesis, Tampere University of Technology, VTT, Espoo (2005), http://www.vtt.fi/inf/pdf/publications/2005/P566.pdf
16. Abowd, G.D., Mynatt, E.D.: Charting past, present, and future research in ubiquitous computing. ACM Trans. Comput.-Hum. Interact 7(1), 29–58 (2000)
17. Baldauf, M., Dustdar, S., Rosenberg, F.: A survey on context-aware systems. International Journal of Ad Hoc and Ubiquitous Computing 2(4), 263–277 (2007)
18. Korpipää, P., Häkkilä, J., Kela, J., Ronkainen, S., Känsälä, I.: Utilising context ontology in mobile device application personalisation. In: MUM 2004: Proceedings of the 3rd international conference on Mobile and ubiquitous multimedia, pp. 133–140. ACM Press, New York (2004)
19. Gu, T., Pung, H.K., Zhang, D.Q.: A service-oriented middleware for building context-aware services. Journal of Network and Computer Applications 28(1), 1–18 (2005)
20. Bechhofer, S., van Harmelen, F., et al. (eds.): OWL Web Ontology Language Reference (2004), http://www.w3.org/TR/2004/REC-owl-ref-20040210/
21. Wang, X., Zhang, D., Gu, T., Pung, H.: Ontology based context modeling and reasoning using owl. In: Proceedings of the Second IEEE Annual Conference on Pervasive Computing and Communications Workshops, pp. 18–22 (2004)
22. Toivonen, S., Kolari, J., Laakko, T.: Facilitating mobile users with contextualized content. In: Artificial Intelligence in Mobile Systems 2003 (AIMS 2003), pp. 124–132 (2003)
23. Dey, A.K., Mankoff, J.: Designing mediation for context-aware applications. ACM Trans. Comput.-Hum. Interact 12(1), 53–80 (2005)
24. Kolari, J., Laakko, T., Hiltunen, T., Ikonen, V., Kulju, M., Suihkonen, R., Toivonen, S., Virtanen, T.: Context-Aware Services for Mobile Users - Technology and User Experiences, number 539. VTT Publications, Espoo (2004)
25. Arbanowski, S., Ballon, P., David, K., Droegehorn, O., Eertink, H., Kellerer, W., van Kranenburg, H., Raatikainen, K., Popescu-Zeletin, R.: I-centric communications: personalization, ambient awareness, and adaptability for future mobile services. IEEE Communications Magazine 42(9), 63–69 (2004)
26. Adomavicius, G., Sankaranarayanan, R., Sen, S., Tuzhilin, A.: Incorporating contextual information in recommender systems using a multidimensional approach. ACM Trans. Inf. Syst. 23(1), 103–145 (2005)
27. Göker, A., Myrhaug, H.: User context and personalization. In: European Conference on Case Based Reasoning, Workshop on Case Based Reasoning and Personalization (2002)

28. Weißenberg, N., Gartmann, R., Voisard, A.: An ontology-based approach to personalized situation-aware mobile service supply. GeoInformatica 10(1), 55–90 (2006)
29. Lee, W.P.: Deploying personalized mobile services in an agent-based environment. Expert Systems with Applications 32(4), 1194–1207 (2007)
30. Sathish, S., Gimson, R., Lewis, R. (eds.): Delivery Context Overview for Device Independence. W3C (2006), http://www.w3.org/TR/2006/NOTE-di-dco-20060320/
31. Kiss, C. (ed.): Composite Capability/Preference Profiles (CC/PP): Structure and Vocabularies 2.0 (2007), http://www.w3.org/TR/2007/WD-CCPP-struct-vocab2-20070430
32. Open Mobile Alliance: User Agent Profile. Approved version 2.0 (2006)
33. Waters, K., Hosn, R.A., Raggett, D., Sathish, S., Womer, M., Froumentin, M., Lewis, R. (eds.): Delivery Context: Client Interfaces (DCCI) 1.0 — Accessing Static and Dynamic Delivery Context Properties. W3C (2007), http://www.w3.org/TR/2007/WD-DPF-20070704/
34. Open Mobile Alliance: Architecture of the Environment using the Standard Transcoding Interface. Approved version 1.0 edn. (2007)
35. Pashtan, A., Kollipara, S., Pearce, M.: Adapting content for wireless web services. IEEE Internet Computing 7(5), 79–85 (2003)
36. Mohan, R., Smith, J., Li, C.H.: Adapting multimedia internet content for universal access. IEEE Transactions on Multimedia 1(1), 104–114 (1999)
37. Hwang, Y., Kim, J., Seo, E.: Structure-aware web transcoding for mobile devices. IEEE Internet Computing 7(5), 14–21 (2003)
38. Lum, W., Lau, F.: A context-aware decision engine for content adaptation. IEEE Pervasive Computing 1(3) (2002)
39. Knutsson, B., Lu, H., Mogul, J.: Architecture and performance of server-directed transcoding. ACM Transactions on Internet Technolog 2(4), 392–424 (2003)
40. Gupta, S., Kaiser, G., Neistadt, D., Grimm, P.: Dom-based content extraction of html documents. In: WWW 2003, Budapest, Hungary (2003)
41. Schilit, B., Trevor, J., Hilbert, D., Koh, T.: Web interaction using very small internet devices. Computer 35(10) (2002)
42. Kurz, B., Popescu, I., Gallacher, S.: Facade - a framework for context-aware content adaptation and delivery. In: Communication Networks and Services Research, 2004. Proceedings. Second Annual Conference, pp. 46–55 (2004)
43. Kaasinen, E., Aaltonen, M., Kolari, J., Melakoski, S., Laakko, T.: Two approaches to bringing Internet services to WAP devices. Computer Networks 33(1–6), 231–246 (2000)
44. Hanrahan, R., Merrick, R. (eds.): Authoring Techniques for Device Independence. W3C (2004), http://www.w3.org/TR/2004/NOTE-di-atdi-20040218/
45. Smith, K.: Device Independent Authoring Language (DIAL). W3C (2007), http://www.w3.org/TR/2007/WD-dial-20070727/
46. Lewis, R., Merrick, R., Froumentin, M. (eds.): Content Selection for Device Independence (DISelect) 1.0, W3C Candidate Recommendation July 25 (2007)
47. Harumoto, K., Nakano, T., Fukumura, S., Shimojo, S., Nishio, S.: Effective web browsing through content delivery adaptation. ACM Transactions on Internet Technology 5(4), 571–600 (2005)
48. Rabin, J., McCathieNevile, C. (eds.): Mobile Web Best Practices 1.0. W3C (2006), http://www.w3.org/TR/2006/PR-mobile-bp-20061102/
49. Palviainen, M., Laakko, T.: The construction and integration of xml editor into mobile browser. In: Creutzburg, R., Takala, J.H. (eds.) Multimedia on Mobile Devices, San Jose, California, SPIE, pp. 231–242 (2005)

50. Lewis, R., Merrick, R. (eds.): Content Selection Primer 1.0. W3C (2007), http://www.w3.org/TR/2007/WD-cselection-primer-20070109/

51. Kaikkonen, A., Roto, V.: Navigating in a mobile xhtml application. In: CHI 2003: Proceedings of the SIGCHI conference on Human factors in computing systems, pp. 329–336. ACM Press, New York (2003)

52. Borodin, Y., Mahmud, J., Ramakrishnan, I.: Context browsing with mobiles - when less is more. In: MobiSys 2007: Proceedings of the 5th international conference on Mobile systems, applications and services, pp. 3–15. ACM Press, New York (2007)

53. Lee, E., Kang, J., Choi, J., Yang, J.: Topic-specific web content adaptation to mobile devices. In: WI 2006: Proceedings of the 2006 IEEE/WIC/ACM International Conference on Web Intelligence, Washington, DC, USA, pp. 845–848. IEEE Computer Society, Los Alamitos (2006)

54. Canali, C., Cardellini, V., Colajanni, M., Lancellotti, R., Yu, P.S.: Cooperative architectures and algorithms for discovery and transcoding of multi-version content. In: WCW 2003, Hawthorne, NY (2003)

55. Palviainen, M., Laakko, T.: mPlaton - Browsing and development platform of mobile applications, number 515. VTT Publications (2003), http://www.vtt.fi/inf/pdf/publications/2003/P515.pdf

5

Knowledge Personalization in Web-Services-Based Environments: A Unified Approach

Alfredo Cuzzocrea

ICAR Institute and DEIS Department
University of Calabria, Italy
cuzzocrea@si.deis.unical.it
http://si.deis.unical.it/~cuzzocrea

Abstract. Inherited from Database and Information Systems research, knowledge representation and management are next challenges for Web Information Systems research. Currently, Web Services are the widely accepted standard for building next-generation Web-based platforms and systems where users and machines are co-operant, and new services can be dynamically composed from a set of pre-existent services. In the context of Web services, due to the enormous size of Web data repositories (e.g., XML-based) and the very large number of available services (e.g., UDDI-based) that can be invoked, the so-called *knowledge personalization problem* plays a critical role, and is currently attracting a lot of attention from research communities. Starting from these considerations, in this chapter we propose models and algorithms for efficiently representing, managing and personalizing knowledge on the Web, while admitting Web-Service-compliant scenarios. These research contributions are synthesized in the *Distributed Knowledge Network* (DKN) framework, which is able to provide personalized knowledge in Web-services-based environments. We complete our analytical contribution with a comprehensive experimental evaluation of the proposed algorithms against significant XML-formatted Web data sets.

5.1 Introduction

Web Services [58] propose a novel way of thinking of Web applications and systems, where data are in different format and nature (i.e., structured, semi-structured, raw, streaming data etc), users access information through heterogeneous connecting devices and with different backgrounds and goals, and services must be (*i*) adaptive, (*ii*) composable, and (*iii*) reliable. "Adaptive" means that users should be able to access and use services according to different profiles and system views. "Composable" means that modern Web applications should be developed by composing distributed-on-the-Internet pre-existent (Web) services, and these services should live together in "orchestrated" and transparent-for-the-user scenarios. "Reliable" means that users should be trusted with services, i.e. trusted users should be authenticated on systems independently from their *current* position and their *current* accessing device, and intelligent middleware should allow them to rapidly access both non-secured services (such as weather forecasting, travelling and tourist information, and so forth) and secured services (such as Web mail, Web banking and ticketing, and so forth).

Without any loss of generality, we can classify Web services on the basis of an unambiguous taxonomy that distinguishes between *Operational Web Services* and

R. Nayak et al. (Eds.): Evolution of the Web in Artificial Intel. Environ., SCI 130, pp. 101–135, 2008.
springerlink.com   © Springer-Verlag Berlin Heidelberg 2008

Data-Intensive Web Services. Data, data size, and data utilization are the main differences between these two classes. Operational Web services are mainly designed to satisfy on-demand weakly-connected-with-data functionalities, and are specialized for particular purposes like building applications for pervasive, ubiquitous, and embedded environments. Data-intensive Web services are the commonly-intended Web services mainly due to the fact that the (traditional) Web is more and more characterized by the presence of heterogeneous and distributed data sources that produce massive data sets in persistent and streaming formats. Due to many *structural* limitations (such as low CPU processing power, low display capabilities, limited memory and battery lifetimes as compared to desktop and laptop systems), the latter scenario is more in conformity with traditional wired environments (like LAN, WAN, Intranet, Internet networks etc), where hosts can produce, process, and manage huge amounts of data. Despite this, WLAN environments are also ready to deal with data-intensive Web services, due to the fact that (*i*) some categories of today's mobile devices, like laptops and PDA, are equipped with powerful CPU and large memories, and (*ii*) today's wireless network protocols, like IEEE 802.11*a* and 802.11*b*, guarantee good performance in both data transmission and reliability.

Among the main challenges of Web services research, i.e. composition and orchestration of Web services, security and privacy issues of Web services etc, the *personalization of presentations and contents* (i.e., their adaptation to user requirements and goals) is becoming a major requirement for this novel Web deployment paradigm, and is converging to the emerging issue of *personalizing knowledge in Web-services-based environments*, which is the main focus of this chapter.

Application fields where personalization can be useful are manifold; they comprise on-line advertising, direct Web-marketing, electronic commerce, *e*-news delivery, online learning and teaching etc. The need for adaptation arises from different aspects of the interaction between users and Web services. For instance, a daily news Web service needs to know what kind of information a user is interested in, as this can be very useful for delivering information which matches user goals. If a certain user is usually interested in sports news and he/she gets, at every access, Web pages containing mostly financial news, probably he/she will not access the Web system and will not subscribe to the service anymore. This means that having potentially the right information for the right user is useless when this information cannot be adaptively delivered to users. The same happens with *e*-commerce systems. In this case, having a wide product warehouse is not enough if proposing the appropriate products based on effective (or probable, at least) user needs is not possible. Analogous situations may be found in the context of *e*-tourism systems. In this case, choosing, composing, and proposing tourist bundles should be performed according to user tastes and past experiences.

Currently, the problem of Web personalization is addressed through approaches based on clustering methodologies which group users based on similarities of navigation through the target Web system. As soon as a new user connects to such system, monitoring components try to understand what he/she needs via simply comparing his/her navigation through the Web system with the clustered sessions of other users. Once the cluster which best represents the current user navigation is located, relevant contents in that cluster are suggested or presented to the current user. Personalization plays a leading role in the context of Web services, as the Web services potentiality in building large-scale Web applications could be reduced by the impersonality of Web information.

The traditional clustering-based approach, coming from the Data Mining research area, results in poor adaptation because it does not consider actual user needs. Instead, it tries to infer them by observing what other users with "similar" behaviour have previously done. Just defining what "similar" means is a basic issue in this context. In fact, there are many, and often very different, proposals defining the similarity concept with respect to the interaction between users and Web systems. For instance, some approaches define the similarity concept in terms of *semantic similarity*; the most popular one among these kinds of proposals is the *Semantic Web* [13]. Other approaches suggest to define the similarity concept in terms of *structural similarity*, and propose designing distance and similarity functions in order to use them as computational tools for making adaptation "measurable" and "computable". Following this direction, there are many interesting papers concerning the detection and the measurement of structural similarity among XML documents, such as [3,33]; they present research results that could be used inside a Web service Provider System for supporting the adaptation task.

To face-off these described knowledge personalization challenges in Web-services-based environments, we believe that a reasonable solution is combining the modern Web services technology with mature results coming from (*i*) the *User and User Behaviour Modelling* research area, which is a traditional well-founded research field mainly devoted to enhance Web systems performance and usability through the analysis of user dynamics, and (*ii*) the *Knowledge Representation, Management, and Reasoning* research area, which recently has benefited from renewed interest thanks to innovative application deployment paradigms like *Knowledge Management Systems* (KMS) [14,31,34,52,53,54], *Enterprise Resource Planning* (ERP) systems [8,35], *Intelligent and Expert Information Systems* (IEIS) [6,23,27], Semantic Web [13].

In the spirit of this research direction, we retain that the knowledge personalization issue in Web-services-based environments must be tackled *by decomposition*, thus selecting and investigating two orthogonal challenges: *data personalization* and *service personalization*. To give background, while there are many proposals and results about the first research issue, the second one is generally recognized as "Web services composition", and, currently, is widely investigated, due to its industrial spin-offs mainly. *Web services interoperability* [48,25,21] is the most investigated topic arising from the latter issue, as it deals with the problem of making heterogeneous Web services communicant and working together. Currently, *Semantic Web Services* [47,32,18] are the widely accepted proposal for addressing Web services interoperability issues. Such a proposal aims at defining *Intelligent Web Services* like those in [1] and [19], i.e. services supporting automatic discovery, composition, invocation, and interoperation.

Novel challenges arising from this way of viewing the personalization problem in the context of Web services (i.e., distinguishing between *data personalization* and *service personalization*) are the following: (*i*) the goal of making Web services *knowledge-aware*, thus enhancing the capability of Web services technology in matching the requirements of modern Web applications; (*ii*) the opportunity of re-visiting and efficiently re-using mature research results about knowledge representation, management, and reasoning; (*iii*) the convenience of treating the two (complex) aspects within a unified framework that aims at solving significant issues introduced by (data-intensive) Web-services-based applications and systems.

This chapter introduces a novel approach whose goal is to increase the accuracy of the general need for adaptation of Web services, by means of fine-grain *OnLine Analytical Processing* (OLAP) [22]-based multidimensional user models and knowledge representation and management techniques. This way, the chapter is also focused on related issues concerning user behaviour tracking and knowledge delivery. In more detail, our approach proposes personalizing knowledge (i.e., data and services) with respect to a set of user profiles, which are clustered on the basis of monitoring the behaviour of users. Therefore, the key idea is to personalize knowledge via continuously observing users, whose behaviour is analysed in order to infer their tastes, preferences and needs. A user profile is built from the user repeated interacting behaviour and is no longer based on what other users with similar behaviour have previously done. In order to improve adaptation performance, we adopt OLAP technology [22] in such a way that each user is not treated as an anonymous user, but instead user registration information is taken into account as the major requirement for accessing Web services.

Following these directions, we propose the *Distributed Knowledge Network* (DKN) framework, whose main goal is to make Web services *knowledge-aware*, according to the above-described guidelines. Such a framework aims at realizing a state of *distributed knowledge*, where users can access and manage useful knowledge via Web services and through different connecting devices. "Distributed knowledge" means that users can retrieve, join, and usefully manage "knowledge chunks" distributed on different *knowledge nodes*, thus defining a sort of *knowledge-oriented pervasive scenario*. In this scenario, the DKN framework implements an abstract knowledge layer that (*i*) sits on top of the heterogeneity of data, services, networks, and devices, and (*ii*) supports a transparent-for-the-user access to knowledge, whose personalized fruition is significantly improved via intelligent knowledge representation and management techniques.

We believe that a reasonable solution to this ambitious goal lies in both the recent progresses made in techniques for knowledge representation and management, but also in the (available) technology, perhaps when the latter is proposed and used in contexts different from traditional ones. In fact, the adoption of already-available *fine-grain user models*, i.e. user models that are able to efficiently map user backgrounds and needs, and to successfully personalize knowledge, allows us to improve the capabilities of service-oriented systems such as Web-services-based (Web) systems in tracking, managing, and supporting users, thus satisfying the emerging need for personalization. To this end, we adopt OLAP technology [22] in order to obtain a multidimensional user model that allows knowledge personalization with a high degree of granularity to be managed. This approach is well founded considering that, usually, Web-services-based systems are built on data domains which are enormous in size (e.g., very large data warehouses) and, consequently, defining a methodology able to provide *the right knowledge to the right user* is very valuable for such systems. Indeed, in data- and service-intensive systems, users could be disoriented when they are interacting with the system front-ends in search for knowledge which is relevant for them and, consequently, they run the risk of retrieving a lot of useless knowledge (for instance, this is highly probable when users interact with *e*-learning systems).

Summary of the Contributions. Overall, in this chapter we make the following contributions:

- We recognize the important need for making Web services knowledge-aware.
- We propose an innovative approach where the knowledge personalization task is decomposed in data personalization and service personalization tasks, which are treated within a unified framework.
- We propose adopting OLAP technology to support fine-grain multidimensional user models, thus significantly improving the degree of granularity of user modelling and management phases.
- We provide formal definitions for our proposed user models, and some properties about them.
- We propose the DKN framework, which implements our approach, and a reference architecture for it.
- We provide experimental results of the effectiveness of our approach.

Outline of the Chapter. The remainder of the chapter is organized as follows. Sect. 5.2 describes related work, focusing the attention on background research areas; Sect. 5.3 presents and investigates definitions, insights, and open issues about supporting knowledge personalization within the currently accepted standard of Web services; in Sect. 5.4, the core user models of our proposal are presented, formalized, and discussed; Sect. 5.5 presents our knowledge personalization process, both for data and services; Sect. 5.6 draws the guidelines of a reference architecture for the DKN framework, which implements our proposal; Sect. 5.7 reports some experimental results confirming the effectiveness of DKN; finally, Sect. 5.8 contains conclusions and ideas for future work.

5.2   Related Work

With respect to our work, there are four major background research areas:

- Web services;
- Knowledge Management Systems;
- User Behaviour Tracking;
- User Profile Modelling.

With respect to the more specific context of *Web services personalization*, we recognize that, to the best of our knowledge, there are very few papers on this important issue. This is due to the fact that currently researchers are mainly involved in other important topics concerning Web services and, more specifically, Web services engineering, like Web services composition, Web services orchestration, and issues about security and privacy of Web services.

In the remaining part of this Section, we provide an overview on related work and discussion about the above-listed research topics.

5.2.1   Web Services

Web services have been defined by the *World Wide Web Consortium* (W3C) [56] as a software application or component with the following properties: (*i*) it is identified by a *Uniform Resource Identifier* (URI); (*ii*) its interface and binding can be described in XML; (*iii*) it can interact directly with other Web services through XML, via Internet-based protocols. Web services are described by means of the XML-based *Web Services Description Language* (WSDL) that specifies the operations and their parameters provided by a Web service, along with its location, its transport protocol, and the invocation style. Web services can be located and integrated within existing Web systems by means of *Universal Description, Discovery, and Integration* (UDDI) functions that are similar to a telephone directory. Web services use SOAP or RPC as communication mechanisms, and dispatching can happen in a synchronous or asynchronous manner. The most important issues in Web services research topics are security and personalization. For the latter, Semantic Web [13] is the currently widely accepted proposal. Semantic Web aims at pushing knowledge inside the link structure of Web applications/services in such a way as to be easily "computable" by both computers and people. In this sense, Semantic Web can be described as an evolving collection of (available) knowledge. Semantic Web can be implemented by using *Ontologies* [17] or *Resource Description Framework* (RDF) [57]. Ontologies are a specification formalism used to describe knowledge in terms of concepts and relationships between concepts. With respect to the Web Engineering research area, ontologies are very often built on XML and used as knowledge inference engine for enabling large scale transactional Web systems on various application fields like digital libraries, skill management systems for corporate enterprises, recommender systems (particularly tailored to support virtual museum tours and thematic *e*-tourism), business intelligence systems, decision-support system etc. RDF is a W3C recommendation about modelling Web resources with XML metadata; like Ontologies, it is usually used to represent knowledge.

5.2.2 Knowledge Management Systems

Briefly, KMS [14,31,34,52,53,54] are usually-provided-with-Web-interface systems that should be able to provide *the right information to the right user at the right time and in the right format*. Application fields of KMS include: *e*-learning systems, health-care systems, *e*-procurement systems, *e*-government systems, industrial building management systems etc. ERP [8,35] represents an effective implementation following the KMS deployment paradigm. In some previous papers, we propose valuable improvements with respect to the commonly accepted architecture for KMS. In more detail, in [11] we propose an innovative reference architecture for *Knowledge Management-based Web Systems* (KM-bWS), which are Web systems that exploit the functionalities of traditional KMS, and make them more general and usable in other application fields. In [12], we propose an extension of the previous work, adding to the reference architecture functionalities useful to improve the Web resource searching and classification phases.

5.2.3 User Behaviour Tracking

A well-known class of systems based on user behaviour tracking are the so-called *Recommender Systems* [38]. These systems learn about user preferences over time, and automatically find things of similar interest, thus reducing the burden of creating explicit queries. They dynamically track users as their interests change, thus user behaviour modelling and tracking are their fundamental research issues. However, such systems require an initial learning phase where behaviour information is built up to form a user profile. During this initial learning phase, performance is often poor due to the lack of user information; this is known as the *cold-start problem* [29]. For these reasons, there has been increasing interest in developing and using tools for creating annotated content and making it available over the Web.

Collaborative Recommender Systems utilize user ratings to recommend items that are of interest for similar people. PHOAKS [46] is an example of collaborative filtering, recommending Web links mentioned in newsgroup articles. Only newsgroups with at least 20 posted Web links are considered by PHOAKS, avoiding the cold-start problem associated with newer newsgroups containing fewer messages. *GroupLens* [24] is an alternative example for recommending newsgroup articles. The authors report two cold-start problems in their experimental analysis. First, users abandon the system before they provide enough ratings to receive recommendations, and, second, early adopters of the system receive poor recommendations until enough ratings are gathered.

Content-based Recommender Systems recommend items with similar content to things the user has shown interest in before. Examples of content-based recommenders are *Fab* [2], and ELFI [42]. Fab recommends Web pages, and needs a few early ratings from each user in order to create a training set. ELFI recommends funding information from a database, observes users using this database and infers both positive and negative examples of interest from their behaviour. Another interesting work focused on knowledge management is [15]; the author proposes a method for finding people that work in similar areas via applying clustering and *Information Retrieval* (IR) technologies.

5.2.4 User Profile Modelling

For our work, methodologies and techniques oriented to define user profile models are also relevant. For instance, Widyantoro *et al.* [50] describe a new scheme to learn dynamic user interests in an automated information filtering and gathering system running on the Internet, using a learning algorithm derived for the particular representation of user interests. In [10], we present a probabilistic model for mapping user behaviour into a weighted digraph, and an algorithm that operates on this digraph in order to dynamically infer the user profile from a set of given profiles. Schiaffino and Amandi [40] propose enhancing the user modelling task in agent-based *e*-service systems via adopting well-known techniques coming from the *Artificial Intelligence* research area like *Case-Based Reasoning* (CBR) and *Bayesian Networks*. An *Incremental Hierarchical Clustering* (IHC) algorithm has been proposed by Widyantoro *et al.* [51]. This algorithm dynamically builds a hierarchy that satisfies the homogeneity and monotonicity properties with respect to user clustering in large-scale Web systems. More recently, specialized applications of the user profile modelling research theme have regarded: *user group modelling and cooperation* [16], *semantic-based user modelling* [37], *multitasking user modelling* [43], *user modelling in mobile and ubiquitous computing* [7].

5.2.5 Web Services Personalization

Web services personalization remains an active area of research and development, mainly because: (*i*) nowadays, Web services run as passive components rather than active components that can be embedded with high-level personalization mechanisms like the one based on *context* modelling and analysing [28]; (*ii*) state-of-the-art approaches (e.g., *Web Services Flow Language* (WSFL) [26] and *Business Process Execution Language for Web Services* (BPEL4WS) [9]) are mainly focused on Web services composition rather than personalization; (*iii*) lack of industrial standards regarding personalization techniques into the Web services initiative.

However, there are some enabling technologies that could be efficiently used to support Web services personalization. Among all, we mention: *profile filling*, which consists in tacking a picture of the user through fill-in forms [30]; *click-stream analysis* and *Web usage mining* [36,45], which consist in collecting data about user movements on a Web site; *collaborative filtering* [4,20], which consists in comparing tastes of a given user with those of other users in order to build up a picture of like-minded people; *cookies*, which continue to be useful for personalization even if they are not particularly new in Internet terms.

With respect to research perspectives, previous experiences addressing the problem of Web services personalization appears in literature in the vest of different alternatives. Riecken [39] propose treating the problem as that of building customer loyalty via meaningful one-to-one relationships. Smith and Cotter [44] propose collecting user preferences during his/her Web interaction and automatically building lists of preferences via analysing variations of user tastes over time. Wells and Wolfers [49] propose that one function for personalization is to help recreate the human element that understands the customer and offers the personalized touch (this approach has been then recognized as very useful in distance learning environments, where personal contact is limited by physical proximity). Schonberg *et al.* [41] propose measuring the effectiveness of a personalization Web service via defining (*i*) what "success" means for the particular Web application, and (*ii*) metrics and feedback techniques for measuring success.

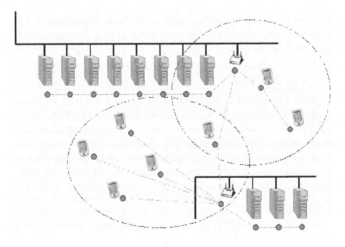

Fig. 5.1. DKN logical model

5.3 A Unified Framework for Making Web Services Knowledge-Aware

5.3.1 Discussion and Insights

Our goal is to delineate an application scenario supporting a *distributed knowledge environment* in which users can access useful knowledge, stored in very large data repositories of different nature (structured, semi-structured, raw, streaming data etc), regardless of their current position and connecting device (see Fig. 5.1). As an example, in such an environment a business analyst can manage trading data both on a desktop computer when he/she is working and on a mobile device when he/she is travelling. The "natural" flexibility offered by the Web services deployment platform allows us to achieve this scenario efficiently, also thanks to the amenity of modelling knowledge by XML, which is an important and popular Web data (and metadata) format among those listed previously. In fact, the same data can be delivered to client devices having different capabilities in terms of computational performance, memory size, and display ability, when they are modelled and formatted with XML[1].

The main goal of DKN consists in supporting the so-delineated distributed knowledge environment. Therefore, without any loss of generality, DKN comprises both wired and wireless domains, and their critical aspect is how to deliver personalized knowledge. A possible limitation for this framework could be represented by computational capabilities of mobile devices, which are usually low, but recent developments of wireless technology make this scenario feasible since current handheld devices are equipped with sufficient capability in terms of available memory and computational power. Of course, the gap with respect to wired devices, such as desktop computers, is still enormous but it has decreased during the last few years. For instance, the display capability is still an important drawback in terms of knowledge visualization, as a handheld device usually can only display few lines of text and small pictures. More generally, this aspect represents a real limitation in the development of very useful digital systems such as *e*-books and *e*-newspapers.

According to our proposal, the overall available knowledge in DKN is the result of single contributions of components belonging to DKN, where each component is specialized in a particular aspect of the knowledge domain. We refer this situation as a state of "distributed knowledge", according to scenarios drawn in Sect. 1. In such an environment, a user can extract knowledge from different distributed sources according to his/her profile, and can use this knowledge in an integrated "transparent-for-the-user" fashion. For instance, a manager can integrate the daily balance plot extracted by using his/her handheld device with other tabular historical data about purchases/sales coming from a remote data server. Moreover, in our vision knowledge delivery is *adaptive* because the underlying system automatically selects the most useful and suitable information for users, and subsequently provides this information by applying the most convenient formatting task for the current device.

The "invariant component" in DKN is the so-called *knowledge node* (see Fig. 5.1: each knowledge node is depicted as a red node), which is a device holding knowledge

---

[1] In the rest of the chapter, we focus our analysis taking into account XML as "dominant" kind of data within the DKN framework, but extending the proposal to other data formats is straightforward.

that is useful for one or more contexts, and all the knowledge nodes form DKN. In other words, knowledge nodes are abstract representations of accessing devices and they, all together, are an abstract representation of DKN with respect to the knowledge instead of the physical network layer. More precisely, we can say that, *with respect to the knowledge*, a knowledge node is a component of the network that holds local useful knowledge, and shares the total knowledge, whereas, *with respect to the network layer*, a knowledge node is a commonly-intended network device (e.g., personal computer, server host, laptop, PDA, third generation-cellular phone etc). A knowledge node is defined as "invariant" meaning that a user can access distributed knowledge via using any node belonging to DKN, and he/she can retrieve useful knowledge independently of the current accessing device, thus taking advantage from system functionalities in charge of formatting knowledge, which are mainly driven by display capabilities of the device.

In this scenario, a user registers on the system, sets an initial, generic profile according to his/her preferences/needs, and subsequently he/she accesses the knowledge available on the system via using wired and wireless devices that adhere to DKN. During the history of interactions between the user and the system, the user profile can be updated and the user can be moved from one profile to another. As a consequence, the middleware layer of the system is a state-aware layer. In fact, because of the complexity of the system, its evolution is described and managed using the *state system abstraction*, based on the *State Design* pattern proposed in the context of software engineering. Therefore, all the interactions of a given user U_i with the system are mapped onto a finite state automaton $A(U_i)$, where each node represents a state of the system and each arc represents a state transaction. For instance, in the context of e-learning systems, a state transaction could capture the situation in which a user reaches a new learning level, more advanced than the previous one. When U_i interacts with the system, the middleware layer of the system updates the involved states of the automaton $A(U_i)$ accordingly, and its topological configuration eventually.

As a consequence, the adaptation process is constituted by two phases. In the first phase, the system adapts the knowledge according to the user profile; in the second phase, the system adapts the knowledge according to technical requirements of the user device. It must be noted that the same knowledge assumes a different format for different devices. For instance, information on stock quotations appear with a rich graphic notation (such as plots having different resolutions and detailed information) on a desktop computer whereas the same information appear with textual format on a handheld device.

5.3.2 Definitions

Formally, the supported scenario is modelled as follows. We introduce: (*i*) a set Φ of N Web services providers $\Phi = \{WSP_0, WSP_1, ..., WSP_{N-1}\}$, (*ii*) a set Θ of P Web services exported by the Web services providers $\Theta = \{WS_0, WS_1, ..., WS_{P-1}\}$, and (*iii*) a set Z of M users $Z = \{U_0, U_1, ..., U_{M-1}\}$. Each user U_i can use Q different devices belonging to the set $\Omega = \{D_0, D_1, ..., D_{Q-1}\}$ for accessing one or more of the exported Web services. Given a user U_i, each session T_m of the system is represented by a tuple $t_{<m,i,W,j>} = <m, U_i, \{WS_h, WS_{h+1}, ..., WS_{h+k}\}, D_j>$. The meaning of $t_{<m,i,W,j>}$ is the following: during the session identified by m, U_i accesses, using the device D_j, Web services belonging to the set $W = \{WS_h, WS_{h+1}, \ldots, WS_{h+k}\}$. As we explain in Sect. 5.4.3 and Sect. 5.4.4, we use this formalism for two fundamental goals: (i) to capture, represent, and manage user behaviours against the DKN framework, and (ii) to reason on such behaviours, thus addressing crucial aspects of user dynamics like *profile change detection*, and so forth.

Furthermore, in the DKN framework a user model $m_{DKN}(UC_i)$, which we present in Sect. 5.4.2, and a user monitoring model $b_{DKN}(UC_i)$, which we present in Sect. 5.4.4, are associated to each user cluster UC_i (and, thus, to each user U_i belonging to the user cluster UC_i^2), and are exploited to manage users for personalization purposes.

5.3.3 Open Issues

As discussed in Sect. 5.1, we believe that *the right knowledge to the right user* paradigm is suitable for facing-off the described requirements, which are very often complex and sometimes conflicting. Furthermore, in a data-intensive Web-services-based environment, distinguishing and treating separately but in a unified framework *data personalization* and *service personalization* can significantly improve modern and successful Web applications and systems like KMS, B2B *e*-commerce systems, Web management systems for corporate data warehouses, *e*-learning systems, and so forth.

There are several open issues about the solution and framework we propose, concerning (i) efficient use of knowledge-based techniques inside a Web-services-based framework, (ii) data personalization, and (iii) service personalization, at both the design level and the technological level.

As regards knowledge-based techniques, we recognize the following open issues:

- *knowledge representation*: currently, there is not any widely accepted standard for knowledge representation; therefore, in many cases, designing and developing ad-hoc representation models is necessary;
- *knowledge inference model*: as for the previous one, often the knowledge inference model (i.e., the model used to extract knowledge from information) is often built as an ad-hoc solution starting from the knowledge representation model.

As regards data personalization, and specifically XML data personalization, we recognize the following open issues:

- *XML querying*: querying XML data is an actual open issue for the database research community as there are many different proposals, some of them based on mapping XML data onto relational databases, and others based on developing specialized query engines independent of the physical representation of XML data – note that this issue strongly influences the performance of DKN;
- *XML updating*: making updates on a very large XML document can be resource-intensive in terms of computational overheads, and this can impact the performance of DKN;
- *XML verification*: verifying the consistency of a very large XML document with respect to a *Document Type Definition* (DTD) or an *XML Schema* can also be a resource-intensive task;

---

[2] In the rest of the chapter, for the sake of simplicity, we restrict our models to the single user U_i as the *representative* of the user cluster UC_i.

- *XML implementation and storing issues*: implementation issues for XML data are still important challenges, which are very often neglected by the research community – nevertheless, this topic has important implications with previous ones and, particularly, becomes a crucial factor in the context of data-intensive Web services.

As regards service personalization, we recognize the following open issues:

- *(Web) services interoperability*: it concerns the fundamental goal of the Web-services-based deployment paradigm, i.e. realizing the interoperability between large-scale distributed applications, overcoming the limitations of more traditional request/response paradigms – in this context, the main issue is how to merge business process and workflow management techniques into today's Web services technology, like in BPEL4WS [9];
- *(Web) services security and privacy*: it represents the most important next-generation challenge in the context of Web services, as modern scenarios, such as *e*-procurement and *e*-government systems, demand more and more for innovative requirements like *confidentiality, authentication, integrity* in data communication and transport, *non-repudiation*; currently, *Cryptography and Public Key Systems* are the most investigated research topics to ensure security and privacy of Web services.

5.4   DKN User and User Monitoring Models

Personalizing knowledge is a critical issue in our approach. In order to tackle this effectively we first provide the description of the multidimensional user model, namely m_{DKN}, we propose to support the DKN framework, and then we present the user monitoring model, namely b_{DKN}, which is in charge of (*i*) capturing and mapping user behaviours, and, above all, (*ii*) detecting profile changes in user behaviours.

5.4.1   Motivations

Because of advances in technology, both in terms of functionalities offered to the end-user and in terms of the degree of complexity of platforms and applications, a need for *multidimensional user models* has arisen recently. Such models describe how users interact with systems via using a *variable set*, usually very large, and, above all, *multiple views* of the model in order to capture characteristics of users. Recent explosion in technology and consequent proliferation of smart, personal devices have made this modelling task very hard, since there are a lot of parameters that must be captured to design "good" user models. Furthermore, another need to be met is that of ensuring the quality of the target framework in efficiently discriminating and managing classes and clusters of users. To this end, the user monitoring model has to be strictly aligned with the user model since the first is in charge of the basic task of intercepting user interaction items and updating values of user model variables accordingly; but, besides this, it could even update the same "current" configuration of the user model. In other words, both the user model and the user monitoring model concur in personalizing knowledge. Note that, usually, many papers appearing in literature are devoted to the two models separately, but we believe that describing the interaction between such models is also very important to knowledge personalization purposes.

5.4.2 The DKN Multidimensional User Model

According to the research direction described in the previous Sections, we introduce a multidimensional user model in which user interaction items are organized in an OLAP fashion, such that (*i*) dimensions represent the observed user variables or *attributes*[3], (*ii*) members of each dimension represent value domains of the corresponding user variables, and (*iii*) measures represent objects to use during the personalization task. Note that, for our context, personalizing knowledge means both personalizing data and services so that each multidimensional entry on the model contains the locations of objects needed to accomplish this task, being such objects dynamically built by the DKN middleware at the beginning of user sessions, on the basis of the methodology we present in Sect. 5.5.

Now, we provide the formal definition for the multidimensional user model we propose. Let $G = \{\rho_0, \rho_1, ..., \rho_{R-1}\}$ be the set of R observed (user) variables about user behaviour (e.g., ρ_r could be the available bandwidth). They represent the dimensions in the OLAP model. For each variable ρ_r, let $E(\rho_r)$ be the set of admissible values for ρ_r (e.g., ρ_r being the available bandwidth, it could be that $E(\rho_r) = \{7.5$ Mbps, 8.0 Mbps, 8.5 Mbps, 9 Mbps$\}$). We denote by $V = \{E(\rho_0), E(\rho_1), ..., E(\rho_{R-1})\}$ the set of all the value domains. Let $l(\rho_r)$ be the level hierarchy defined on the variable ρ_r and let $\Gamma(\rho_r,h)$ be a function that takes the hierarchy $l(\rho_r)$ and an integer h, and returns the level h of $l(\rho_r)$, if it exists. We denote by $L = \{l(\rho_0), l(\rho_1), ..., l(\rho_{R-1})\}$ the set of all the hierarchies. Therefore, the multidimensional user model m_{DKN} is defined as follows:

$$m_{DKN} = <G, V, L, \Gamma> \qquad (5.1)$$

According to *the right knowledge to the right user* paradigm described in Sect. 1, each multidimensional entry in m_{DKN} contains the locations of the so-called *personalizing objects* (mentioned before), i.e. resources computed and used at run-time to personalize knowledge. These objects are: (*i*) an XML Schema file f_{XSD} modelling the scheme of personalized data to be presented to users (note that f_{XSD} supports the *data personalization* task, which we describe in Sect. 5.1); (*ii*) a set of WSDL files $\mathfrak{I}_{WSDL}$ describing Web services to be invoked on so-personalized data (note that $\mathfrak{I}_{WSDL}$ supports the *service personalization* task, which we describe in Sect. 5.2); (*iii*) an XSL file f_{XSL} containing the formatting rules to be applied on data that are personalized by means of the scheme stored in f_{XSD}, and processed by services described in $\mathfrak{I}_{WSDL}$.

This model is then specialized for each user U_i, i.e. $m_{DKN}(U_i)$, via exploiting OLAP technology and, particularly, the multidimensional view mechanism [22]. In order to compute personalizing objects, a collection of sets is introduced. We define these components of the model we propose as "latent" because they are transparent-for-the-user components that are exploited by the DKN middleware for supporting the

---

[3] In the rest of the chapter, we use the terms "observed user variable" and "attribute" interchangeably.

dynamic building of personalizing objects. To this end, for each multidimensional entry $m_{DKN}(U_i)$, the following sets are defined:

- a set of concepts C^D supporting data personalization;
- a set of knowledge nodes Z^D supporting data personalization;
- a set of services W^S supporting service personalization;
- a set of knowledge nodes Z^S supporting service personalization.

The meaning and the use of such sets will be described in Sect. 5.1 and Sect. 5.2. Thanks to intelligent personalization techniques that sequentially elaborate at first the pair $\{C^D, Z^D\}$ and then the pair $\{W^S, Z^S\}$, a multidimensional relation between a given user profile and personalizing objects is computed at run-time, and used to personalize knowledge. To give an idea, let U_i be a user, a possible entry of the model we propose could be the following:

$$m_{DKN}(U_i)[\texttt{tourist, expert, PDA, 9 Mbps,}$$

$$\texttt{MS Windows CE .NET,}$$

$$\texttt{MS Internet Explorer CE .NET}] =$$

$$\{\backslash\backslash\texttt{dynContent}\backslash\texttt{scheme.xsd,}$$

$$\backslash\backslash\texttt{dynService}\backslash\texttt{service.wsdl,}$$

$$\backslash\backslash\texttt{dynPresentation}\backslash\texttt{presentation.xsl}\}$$

(5.2)

This entry means that when U_i belongs to the cluster representing people of the class `tourist` and profile `expert` profile, uses a `PDA` as connecting device with a bandwidth equal to `9 Mbps`, adopts `MS Windows CE .NET` as OS platform and `MS Internet Explorer CE .NET` as Web browser, the DKN middleware provides access to the overall knowledge for U_i via (*i*) accessing and loading personalized data modelled by the XML Schema file `scheme.xsd`, located in the directory `\\dynContent\`, (*ii*) interacting with the Web service described in the WSDL file `service.wsdl`, located in the directory `\\dynService\`, which runs on personalized data, and (*iii*) formatting the processed (personalized) data on the basis of the presentation rules stored in the XSL file `presentation.xsl`, located in the directory `\\dynPresentation\`.

This approach allows us to manage users with a very high level of granularity: for instance, the same user U_i of the running example could have *other* entries on the *same* multidimensional model $m_{DKN}(U_i)$ in such a way that such entries state how knowledge is personalized when U_i is positioned in clusters represented by configurations different from the one of the running example because of one or more attributes. Therefore, another entry in $m_{DKN}(U_i)$ could establish that U_i must interact with DKN through the service described by the WSDL file `service1.wsdl`, located in the directory `\\dynService\`, when U_i accesses DKN adopting `WebToGo` as Web browser instead of `MS Internet Explorer CE .NET`, maintaining the other attributes and the other personalizing objects unchanged. Note that this mechanism allows us to capture thin variations in the attribute domains, thus to manage in different ways users whose behaviours are different for very few observed variables.

We highlight that we are assuming that personalizing objects are stored in local directories (i.e., directories of hosts where the DKN middleware is running), but extending the actual framework to deal with more sophisticated resource locators that also take into account distributed and decentralized objects, like URI, or semantics-based resource locators that allow to make the resource accessing process "computable", like RDF, is an exciting and promising research goal.

The considerations above give raise to the following definition.

Definition 5.1: DKN Multidimensional User Model. *The DKN multidimensional user model m_{DKN} is a relation between tuples $<G, V, L, \Gamma> \rightarrow <f_{XSD}, \Im_{WSDL}, f_{XSL}>$, such that: (i) $G = \{\rho_0, \rho_1, ..., \rho_{R-1}\}$ is the set of observed (user) variables; (ii) $V = \{E(\rho_0), E(\rho_1), ..., E(\rho_{R-1})\}$ is the set of admissible values for variables in G, being $E(\rho_r)$ the value domain of ρ_r; (iii) $L = \{l(\rho_0), l(\rho_1), ..., l(\rho_{R-1})\}$ is the set of level hierarchies defined on variables in G, being $l(\rho_r)$ the level hierarchy defined on ρ_r; (iv) Γ is a function that selects levels from L; (v) f_{XSD} is an XML Schema file used to personalize data; (vi) $\Im_{WSDL}$ is a set of WSDL files used to personalize services; (vii) f_{XSL} is an XSL file used to personalize presentations.*

5.4.3 OLAP-Based User Profile Fine-Grain Modelling and Management

OLAP technology was first proposed with the aim of managing massive data sets in the context of Data Warehousing systems. From a practical point of view, currently there are many efficient commercial implementations of OLAP technology that ensure good performance, both in the physical representation of the multidimensional data and in the query evaluation. For the purpose of our framework, these features allow us to achieve the following amenities: (i) many user classes can be managed with a high degree of granularity; (ii) the overall throughput of the system is significantly improved.

In more detail, we exploit the OLAP logic model and use the well-known mechanism of *aggregations* on levels [22], i.e. the amenity of aggregating the same (multidimensional) data according to different levels of resolution. This allows us to represent and manage user variables with a *very high level of granularity*, and, as a consequence, to significantly improve the knowledge personalization task. In order to better understand this claim, we first provide the definition of *user profile* in the multidimensional user model m_{DKN}.

Definition 5.2: DKN User Profile. *Given the DKN multidimensional user model $m_{DKN} = <G, V, L, \Gamma>$, a DKN user profile P is a multidimensional view $\Pi^M(k_0, k_1, ..., k_{R-1})$ on V obtained by aggregating values in V according to levels provided by $\Gamma(\rho_0, k_0), \Gamma(\rho_1, k_1), ..., \Gamma(\rho_{R-1}, k_{R-1})$, respectively.*

Among all the possible user profiles, the "default" user profile $\Pi^M(\Gamma(\rho_0, 0), \Gamma(\rho_1, 0), ..., \Gamma(\rho_{R-1}, 0))$ (i.e., the user profile for which any personalization is done) is the same of the value set V. This gives raise to the following lemma.

Lemma 5.1. *Given the DKN multidimensional user model $m_{DKN} = <G, V, L, \Gamma>$, $V \equiv \Pi^M(\Gamma(\rho_0, 0), \Gamma(\rho_1, 0), ..., \Gamma(\rho_{R-1}, 0))$.*

With respect to the user variable domain, a profile P is defined as a set of admissible value sets for each observed parameter (i.e., dimension) ρ_r belonging to the multidimensional model m_{DKN}, as follows:

$$P = \{E_P(\rho_0), E_P(\rho_1), ..., E_P(\rho_{R-1})\} \tag{5.3}$$

such that $E_P(\rho_r) \subseteq E(\rho_r) \; \forall \; r \in \{0, 1, ..., R-1\}$. More precisely, (5.3) represents a *stereotypical* user profile [5] with respect to the overall knowledge available in the system. At the boot phase, the system can start with a set of pre-defined user profiles that could be created by the system administrator on the basis of his/her expertise on the particular application domain. At the run-time phase of the system, the profile set can be updated according to outputs of the user monitoring model. Similarly, profiles that no longer match the current users can be deleted from the profile set. To adequately support the *strong* need for efficiently managing user profiles arising from such a framework, a relation database *PDB* stores the mapping between user profiles as set of attributes (i.e., $k_0, k_1, ..., k_{R-1}$ in Def. 5.2) and user profiles as OLAP views (i.e., $\Pi^M(k_0, k_1, ..., k_{R-1})$ in Def. 5.2).

The OLAP-based user model allows us to capture sophisticated aspects concerning both the profile modelling and the profile management tasks. For instance, we can efficiently model profiles defined starting from the "intersection" of two or more stereotypical profiles. We highlight that this is a very frequent condition in the context of modern Web applications and systems, as Web users very often indirectly define overlapping clusters, being features of some clusters in some others, and so forth. As an example, in Fig. 5.2 the profile P_c is obtained from the profiles P_a and P_b: therefore, profile P_c can effectively represent users having features that are in P_a and P_b separately, but not in both profiles.

User Model

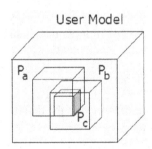

Fig. 5.2. Advanced profile modelling in DKN

The indirect advantage of this approach is that we can build formal models, like *hierarchies of profiles*, and make computation over them, perhaps by resorting to algorithms and techniques coming from the Data Mining and Artificial Intelligence research areas. All considering, we can claim that the adoption of the OLAP logic model inside the DKN proposal allows us to improve the semantics and the computational power of the user profile management.

The OLAP-based user model also supports *multi-resolution management of profiles*, i.e. the amenity of managing profiles according to different levels of resolution.

In other words, our proposed framework is able to provide personalized knowledge to a *same* user according to *different* levels of resolution defined on the *same* user variables via taking into account *ranges of attributes* instead of single attributes like those in the previous running example. With respect to the previous example, we can also express multidimensional entries like the following:

$m_{DKN}(U_i)$[{tourist..analyst},

{beginner..expert},

{PDA..PC},{4 Mbps..12 Mbps},

{MS Windows CE .NET..Palm OS},

{MS Internet Explorer CE   (5.4)
.NET..WebToGo}]=

{\\dynContent\scheme.xsd,

\\dynService\service.wsdl,

\\dynPresentation\presentation.xsl}

that is, given a user U_i, we can specify *ranges* of attributes in the attribute domains, and augment the power of the personalization task via setting, for a unique multidimensional entry $m_{DKN}(U_i)$, *different* personalizing objects that are valid for *sub-ranges* of ranges of attributes defined in $m_{DKN}(U_i)$. We highlight that this approach is very useful in many real-life Web dynamics: for instance, an analyst that accesses business data through a handheld device could retrieve these data in a graphical fashion if the current bandwidth b belongs to the range {10..15} Mbps, or in a textual fashion if b belongs to the range {4..8} Mbps. This leads to the definition of the *DKN multi-resolution user model* m_{DKN}^{MR}, which extends the DKN multidimensional user model m_{DKN} (see Def. 5.1).

Definition 5.3: DKN Multi-Resolution User Model. *The DKN multi-resolution user model* m_{DKN}^{MR} *is a tuple* $<m_{DKN}, R, \Gamma^+>$, *such that: (i)* m_{DKN} *is the DKN multidimensional user model; (ii) R is a set of ranges defined on the observed (user) variable set* $G \in m_{DKN}$; *(iii)* Γ^+ *is a function that, for each assignment* $\{r_0, r_1, ..., r_{R-1}\}$ *in R, returns a tuple* $<f_{XSD}, \mathcal{S}_{WSDL}, f_{XSL}>$ *in the co-domain of* m_{DKN}.

Lemma 5.2 asserts the equivalence between m_{DKN}^{MR} and m_{DKN} when $|R| = 0$, i.e. when there not exist ranges defined on G (note that this depends on the particular application domain on which the DKN framework is modelled).

Lemma 5.2. *Under the condition that* $|R| = 0$, $m_{DKN}^{MR} \equiv m_{DKN}$.

Finally, as regards computational issues concerning the described profile management, we recall that OLAP technology allows us to efficiently manage views (i.e., profiles), and, thus, we can stress this aspect of the proposed framework and fit the need for flexibility in managing users and user clusters, as required by modern Web applications and systems.

5.4.4 The DKN Multidimensional User Monitoring Model, and the Profile Change Detection Process

Given a user U_i and M different subsequent sessions $T = \{T_0, T_1, ..., T_{M-1}\}$, at session T_j the proposed approach provides the right knowledge (i.e., data and services) to U_i on the basis of his/her current (user) profile P_j, which is computed, at the *beginning* of T_j, by evaluating his/her behaviour from session T_0 until session T_{j-1}. During the interaction with the DKN framework at session T_j, the monitoring components collect parameters for computing the "next" profile of U_i at session T_{j+1}. Note that M is an input parameter for our proposed framework, and it must be empirically set, mainly taking into account (*i*) performance of hosts where the DKN middleware is running, and (*ii*) load-balancing issues.

In our formal model, given a user U_i and a session T_k, we define the *behaviour* $B_{i,k}$ of U_i at session T_k as follows:

$$B_{i,k} = \{b_0, b_1, ..., b_{R-1}\} \tag{5.5}$$

such that $b_0 \in E(\rho_0)$, $b_1 \in E(\rho_1)$, ..., $b_{R-1} \in E(\rho_{R-1})$. In other words, $B_{i,k}$ is an R-wide tuple representing values of the observed variables that are automatically detected and captured by the monitoring components for U_i during T_k, on the basis of his/her interaction with the DKN framework. For instance, with respect to (5.2) and (5.4), a possible value for $B_{i,k}$ could be the following:

$$\begin{aligned} B_{i,k} = \{\texttt{analyst}, \texttt{beginner}, \texttt{PC}, \texttt{10 Mbps},\\ \texttt{MS Windows XP}, \texttt{Netscape Navigator}\} \end{aligned} \tag{5.6}$$

From a practical point of view, user behaviours are stored in the relational database *BDB* and, when a user U_i accesses the system at session T_k, the DKN framework automatically loads and associates to U_i the "previous" behaviour $B_{i,k-1}$ (i.e., the behaviour of U_i during the session T_{k-1}), which is used to provide personalized knowledge for U_i during T_k. The "current" behaviour $B_{i,k}$ is computed during T_k, and used to update *BDB* at the *end* of T_k, thus being available for the session T_{k+1}.

We define the DKN multidimensional user monitoring model b_{DKN} as a relation between the domain $G \in m_{DKN}$ and the domain $V \in m_{DKN}$ (we recall that $V = \{E(\rho_0), E(\rho_1), ..., E(\rho_{R-1})\}$). Therefore, $B_{i,k} \in b_{DKN}$.

Definition 4: DKN Multidimensional User Monitoring Model. *The DKN multidimensional user monitoring model b_{DKN} is a relation between domains $G \rightarrow V$, such that (i) $G = \{\rho_0, \rho_1, ..., \rho_{R-1}\}$ is the set of observed (user) variables, and (ii) $V = \{E(\rho_0), E(\rho_1), ..., E(\rho_{R-1})\}$ is the set of admissible values for the variables in G, being $E(\rho_r)$ the value domain of ρ_r.*

This model is then specialized for each user U_i and for each session T_k, i.e. $b_{DKN}(U_i, T_k)$, in such a way that users are monitored with a high level of accuracy, also tracking possible variations in their behaviours.

Furthermore, we denote by $W_1(T_{k-1}, T_k)$ the *1-step transaction* from session T_{k-1} to session T_k. More generally, we denote by $W_m(T_{k-m}, T_k)$ the *m-step transaction* from session T_{k-m} to session T_k. These definitions are useful to model an important process of the DKN framework, i.e. the *profile change detection*, which is the basic operation to cluster users into profiles.

To this end, we define the behaviour $B_{i,k-1 \rightarrow k}$ of U_i across a *1*-step transaction $W_l(T_{k-1}, T_k)$ as follows:

$$B_{i,k-1 \rightarrow k} = \{<b_{0,k-1}, b_{0,k}>, <b_{1,k-1}, b_{1,k}>, ..., <b_{R-1,k-1}, b_{R-1,k}>\} \qquad (5.7)$$

and the behaviour $B_{i,k-m \rightarrow k}$ of U_i across an *m*-step transaction $W_m(T_{k-m}, T_k)$ as follows:

$$B_{i,k-m \rightarrow k} = \{<b_{0,k-m}, .., b_{0,k}>, <b_{1,k-m}, .., b_{1,k}>, ..., <b_{R-1,k-m}, .., b_{R-1,k}>\} \qquad (5.8)$$

Profile change detection is the most important issue for our work on the user monitoring model. With respect to a *1*-step transaction $W_l(T_{k-1}, T_k)$, we introduce the *variation* of the behaviour $\Delta B_{i,k-1 \rightarrow k}$ for a given user U_i during $W_l(T_{k-1}, T_k)$ as follows:

$$\Delta B_{i,k-1 \rightarrow k}(B_{i,k-1}, B_{i,k}) = B_{i,k} - B_{i,k-1} \qquad (5.9)$$

Therefore, we can model a profile change for U_i belonging to the profile P_j during $W_l(T_{k-1}, T_k)$ as follows. If $\Delta B_{i,k-1 \rightarrow k}(B_{i,k-1}, B_{i,k}) = \varnothing$, then U_i still belongs to the profile P_j. If $\Delta B_{i,k-1 \rightarrow k}(B_{i,k-1}, B_{i,k}) = \{b_p, b_{p+1}, ..., b_{p+q}\}$, such that $p > 0 \wedge q > 0$, and $b_h \in P_j \forall h \in \{p, p+1, ..., p+q-1\}$, then U_i still belongs to the profile P_j; otherwise (i.e., $\exists\, b_h \notin P_j \wedge \exists\, P_z: b_h \in P_z$) U_i belongs to another profile P_z. If P_z does not exist, then a new profile is created and added to the database *PDB*.

Note that assigning a given user U_i to a pre-existent profile or creating a new profile are critical aspects of our framework. In fact, we highlight that, as regards the first issue, many profiles satisfying the above-described condition could exist, i.e. there could exists a set of profiles $Y = \{P_z, P_{z+1}, ..., P_{z+t-1}\}$, with $t > 0$, such that $b_h \in P_l \forall h \in \{p, p+1, ..., p+q-1\}$, with $p > 0 \wedge q > 0$, $\forall\, l \in \{z, z+1, ..., z+t-1\}$, and, thus, the consequential problem is to determine, among all the possibilities, which profile $P_z \in Y$ has to be selected in order to assign U_i to it. To solve this problem, we devise the following criterion: given a user U_i, an input observed variable set $\Delta B_{i,k-1 \rightarrow k}(B_{i,k-1}, B_{i,k}) \neq \varnothing$ (i.e., such that a profile change has happened), and a set of eligible profiles Y, we select the profile $P_z \in Y$ having (*i*) the *biggest* number of "changed" variables (or, equally, the biggest value of $|\Delta B_{i,k-1 \rightarrow k}(B_{i,k-1}, B_{i,k})|$), and (*ii*) the *smallest* ranges defined on the involved (observed) variables $G_{i,k} = \{\rho_p, \rho_{p+1}, ..., \rho_{p+q}\}$ (see Def. 5.3). This approach aims at classifying U_i in the most "specific" one among all the available profiles in Y, in such a way as to accurately fit his/her needs and objectives. Augmenting the granularity of the DKN framework in managing users is the underlying goal of this approach.

As regards the second issue, i.e. the creation of a new profile, we exploit the "expressive power" and the flexibility offered by the DKN multi-resolution user model m_{DKN}^{MR} once again. In this case, we build the new profile P_z in such a way that it has (*i*) the *smallest* number of "changed" variables (or, equally, the smallest value of $|\Delta B_{i,k-1 \rightarrow k}(B_{i,k-1}, B_{i,k})|$), and (*ii*) the *biggest* ranges defined on the involved (observed) variables $G_{i,k} = \{\rho_p, \rho_{p+1}, ..., \rho_{p+q}\}$ (see Def. 5.3). This approach aims at creating P_z as a "general" profile that should be able to fit the requirements of unusual users (i.e., users that do not match any of the pre-existent profiles) through a sort of "best-effort" strategy. This new profile P_z will be then consolidated and refined during the next iterations, by means of the above-defined mechanism for assigning new users to pre-existent profiles.

Without any loss of generality, we introduce the variation of the behaviour $\Delta B_{i,k\text{-}m\rightarrow k}$ for a given user U_i during an m-step transaction $W_m(T_{k\text{-}m}, T_k)$ as follows:

$$\Delta B_{i,k\text{-}m\rightarrow k}(B_{i,k\text{-}m}, ..., B_{i,k}) = \Delta B_{i,k\text{-}m\text{-}1\rightarrow k\text{-}m} \cup ... \cup \Delta B_{i,k\text{-}1\rightarrow k} \qquad (5.10)$$

and the change profile detection on an m-step transaction is modelled via generalizing the task given for the change profile detection on a 1-step transaction.

5.5 Personalizing Knowledge in DKN

In this Section, we first provide a formal theoretical framework for representing and extracting personalized data in DKN; after that, we will describe how services are personalized in DKN, thus achieving the global goal of modelling knowledge personalization by means of data and service personalization.

The data personalization framework is *data-centric*, meaning that it takes into account knowledge representation and management techniques on XML repositories without looking at the particular deployment paradigm (i.e., Web services). In other words, we can claim that our data personalization proposal is "orthogonal" with respect to the deployment paradigm. As a consequence, such a proposal can be used along with different deployment paradigms like the traditional *Client/Server* paradigm, the *Parallel Computing* paradigm, and so forth.

5.5.1 Data Personalization in DKN

In our framework, the set of knowledge nodes are denoted by $K = \{n_0, n_1, ..., n_{Q\text{-}1}\}$. As described in Sect. 5.3, each knowledge node n_q can represent both a wired device or a wireless device, belonging to the set Ω (see Sect. 5.3.2). Furthermore, for each $n_q \in K$, we denote by $\Sigma(n_q)$ the representation model for the knowledge held in n_q. $\Sigma(n_q)$ is modelled as a *network of inter-related concepts*, exploiting the XLink excellent functionalities for representing "reticular" information. As a consequence, $\Sigma(n_q)$ is a direct graph where (*i*) nodes represent concepts, and (*ii*) each edge $C_i \rightarrow C_j$ represents the fact that the concept C_j is "related" to the concept C_i (i.e., a certain relationship r_{ij} between C_i and C_j exists). We believe this kind of representation as the most convenient way to model knowledge, taken into consideration characteristics of the application scenario described by the DKN framework. We highlight that $\Sigma(n_q)$ is different from the proper XML repository, denoted by $R(n_q)$, which stores knowledge contents. In other words, we can say that $\Sigma(n_q)$ is a "description", with respect to the knowledge level, of $R(n_q)$. For instance, in an *e*-learning context $\Sigma(n_q)$ could represent the set of topics (i.e., concepts, in the learning domain) belonging to an *e*-course provided by a server n_q in a university, whereas $R(n_q)$ could represent the proper didactical material. The network of concepts (i.e., $\Sigma(n_q)$) can in turn be stored within XML documents,

```
<teaching_course>
  <name>Database Systems</name>
  <class>Computer Science</class>
  <topics>
    <topic ID="CS_DB_0001">
      <name>
        Entity Relationship Model
      </name>
      <location xlink:type="simple"
      link:href=
      "file://topicsRepository/dataBase/
      topics/ER.xml"/>
      [...]
    </topic>
    [...]
  </topics>
  <related_topics>
    <name ID="CS_DM_0001">
      Domain Modelling
    </name>
    <name ID="CS_CM_0009">
      Constraint Management
    </name>
    [...]
  </related_topics>
  [...]
</teaching_course>
```

Fig. 5.3. $\Sigma(n_q)$ for a CS *e*-course on database systems

since the well-known properties of XML effectively support the need for flexibility
and frequent updates of data as required by knowledge management. Fig. 5.3 shows
the $\Sigma(n_q)$ for a Computer Science *e*-course on database systems.

As shown in Fig. 5.3, the modularity of the XML namespace-based distributed data
framework and the XLink facilities can perfectly match the need for dynamism in the
knowledge management. Note that contents stored in $R(n_q)$ and "described" by $\Sigma(n_q)$
are located through XLink pointers. This important feature is exploited by the data
personalization task in DKN, as described next.

It should be noted that, starting from the data-centric approach illustrated above, it
is easy to build a network of inter-related concepts on the basis of filtering processes
running on the model of the current user. Also, we highlight that, in order to provide
data personalization, the DKN middleware processes the description $\Sigma(n_q)$ instead of
the repository $R(n_q)$ for each involved knowledge node n_q, thus introducing low computational overheads because the size of $\Sigma(n_q)$ is always many orders of magnitude
smaller than the size of $R(n_q)$.

The DKN data personalization scenario we are drawing works follows. Given a set
of very large XML repositories $R(n_q)$ distributed on DKN (see Fig. 5.4), the goal is to

provide personalized data to users by means of knowledge management techniques performing on the descriptions of repositories $R(n_q)$, $\Sigma(n_q)$. To this end, we introduce the knowledge extraction task performed on the node n_q, denoted by $\Pi(n_q)$, being Π the so-called *knowledge extraction operator*. In contrast with the previous case, this formalization is tightly tied to the specific application domain. Without any loss of generality, we can think of this task as the selection of a subset of $\Sigma(n_q)$, that is, $\Pi(n_q) = \prod_D \Sigma(n_q)^4$, such that D is the selecting (concept) set, and a concept C_i be-

longing to the set $\Sigma(n_q)$ is projected into the set $\Pi(n_q)$ iff $C_i \cap D \neq \varnothing$. Essentially, building the selecting set D is the fundamental operation, as $\Pi(n_q)$ is implemented through the well-known projection operation coming from Set Theory. The obvious advantage of this approach is just being data-centric, as mentioned above, i.e. it directly works on data through very simple procedures like set projection, without invoking the execution of complex and resource-intensive algorithms that could mine the efficiency and, above all, the scalability of the proposed framework.

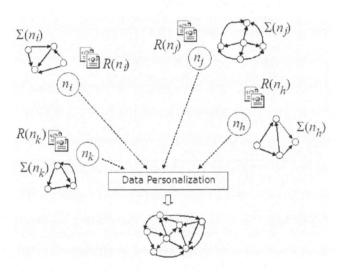

Fig. 5.4. Data personalization in DKN

As mentioned in Sect. 5.4.2, in order to support the data personalization task, for each user U_i a collection of concepts C^D and a collection of knowledge nodes Z^D are stored in the model $m_{DKN}^{MR}(U_i)$. C^D holds concepts that are retrieved by U_i from the knowledge nodes held in Z^D. In order to better understand the details of the data personalization task, the definition of *k-degree related concept* $C_{i,j}^k$ of a given concept C_i is needed.

---

[4] Note that, besides the knowledge extraction operator, $\Pi(n_q)$ also denotes the set of concepts extracted from $\Sigma(n_q)$.

Definition 5.5: *k*-Degree Related Concept. *Given a knowledge node n_q and a concept C_i belonging to the description $\Sigma(n_q)$, a k-degree related concept $C_{i,j}^k$ of C_i is a node in $\Sigma(n_q)$ that can be reached starting from C_i through a k-wide direct path.*

Lemma 5.3 and Lemma 5.4 report some inferred proprieties driven by considerations about the semantics of concepts and topology of direct graphs.

Lemma 5.3. *Given a knowledge node n_q and a concept C_i belonging to the description $\Sigma(n_q)$, any "semantically-intended" related concept C_j of C_i is a 1-degree related concept of C_i with respect to the DKN framework.*

Lemma 5.4. *Since each concept C_i is semantically related to itself, any concept C_i is a 0-degree related concept of itself with respect to the DKN framework, i.e. $C_{i,i}^0 \equiv C_i$ $\forall\ C_i \in \Sigma_{DKN} = \{\Sigma(n_0), \Sigma(n_1), ..., \Sigma(n_{Q-1})\}$.*

When the user U_i with behaviour $B_{i,k-1}$ accesses the system at session T_k, the DKN middleware performs, at the beginning of session T_k, the following data personalization operations:

1. extract from $m_{DKN}^{MR}(U_i)$ the set of concepts $C_{i,k}^D$ via applying the multidimensional entry $B_{i,k-1}$;

2. extract from $m_{DKN}^{MR}(U_i)$ the set of knowledge nodes $Z_{i,k}^D$ via applying the multidimensional entry $B_{i,k-1}$;

3. for each knowledge node n_z belonging to the set $Z_{i,k}^D$, process the description $\Sigma(n_z)$ and retrieves (*i*) the set of concepts that match the set $C_{i,k}^D$, namely $E_{i,k}(n_z)$, and (*ii*) the set of *1*-degree related concepts of the concepts in $E_{i,k}(n_z)$, namely $E_{i,k}^1(n_z)$;

4. execute procedure `personalizeData`, described below, and build the XML Schema file f_{XSD} that models the scheme of personalized data.

Procedure `personalizeData` takes as input: (*i*) the set of knowledge nodes $Z_{i,k}^D$, (*ii*) the set of concepts $E_{i,k}^T = \bigcup\limits_{z=0}^{|Z_{i,k}^D|-1} E_{i,k}^T(n_z)$, such that $E_{i,k}^T(n_z) = E_{i,k}(n_z) \cup E_{i,k}^1(n_z)\ \forall\ n_z$, and (*iii*) the name of the output XML schema file f_{XSD}, *filename*, and returns as output the file `filename.xsd` that models the mapping between concepts belonging to $E_{i,k}^T$ and resources (i.e., the XML repositories $R(n_z)$) distributed across DKN, thus realizing the entity f_{XSD} in the model m_{DKN}^{MR} (see Sect. 5.4.3).

During session T_k, the DKN middleware performs other management operations, which concur to form the "new" contents of sets $C_{i,k+1}^D$ and $Z_{i,k+1}^D$, which will be used by the DKN middleware to provide personalized data to U_i at the next session T_{k+1}. These operations are the following:

1. if some of the concepts belonging to the set $C_{i,k}^D$ are no longer retrieved by U_i, remove them from $m_{DKN}^{MR}(U_i)$;

2. if some of the knowledge nodes belonging to the set $Z_{i,k}^D$ are no longer browsed by U_i, remove them from $m_{DKN}^{MR}(U_i)$;

3. if new concepts (i.e., concepts that were not in $E_{i,k}^T$) are retrieved by U_i, add them to $m_{DKN}^{MR}(U_i)$;

4. if new knowledge nodes (i.e., knowledge nodes that were not in $Z_{i,k}^D$) are browsed by U_i, add them to $m_{DKN}^{MR}(U_i)$.

In our proposal, the knowledge extraction task $\Pi(n_q)$ is improved via defining a *semantic metrics m* on concepts, so that a concept C_i not satisfying the non-null intersection condition can be added to the set $E_{i,k}$ if another concept C_j such that $m(C_i,C_j) < V$ exists, i.e. the semantic distance between C_i and C_j is bounded by an empirically-determined threshold V. This allows us to efficiently manage the uncertainty that could be due to the proposed approach. In more detail, with regard to this aspect of our approach, we propose a very simple solution based on *WordNet* [55], a lexical database for the English language, in order to detect terms as synonyms and related-words that allow us to improve the task $\Pi(n_q)$. As for the set $\Sigma(n_q)$, the context of *e*-learning systems is a clarifying example about this amenity of the operator $\Pi(n_q)$. Here, a student, or a teacher too, can assemble an *e*-course via selecting a set of topics, even from different scientific fields of study. Therefore, when the user composes the *e*-course in a transparent manner, the underlying application logic should be able to define and execute the right data personalization task (i.e., the right instantiation for $\Pi(n_q)$). Despite the previous guidelines, defining ad-hoc data personalization engines (perhaps as specialized instances of this general framework) for different application domains, such as people-finding systems (i.e., systems devoted to find people skilled in a certain role for very complex organizations), is a must in order to significantly improve the overall throughput of the target knowledge personalization system.

5.5.2 Service Personalization in DKN

Personalizing services is a task more difficult than personalizing data. At present, we use a particular service for one or more multidimensional entries on m_{DKN} (see Fig. 5.5), and we manage services with a similar spirit that data. As mentioned in Sect. 5.4.2, in order to support the service personalization task, for each user U_i a collection of services W^S and a collection of knowledge nodes Z^S are stored in the model $m_{DKN}^{MR}(U_i)$. W^S holds services that are invoked from U_i on the knowledge nodes held in Z^S. Note that this formalization is "parallel" to that provided for data personalization.

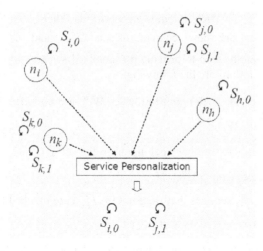

Fig. 5.5. Service personalization in DKN

We highlight that in the case of data personalization, the output personalized data are *related* among them, and, for this reason, they are represented as a network of concepts. On the contrary, in the case of service personalization, the output of the task is a collection of *selected* services, without any relationship among them.

When the user U_i with behaviour $B_{i,k-1}$ accesses the system at session T_k, the DKN middleware performs, at the beginning of session T_k, the following service personalization operations:

1. extract from $m_{DKN}^{MR}(U_i)$ the set of services $W_{i,k}^{S}$ via applying the multidimensional entry $B_{i,k-1}$;

2. extract from $m_{DKN}^{MR}(U_i)$ the set of knowledge nodes $Z_{i,k}^{S}$ via applying the multidimensional entry $B_{i,k-1}$;

3. for each knowledge node n_z belonging to the set $Z_{i,k}^{S}$, retrieve the set of services that match the set $W_{i,k}^{S}$, namely $J_{i,k}(n_z)$;

4. execute procedure `personalizeServices`, described below, and build the set of WSDL files $\mathcal{I}_{WSDL}$ that collects the personalized services.

Procedure `personalizeServices` takes as input: (*i*) the set of knowledge nodes $Z_{i,k}^{S}$, and (*ii*) the set of services $J_{i,k}^{T} = \bigcup_{z=0}^{|Z_{i,k}^{S}|-1} J_{i,k}(n_z)$, and returns as output a set of personalized services running on DKN, thus realizing the entity $\mathcal{I}_{WSDL}$ in the model m_{DKN}^{MR} (see Sect. 5.4.3).

We highlight that services belonging to $\mathcal{I}_{WSDL}$ run on data that have been previously personalized according to the technique given in Sect. 5.5.1.

During session T_k, the DKN middleware performs other management operations, which concur to form the "new" contents of sets $C_{i,k+1}^{S}$ and $Z_{i,k+1}^{S}$, which will be used by the DKN middleware to provide personalized services to U_i at the next session T_{k+1}. These operations are the following:

1. if some of the services belonging to the set $W_{i,k}^{S}$ are no longer invoked from U_i, remove them from $W_{i,k}^{S}$;

2. if some of the knowledge nodes belonging to the set $Z_{i,k}^{S}$ are no longer browsed by U_i, remove them from $m_{DKN}^{MR}(U_i)$;

3. if new services (i.e., services that were not in $J_{i,k}^{T}$) are invoked from U_i, add them to $m_{DKN}^{MR}(U_i)$;

4. if new knowledge nodes (i.e., knowledge nodes that were not in $Z_{i,k}^{S}$) are browsed by U_i, add them to $m_{DKN}^{MR}(U_i)$.

This methodology can be improved via specializing the same service on the basis of its input values, or better, its *context of execution*. This allows us to improve the semantics of services, and, consequently, augment their expressive power and the effectiveness of reasoning on them. For instance, two releases could be defined for the same GIS Web service WS_j depending if GIS data are more or less fine-grained, and are equipped with metadata or not. Note that, in such described cases, WS_j executes different runs as it could offer many or few functionalities to users, according to the nature of GIS data. Such a context-oriented framework could be further improved via taking into account the computational resources of the host on which services run. Currently, we are designing and developing a knowledge-aware and cost-based methodology for dynamically selecting *the right (Web) service to the right user*. We are working in this direction not only considering the signature of the Web service (as many recent approaches have proposed) but also by taking into consideration the needs of the Web service in terms of spatial and temporal resources, in order to define a computational framework able to segment and allocate different computational tasks across a set of computational nodes according to cost-based functions. We argue that, looking at the novel *aspect-oriented computing paradigm* could allow us to achieve significant advantages in this research effort.

5.5.3 Formatting Personalized Knowledge

In order to support the *presentation personalization task*, we devise a template-oriented technique that generates the right XSL formatting rules in dependence on the display capabilities of user devices. To this end, an XSL template database *TDB* is defined inside the DKN framework. *TDB* stores XSL rules stating how GUI components (such as frames, lists, check boxes, combo boxes etc) must be presented on a given device (such as desktop computers, laptop computers, PDA etc).

Given a user U_i at session T_k, to format (personalized) knowledge, the DKN middleware retrieves the features of the user device $D_{i,k}$, and automatically builds an output XSL file via (i) querying the database *TDB* to extract XSL rules tailored for the features of $D_{i,k}$, and (ii) putting together such rules, thus realizing the entity f_{XSL} in the model m_{DKN}^{MR} (see Sect. 5.4.3).

5.5.4 Knowledge Personalization in Action

In this Section, we describe how the system performs the knowledge personalization task. According to the formal models presented in Sect. 5.5.1 and Sect. 5.5.2, in a given session T_k, a user U_i belonging to a (previous) profile P_{k-1} has a (previous) behaviour $B_{i,k-1}$. Therefore, when U_i accesses the system, session T_k starts and the DKN middleware (i) executes the data personalization task, (ii) executes the service personalization task, (iii) executes the presentation personalization task, (iv) starts the management operations for both data and services, (v) begins to collect (new) information about U_i. Note that when T_k starts, the DKN middleware provides personalized knowledge for U_i via accessing both the user model $m_{DKN}^{MR}(U_i)$ and the user monitoring model $b_{DKN}(U_i,T_k)$ using $B_{i,k-1}$, retrieved from the database *BDB* as a multidimensional entry. During T_k, the DKN middleware computes the "new" behaviour $B_{i,k}$. When T_k ends, the new behaviour $B_{i,k}$ is defined and the DKN middleware computes the value of $\Delta B_{i,k-1 \to k}(B_{i,k-1}, B_{i,k})$ in order to determine if U_i still belongs to the same profile P_{k-1} (i.e., $P_k = P_{k-1}$) or to another one (i.e., $P_k \neq P_{k-1}$). Eventually, user profiles defined on the system are updated according to the method described in Sect. 5.4.4. In any case, a *current* entry $B_{i,k}$ for both the user models is defined, and it will be used to provide personalized knowledge to U_i at the next session T_{k+1}. The knowledge personalization task for an m-step transaction is the generalization of the latter task provided for a *1*-step transaction.

To face-off the so-called *cold-start problem* (see Sect. 5.2) that could degrade the effectiveness of the above-described mechanism, we propose submitting to the users, at the first log-in, a sort of "setting-up questionnaire" that allows us to (i) define initial user profiles, (ii) position users into these profiles, and (iii) compute the starting configurations for user behaviours and for the "latent" sets C^D, Z^D, W^S, and Z^S.

5.6 The Proposal of a Reference Architecture Implementing the DKN Framework

As described in Sect. 5.3, the DKN framework is composed of different kinds of nodes, which can belong to a wired or a wireless environment. In this heterogeneous domain, each node provides knowledge, represented as XML data and WSDL descriptions, and indexes other nodes which hold correlated knowledge. This approach allows the status of distributed knowledge to be maintained, in such a way that a user can access useful knowledge wherever he/she is located.

A basic issue in this context is how to make Web services knowledge-aware, that is, how to push knowledge management into Web services technology, or, in other

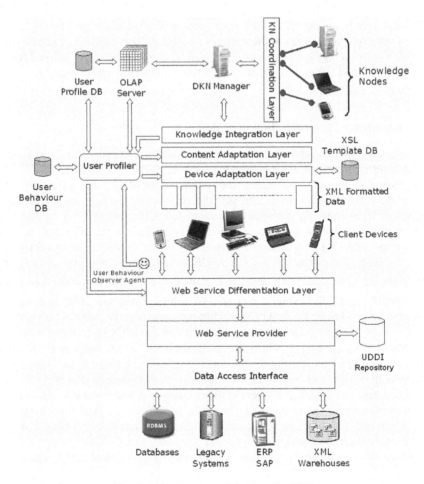

Fig. 5.6. A reference architecture for DKN

words, how to personalize Web services with respect to users. As a consequence, having dynamic, lightweight software modules is mandatory, because of scalability and reliability are the most critical design requirements in a system in which there is the need for activating or deactivating software components quickly. However, as mentioned above, with respect to the specific service personalization requirement, in the current DKN framework, we implement the selection mechanism of *the right (Web) service to the right user*, via giving meaningful support based on multidimensional reasoning. On the contrary, as regards the data personalization requirement, we devise an innovative strategy based on representing the knowledge of very large XML repositories through networks of concepts, and processing such networks via knowledge management techniques.

The reference architecture for DKN we propose is depicted in Fig. 5.6. From a technological point of view, the DKN framework is a component-based system in which each component is specialized in the execution of a particular task. Agent technology is also integrated inside the DKN framework, properly for implementing client functionalities like user behaviour tracking. Of course, the DKN middleware is running in each knowledge node. It supports: (*i*) connection and authentication functionalities, (*ii*) data access functionalities, (*iii*) proxy server functionalities, (*iv*) data, service, and presentation personalization functionalities in cooperation with other intelligent components running in the server side of the framework.

As shown in Fig. 5.6, the components of the proposed reference architecture for DKN are the following.

The *DKN Manager* is the application server that is in charge of coordinating and managing all the knowledge nodes. Therefore, its goal is to handle and personalize knowledge on the basis of the above-described models and techniques.

The *KN Coordination Layer* implements XML-based coordination/synchronization protocols between two or more nodes.

The *Knowledge Integration Layer* deals with the integration of the different knowledge nodes on the basis of a model-oriented schema integration.

The *Content Adaptation Layer* adapts extracted contents taking into account the current user profile (i.e., it deals with personalizing knowledge with respect to data).

The *Device Adaptation Layer* is the software layer that adapts data coming from the Content Adaptation Layer on the basis of the current user device (i.e., it deals with formatting knowledge).

The *XSL Template DB* is a relational database that stores XSL formatting rules, and gives support to the personalization of presentations.

The *Web Services Differentiation Layer* aims at differentiating the exported Web services according to the current user profile (i.e., it deals with personalizing knowledge with respect to services).

The *Data Access Interface* is a "transparent-for-the-format" data access component that provides uniform access to highly heterogeneous data sources, such as: RDBMSs, XML Warehouses, Legacy Systems, ERP systems, SAP systems etc.

The *Web Services Provider* is a collection of (Web services) providers that interact with a large UDDI repository and provide services for different application contexts.

The *OLAP Server* gives support to the implementation of both the multi-resolution user model m_{DKN}^{MR} and the multidimensional user monitoring model b_{DKN}.

The *User Profile DB* is a relational database that maps the matching between (user) profiles and multidimensional OLAP views (i.e., profiles, in the model m_{DKN}^{MR}), and gives support to the profile management task.

The *User Behaviour Observer Agent* is devoted to (*i*) monitor the user behaviour during the user/DKN interaction, and (*ii*) collect the variables that will be used for clustering the user and assigning him/her to the appropriate user profile.

The *User Behaviour DB* is a relational database that stores user behaviours, and gives support to the personalization of knowledge.

The *User Profiler* is the middleware software that runs on both client and server components of the proposed architecture; it finally personalizes knowledge both with respect to data and services, and computes the right XSL formatting rules to delivery (so-personalized) knowledge.

5.7 Experimental Assessment

In this Section, we present results of some experiments that give an insight into the effectiveness of our proposed implementation of DKN. In these experiments, we designed and developed a simulator of DKN for supporting *e*-learning environments. System contents have been represented as a very large XML repository of *e*-courses, topics, links pointing to the related topics, and so forth. Besides this, the synthetic system has been organized in thematic sections according to a pre-defined hierarchy of themes, so that several topics belonging to different *e*-courses can also stay in the same thematic section (for instance, *e*-course Web Systems Design could be related to the topic Data Modelling belonging to the *e*-course Database Systems). Furthermore, we designed a Web service allowing to ensemble study plans via selecting different *e*-courses. This Web service is specialized as follows: for students, it dynamically checks the referential constraints among the selected *e*-courses; for teachers, it allows new *e*-courses to be created by composing topics coming from different pre-existing *e*-courses. We delineated an application scenario in which users can access the knowledge with a desktop workstation or with a PDA. A traffic generator providing user behaviours according to different data distributions completes our simulation environment.

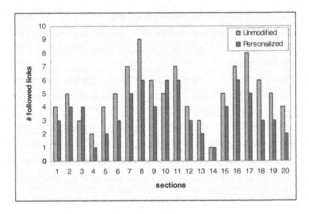

Fig. 5.7. Followed links from the experimental assessment focused on *e*-learning environments

We measured user effort in terms of both the amount of navigation (i.e., number of followed links) and the time to attain goals inside thematic sections. Fig. 5.7 and Fig. 5.8 report our results comparing the "personalized" system, i.e. DKN equipped with components for personalized knowledge delivery (see Fig. 5.6), versus the "unmodified" system, i.e. DKN enabling the access to the information only, without personalization. For both parameters (i.e., number of followed links and time required) we observe that performance is decidedly better when using our knowledge personalization framework.

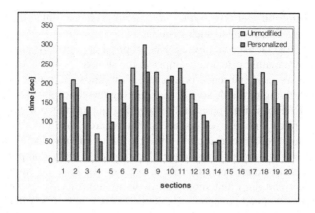

Fig. 5.8. Time required from the experimental assessment focused on *e*-learning environments

5.8 Conclusions and Future Work

In this chapter, we presented the definition of the DKN framework, and a proposal for its implementation. DKN is a network of devices (wired or wireless) that provides the right knowledge to the right user at the right time and in the right format, taking into account (*i*) user profile, and (*ii*) user device features. A very promising approach is the adoption of OLAP technology in order to improve the capabilities and granularities of user models, both for the profiling phase and for the tracking/observing phase. Rigorous multidimensional user and user monitoring models have been presented, and formally defined. Some lemmas about them have also been derived. Mechanisms and techniques for personalizing data and services (i.e., knowledge, in our feeling) have been proposed, formalized, and discussed in detail. A template-oriented technique for formatting personalized knowledge via XSL rules has also been proposed. To the best of our knowledge, there are no other similar approaches in this research area, and consequently, our proposal can be considered innovative. Experimental results seem to confirm the correctness and the efficiency of our approach. However, there is still a lot of work to be done in order to develop a complete prototype of the proposed reference architecture for DKN. Moreover, due to the interdisciplinary nature of this architecture, currently there are also some open issues of technological kind such as XML querying, updating and verification, and Web services interoperability, security and privacy.

Future work is focused on two directions. On one hand, we will stress personalization capabilities of the proposed architecture by means of testing its performance against more probing real-life settings, which, beyond the "natural" context of *e*-learning environments, like the one investigated in this chapter, could also consider other interesting "knowledge-intensive" environments such as those of *e*-procurement and *e*-government systems. On the other hand, we will concentrate on the exploitation of the fundamental distributed knowledge concept introduced in this chapter via directing our research effort towards the integration of this concept with recent proposals focusing on *pervasive and ubiquitous systems*, thus trying to add a "knowledge-oriented fashion" to such systems.

Some other particular aspects of the DKN proposal are currently in progress of development, and are mainly oriented to improve the semantics of the knowledge representation, discovery, and inference processes. As regards the multidimensional user model, we are studying new solutions for integrating the concept of *Quality of Service* (QoS) in our OLAP-based proposal, thus dowering our framework with the amenity of managing applications as well as users. As regards the multidimensional user monitoring model, we are considering the possibility of introducing the notion of *ageing of user behaviour*, and, consequently, capturing more complex (Web) user dynamics. As regards data personalization, we are working on innovative *graph-based reasoning techniques* that should allow us to significantly improve the definition and the use of "related concepts" inside the DKN framework. Finally, as regards service personalization, we are developing intelligent schema taking into account the definition of *invocation chains* of Web services, thus enforcing services dissemination across the DKN framework.

References

1. Anyanwu, K., Sheth, A.P.: P-Queries: Enabling Querying for Semantic Associations on the Semantic Web. In: Proceedings of the 12th International World Wide Web Conference, pp. 690–699 (2003)
2. Balabanovi, M., Shoham, Y.: Fab: Content-based, Collaborative Recommendation. Communication of the ACM 40(3), 66–72 (1997)
3. Bertino, E., Guerrini, G., Mesiti, M.: A Matching Algorithm for Measuring the Structural Similarity between an XML Document and a DTD and its Applications. Information Systems 29(1), 23–46 (2004)
4. Breese, J., Heckerman, D., Kadie, C.: Empirical Analysis of Predictive Algorithms for Collaborative Filtering. In: Proceedings of the 40th International Conference on Uncertainty in Artificial Intelligence, pp. 43–52 (1998)
5. Brusilovsky, P.: Methods and Techniques of Adaptive Hypermedia. User Modelling and User-Adapted Interaction 6(2-3), 87–129 (1996)
6. Buchanan, B.G., Smith, R.G.: Fundamentals of Expert Systems. In: The Handbook of Artificial Intelligence, vol. 4, pp. 149–192. Addison-Wesley, Reading (1989)
7. Bull, S.: User Modelling and Mobile Learning. In: Proceedings of the 9th International Conference on User Modelling, pp. 383–387 (2003)
8. Burn, J.M., Ash, C.G.: Managing Knowledge in an ERP Enabled Virtual Organization. In: Internet-based Organizational Memory and Knowledge Management, pp. 222–240. Idea Group Publishing (2000)
9. Business Process Execution Language for Web Services, http://www-106.ibm.com/developerworks/library/ws-bpel/
10. Cannataro, M., Cuzzocrea, A., Pugliese, A.: XAHM: an Adaptive Hypermedia Model based on XML. In: Proceedings of the 14th ACM International Conference on Software Engineering & Knowledge Engineering, pp. 627–634 (2002)
11. Cuzzocrea, A., Mastroianni, C.: A Reference Architecture for Knowledge Management-based Web Systems. In: Proceedings of the 4th IEEE International Conference on Web Information Systems Engineering, pp. 347–354 (2003)

12. Cuzzocrea, A., Mastroianni, C.: Pushing Knowledge Management in Web Information Systems Engineering. In: Proceedings of the 8th International Symposium on Database Engineering and Applications, pp. 480–485 (2004)
13. Daconta, M.C., Obrst, L.J., Smith, K.T.: The Semantic Web: A Guide to the Future of XML, Web Services, and Knowledge Management. John Wiley & Sons, Chichester (2003)
14. Davenport, T., Prusak, L.: Working Knowledge: Managing What Your Organisation Knows. Harvard Business School Press (1997)
15. Dunlop, M.D.: Development and Evaluation of Clustering Techniques for Finding People. In: Proceedings of the 3rd International Conference on Practical Aspects of Knowledge Management (2000)
16. Feinman, A., Alterman, R.: Discourse Analysis Techniques for Modelling Group Interaction. In: Proceedings of the 9th International Conference on User Modelling, pp. 228–237 (2003)
17. Fensel, D.: Ontologies: A Silver Bullet for Knowledge Management and Electronic Commerce. Springer, Heidelberg (2001)
18. Florescu, D., Grünhagen, A., Kossmann, D.: XL: an XML Programming Language for Web Service Specification and Composition. In: Proceedings of the 11th International World Wide Web Conference, pp. 65–76 (2002)
19. Gibbins, N., Harris, S., Shadbolt, N.: Agent-based Semantic Web Services. In: Proceedings of the 12th International World Wide Web Conference, pp. 710–717 (2003)
20. Goldberg, D., Nichols, D., Oki, B.M., Terry, D.: Using Collaborative Filtering to Weave an Information Tapestry. Communications of the ACM 35(12), 61–70 (1992)
21. Hull, R., Benedikt, M., Christophides, V., Su, J.: E-Services: a Look behind the Curtain. In: Proceedings of the 22th ACM Symposium on Principles of Database Systems, pp. 1–14 (2003)
22. Gray, J., Bosworth, A., Layman, A., Pirahesh, H.: Data Cube: a Relational Operator Generalizing Group-By, Cross-Tab and Sub-Totals. In: Proceedings of the 12th International Conference on Data Engineering, pp. 152–159 (1996)
23. Jackson, P.: Introduction to Expert Systems. Addison-Wesley, Reading (1998)
24. Konstan, J.A., Miller, B.N., Maltz, D., Herlocker, J.L., Gordon, L.R., Riedl, J.: GroupLens: Applying Collaborative Filtering to Usenet News. Communication of the ACM 40(3), 77–87 (1997)
25. Kreger, H.: E-Services: Fulfilling the Web Services Promise. Communication of the ACM 46(6), 29–34 (2003)
26. Leymann, F.: Web Services Flow Language (WSFL 1.0). IBM Technical Report (2001), http://www4.ibm.com/software/solutions/webservices/pdf/WSFL.pdf
27. Liebowitz, J.: The Handbook of Applied Expert Systems. CRC Press, Boca Raton (1997)
28. Maamar, Z., Al Khatib, G., Kouadri Mostefaoui, S., Lahkim, M.B., Mansoor, W.: Context-based Personalization of Web Services Composition and Provisioning. In: Proceedings of the 30th EUROMICRO International Conference, pp. 396–403 (2004)
29. Maltz, D., Ehrlich, E.: Pointing the Way: Active Collaborative Filtering. In: Proceedings of the Conference on Human Factors in Computing Systems, pp. 202–209 (1995)
30. Manher, U., Patel, A., Robison, J.: Experience with Personalization on Yahoo! Communications of the ACM 43(8), 31–34 (2000)
31. Milton, N., Shadbolt, N., Cottman, H., Hammersley, M.: Towards a Knowledge Technology for Knowledge Management. International Journal Human-Computer Studies 53(3), 615–664 (1999)

32. Narayanan, S., Mc Ilraith, S.A.: Semantic Web Services: Simulation, Verification and Automated Composition of Web Services. In: Proceedings of the 11th International World Wide Web Conference, pp. 77–88 (2002)
33. Nierman, A., Jagadish, H.V.: Evaluating Structural Similarity in XML Documents. In: Proceedings of the 5th International Workshop on the Web and Databases, pp. 61–66 (2002)
34. Nonaka, I.: The Knowledge-Creating Company. Harvard Business Review 69, 96–104 (1991)
35. O'Leary, D.: Knowledge Management in Enterprise Resource Planning Systems. In: Proceedings of the AAAI Workshop on Exploring Synergies of Knowledge Management & Case-Based Reasoning, pp. 70–72 (1999)
36. Pierrakos, D., Paliouras, G., Papatheodorou, C., Spyropoulos, C.D.: Web Usage Mining as a Tool for Personalization: a Survey. User Modelling and User-Adapted Interaction 13(4), 311–372 (2003)
37. Razmerita, L., Angehrn, A.A., Maedche, A.: Ontology-Based User Modelling for Knowledge Management Systems. In: Proceedings of the 9th International Conference on User Modelling, pp. 213–217 (2003)
38. Resnick, P., Varian, H.R.: Recommender Systems. Communication of the ACM 40(3), 56–58 (1997)
39. Riecken, D.: Personalized Views of Personalization. Communications of the ACM 43(8), 27–28 (2000)
40. Schiaffino, S.N., Amandi, A.: User Profiling with Case-Based Reasoning and Bayesian Networks. In: Proceedings of the International Joint Conference, 7th Ibero-American Conference on Artificial Intelligence and 15th Brazilian Symposium on Artificial Intelligence, pp. 12–21 (2000)
41. Schonberg, E., Cofino, T., Hoch, R., Podlaseck, M., Spraragen, S.L.: The Human Element: Measuring Success. Communications of the ACM 43(8), 53–57 (2000)
42. Schwab, I., Pohl, W., Koychev, I.: Learning to Recommend from Positive Evidence. In: Proceedings of the 2000 International Conference on Intelligent User Interfaces, pp. 241–247 (2000)
43. Slaney, M., Subrahmonia, J., Maglio, P.P.: Modelling Multitasking Users. In: Proceedings of the 9th International Conference on User Modelling, pp. 188–197 (2003)
44. Smith, B., Cotter, P.: A Personalized Television Listing Service. Communications of the ACM 43(8), 107–111 (2000)
45. Srivastava, J., Cooley, R., Deshpande, M., Tan, P.-N.: Web Usage Mining: Discovery and Applications of Usage Patterns from Web Data. SIGKDD Explorations 1(2), 12–23 (2000)
46. Terveen, L., Hill, W., Amento, B., MacDonald, D., Creter, J.: PHOAKS: a System for Sharing Recommendations. Communications of the ACM 40(3), 59–62 (1997)
47. Trastour, D., Bartolini, C., Preist, C.: Semantic Web Services: Semantic Web Support for the Business-To-Business E-Commerce Lifecycle. In: Proceedings of the 11th International World Wide Web Conference, pp. 89–98 (2002)
48. Van der Aalst, W.M.P.: Don't go with the Flow: Web Services Composition Standards Exposed. IEEE Intelligent Systems 18(1), 72–85 (2003)
49. Wells, N., Wolfers, J.: Finance with a Personalized Touch. Communications of the ACM 43(8), 31–34 (2000)
50. Widyantoro, D.H., Ioerger, T.R., Yen, J.: An Adaptive Algorithm for Learning Changes in User Interests. In: Proceedings of the 8th Conference on Information and Knowledge Management, pp. 405–412 (1999)

51. Widyantoro, D.H., Ioerger, T.R., Yen, J.: An Incremental Approach to Building a Cluster Hierarchy. In: Proceedings of the 2002 IEEE International Conference on Data Mining, pp. 705–708 (2002)
52. Wiig, K.: Knowledge Management Foundations: Thinking about Thinking-how People and Organizations Create, Represent and Use Knowledge. Schema Press (1994)
53. Wiig, K.: Knowledge Management: Where Did It Come From and Where Will It Go? Expert Systems With Applications 13(1), 1–14 (1997)
54. Wilkins, J., Van Wegen, B., De Hoog, R.: Understanding and Valuing Knowledge Assets: Overview and Method. Expert Systems With Applications 13(1), 55–72 (1997)
55. WordNet: a Lexical Database for the English Language, http://www.cogsci.princeton.edu/~wn/
56. The World Wide Web Consortium, http://www.w3.org/
57. W3C Resource Description Framework (RDF), http://www.w3.org/RDF/
58. W3C Web Services Activity, http://www.w3.org/2002/ws/

6

Model-Based Data Engineering for Web Services

Andreas Tolk and Saikou Y. Diallo

Old Dominion University, Norfolk, VA 23529, USA
atolk@odu.edu, sdiallo@odu.edu

The application of the Extensible Mark-up Language (XML) and Web services enabled a new level of interoperability for heterogeneous IT systems. However, although XML enables separation of data definition and data content, it doesn't ensure that data exchanged are interpreted correctly by the receiving system. This motivates data management to support unambiguous definition of data elements for information exchange. Using a common reference model improves this process leading to Model-based Data Engineering (MBDE). The results can be used immediately to configure mediation layers integrating services into an overall service oriented architecture. Ultimately, the objective must be to describe the information exchange requirements, contexts, and constraints of web services in metadata allowing intelligent agents to conduct these engineering steps without human support. This chapter describes the current state of the art of MBDE and how it relates to Service Oriented Architectures (SOA) in general and Web Services in particular.

6.1 Introduction

Independently developed and distributed applications each have internal representations of their data. Therefore a transformation layer translating the internal representations into each other has to be created to make information exchange possible between these systems. The traditional approach is to utilize the Extensible Mark-up Language (XML) to enable data exchange between any two systems; nonetheless XML doesn't ensure that data exchanged are interpreted correctly by the receiving system. Furthermore, XML does not cope with the problem of semantic information exchange. This motivates data engineering to support the unambiguous definition of data elements for information exchange and the definition of a standard approach to mapping heterogeneous data models. This chapter presents an algorithm that can be applied to the data engineering process to ensure correctness in the exchange of bits and bytes but more importantly correctness in the conceptual and semantic exchange.

Practical applications have shown that using a common reference model improves this process leading to Model-based Data Engineering (MBDE). In addition, in order to support operations with rapidly changing requirements, service oriented architectures are needed instead of the traditional solutions, which are often too inflexible. As an alternative to having a system fulfilling a set of predefined requirements, services fulfilling requirements are identified, composed and orchestrated to meet the current users' needs in an ongoing operation.

R. Nayak et al. (Eds.): Evolution of the Web in Artificial Intel. Environ., SCI 130, pp. 137–161, 2008.
springerlink.com

The ideas of MBDE are rooted in federated databases Spaccapietra and colleagues identify the following four classes of conflicts to be solved by data engineering [1]:

- *Semantic Conflicts* occur when concepts of the different local schemata do not match exactly, but have to be aggregated or disaggregated. They may only overlap or be subsets of each other, etc.
- *Descriptive Conflicts* describe homonyms, synonyms, and different names for the same concept, different attributes or slot values for the same concept, etc.
- *Heterogeneous Conflicts* result from substantially different methodologies being used to describe the concepts.
- *Structural Conflicts* results from the use of different structures describing the same concept.

Spaccapietra et al. concluded that a generic meta data model comprising only objects and attributes for values and references is needed to support efficient data management. This generic data model would describe all information exchange requirements and constraints for participating systems, allowing to construct a federated database schema in which (1) all information elements to be exchanged between the systems are modelled in an unambiguous way, and (2) the mapping of these infromation elements to representing entities in the participating systems is captured unambiguously as well. In order to enable composable services, these ideas must be captured in a framework that consistently captures the required information in the form of metadata.

MBDE was developed in support of integrating Modeling and Simulation (M&S) applications and operational Information Technology (IT) systems to enable decision support – such as alternative simulation and evaluation –, training applications – such as using simulation as a synthetic environment for the trainee using the operational system he is used to –, and testing applications. The second part of this chapter will focus on some special challenges of M&S web services that may go beyond the general integration challenges of web services. Most of this chapter, however, will deal with general challenges of composing web services based on MBDE.

6.2 Model-Based Data Engineering for Composable Services

Most Information Technology (IT) specialists consider the ability to merge heterogeneous data models as the single most important challenge in the business world today [1, 2]. The proliferation of independently developed data models and the diversity of data sources and potential consumers, motivate the need for a clearly defined Data Engineering process. Some of the most immediate challenges are:

- *Multiple sources with different formats*: Data are no longer only located in databases or text files. The advent of XML, Web services and Real Simple Syndication (RSS) Streams has added to the variety of formats that need to be handled.
- *Structured, Semi Structured, Unstructured data*: Based on the definitions given in [3], data are considered unstructured when they can be of any type, do not necessarily follow any sequence, format or rule and are not predictable. Data are semi-structured if they are organized in semantic entities and similar entities are grouped together. Entities in the same group may not have the same attributes however, and the order

of attributes is not necessarily important. Additionally, all attributes may not be required and the size and type of the same attributes in a group may differ. Data are considered structured when they are organized in semantic chunks (entities). Similar entities are grouped together (relations or classes) and entities in the same group have the same descriptions (attributes). Descriptions for all entities in a group (Schema) must be all present and follow the same order. Additionally, they must have the same defined format and predefined length. Disparity in data format further complicates the task of model interoperability.

- *Data Quality and Reliability*: Making the right decision or getting the right overall picture depends heavily on the quality and reliability of data. It is therefore crucial to have dependable sources providing quality information to the decision maker. Modelers have to devise a mechanism rating the quality of sources in order to provide the most reliable response to user's queries.
- *Transition from Data to Information*: It is not sufficient to have data; the goal is to gain insight from the data and obtain information. Data have to be transformed in a format that makes that process obvious. The decision maker, whether it is a commander or a business leader, should be able to rapidly gain information from data by the way it is presented.

Data integration is the most expensive part of any business intelligence or simulation interoperability project. According to Kamal Hathi [2] from Microsoft, some of the key factors adding to the cost include:

- Getting the data out in the format that is necessary for data integration ends up being a slow and torturous process fraught with organizational power games.
- Cleansing the data and mapping the data from multiple sources into one coherent and meaningful format is extraordinarily difficult.
- More often than not, standard data integration tools don't offer enough functionality or extensibility to satisfy the data transformation requirements for the project. This can result in the expenditure of large sums of money in consulting costs to develop special ETL code to get the job done.
- Different parts of the organization focus on the data integration problem in silos.

Currently, mapping heterogeneous data models is a complex, time consuming and error-filled endeavor. The traditional approach is to have experts on both sides establish correspondences between concepts, entities and relationships within the two models. However this is not always possible, simply because the original creators of the model have changed jobs, moved on to other projects or the model has greatly evolved since its inception. Most often, it is left to modelers to figure out what the mappings should be and the lack of documentation greatly complicates the issue. Some of the difficulties facing modelers include:

- *The lack of an explicit ontology:* For practical use an ontology can be defined as "a controlled vocabulary that describes objects and the relations between them in a formal way, and has a grammar for using the vocabulary terms to express something meaningful within a specified domain of interest. The vocabulary is used to make queries and assertions" [4]. Even though a model necessarily has an implicit ontology, modelers seldom take the time to formally establish and record it, thus introducing an element of ambiguity and interpretation when it has to be mapped to another model.

- *The complexity of models:* Models vary in size and complexity. For relatively simple models that have similar ontologies the mapping effort is more or less trivial. This is the case in business applications where most models are developed to fulfill somewhat similar needs (payroll models, personnel models etc...) The difficulty in this case lies in reconciling disparate formats and data sources, not in matching concepts and relationships. In some areas however, the domain space is so complex and heterogeneous that the mapping effort is exponentially more difficult. This is the case in the military domain where there is a wide range of entities and relationships that can be modeled in significantly different ways depending on the intended use and audience. Other factors such as Terrain, Weather and Civilian entities can further complicate the task.
- *A disparity in Domain space:* The common assumption is that any two models can be mapped or at worst merged. This is not always the case and modelers might not realize it until it is too late.

Even in successful cases there is very little documentation regarding the mapping process and as a result it is nearly impossible to measure the effectiveness of one approach compared to another. The authors believe that the main issue lies in the fact that there is no standard or framework to help modelers anticipate the problems they might face.

All of these factors highlight the need for an engineering process that will define a framework for data integration. This framework will be presented in detail in the remainder of this chapter.

6.2.1 The Example

The example presented here will serve as an illustration of the challenges inherent to integration projects in real life. The EZ rental car rental company has a data model that keeps track of cars and customers that rent them. EZ rental gets some of its cars either indirectly from a local dealer which has a data model to monitor cars, customers and parts from their internal garage, or directly from CheapCar a local U.S manufacturer. The local dealer also gets its cars and parts from CheapCar. The manufacturer has its own internal model that records and monitors cars, parts and customers among other things. In order to better manage the rental fleet, the management at EZ rental needs the ability to interface with both the dealership and the manufacturer. The dealership would be able to monitor cars and know which ones to buy and resell as used as well as provide new cars to EZ rental. The manufacturer can update the fleet either directly or through the dealership. Additionally, in an effort to bolster the amount of parts sold each year, CheapCar wants a limited instantaneously access to dealer's models to monitor the amount and type of parts that are used. They want to use this information to predict the needs of their customers and reduce stocking time. Furthermore, CheapCar wants to use comments from A&C's customers as instant feedback about the quality, faults and problems related to their product. After several meetings, the businesses agree to expose part of their model to each other. However, each business retains control of its internal model and

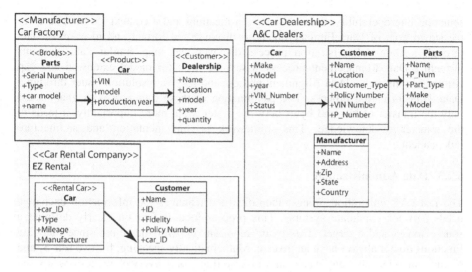

Fig. 6.1. Conceptual Model of the Example

communicates only parts of its data. To fulfill this task IT specialists for each business must meet and agree on a common information exchange framework. The remainder of this chapter will use this scenario to illustrate the problems inherent to the integration process. In addition, it will be used to show the applicability of the steps in MBDE.

Figure 6.1 shows a sample representation of the data model in each business. Even though the same concepts are represented, the focus and goal of the business dictates the structure of the model. The car rental company cares about the mileage of a rental while the manufacturer and the dealer do not. In terms of syntax, the manufacturer identifies a car by its unique Vehicle Identification Number (VIN), while the rental business assigns a unique ID number that might or might not be the VIN. The dealer on the other hand has an attribute representing the VIN by calls it "VIN_Number". Two models (Dealer and Manufacturer) have a parts object while the Rental does not really care about it. This means that the model does not record customer's complaints about parts for example. These are only some of the obvious differences between these three models. Each specific problem will be examined in detail in the discussion of MDBE.

6.2.2 Data Engineering

The process of mapping heterogeneous data models has been extensively addressed over the last twenty-five years. Several tools and approaches have been proposed but there is no consistent coherent framework upon which modelers can rely on to successfully map heterogeneous models. Data Engineering is based upon a simple observation that holds true regardless of the size and complexity of the models involved. Simply stated, data has a format (structured, unstructured, or semi-structured) and a physical location (text file, relational Database etc...). In order to transition from data exchange to information exchange-which is the stated goal of semantic interoperability, data must have a meaning and a context [5, 6]. Therefore the stated goal of Data Engineering is to discover the format and location of data through a Data Administration process discover and map similar data elements through a Data Management process, assert the need for model extension and gap elimination through a Data Alignment process and resolve resolution issues through a Data Transformation process. The combination of these four processes enables not only the transfer of bits and bytes between systems but more importantly, it leads to the transfer of knowledge. This framework is implementation and architecture independent.

6.2.3 Data Administration

The Data Administration process identifies and manages the information exchange needs between candidate systems. This process focuses first on clearly defining a source model and a target. This is an important step for the future since mapping functions do not always have an inverse. Mathematically speaking, for two sets S_1 and S_2, any mapping function f has a valid inverse if and only every element of S_1 has one and only one counterpart in S_2 [7]. Simply stated f must be bijective (see figure 6.2). This is clearly not the case and in fact research shows that while *1:1* mappings do exist, *n: m* mappings are more prevalent [8]. The chapter will address the issue of complex mapping during Data Management.

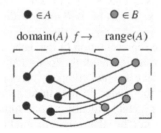

Fig. 6.2. Bijective Function

Data Administration also aims at aligning formats and documentation, examining the context of validity of data and asserting the credibility of its sources. A special emphasis is put on format alignment. Format alignment implies that modelers should not only agree on a common format (XML, text file) for data exchange but also that semi-structured and unstructured data be enriched semantically and syntactically. Data administration is the first step to ensuring that systems communicate in a complete and meaningful manner.

2.4 Applying Data Administration to the Example

In terms of the example provided earlier, since Data Administration requires a clear definition of source and target, the team agreed that:

- From EZ rental to A&C Dealerships: For car information the rental company is the source and the dealership is the target.

- From A&C to CheapCar: For customer information, the dealer is the source and the manufacturer is the target.
- From CheapCar to A&C Dealership: For parts information the dealership is the source and the manufacturer is the target.

These definitions highlight the fact that source and target identification is not a one time process. It is a consensus human decision based on the exchange requirements.

The next step during Data Administration is to agree on a common exchange format and perform semantic enrichment to eliminate assumptions embedded within the models. The modeling team decides that XML is the common format that they will use and each model should publish a XML Schema Definition (XSD) encompassing the objects that will be exchanged. Figure 6.3 shows the XSD of each of the models.

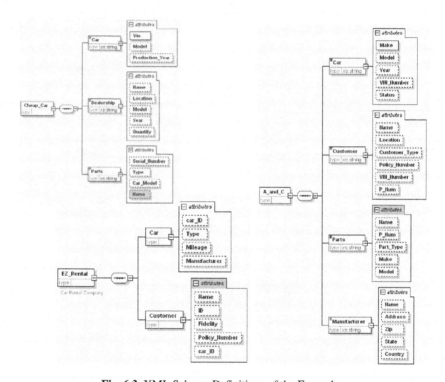

Fig. 6.3. XML Schema Definitions of the Example

The team observes that the dealership and the manufacturer expect string values for their elements while the rental does not specify a type. For the sake of simplify, they agree that any value exchanged will be of type "String". Each model must add a conversion layer to align its internal type to the agreed type. They further observe that there is a strong possibility of error due to the occurrence of homonyms and synonyms within models and across models. A&C for example has an attribute

"Name" for both the "Car" and "Part" element. The dealership refers to the make and model of a car while the rental company has an attribute manufacturer. This begs the question as to whether the manufacturer refers to the make, the model or to both. As a result the team decides to make all of the assumptions and other documentation explicit within the XSD.

Figure 6.4 shows the XSD of the car rental company. It has been augmented with a definition of type, a description of elements and constraints such as unique keys. The manufacturer and the dealer have a similar schema. This XSD shows how XML can be used to better serve the mapping effort. Other models can now use this schema in the Data Management process. It is worth noting that further enhancements are needed in this XSD. The documentation remains a little bit ambiguous (the car type definition does not specify what the enumeration values are for example).

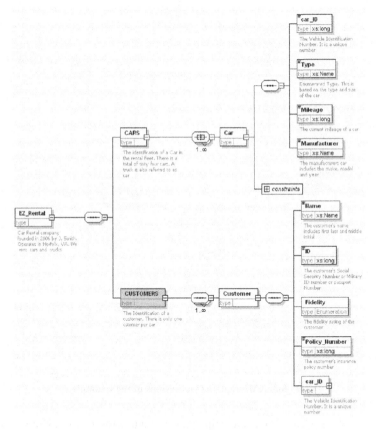

Fig. 6.4. Augmented XSD of EZ Rental

The enhancement process is repeated until everyone is satisfied that the schema is clear and unambiguous. As stated before, a good Data Administration process is the key to an easier, more accurate integration process.

6.2.5 Data Management

The goal of Data Management is to map concepts, data elements and relationships from the source model to the target model. Data Management is the most time consuming and difficult area of Data Engineering. As the literature has shown, mapping can be done either manually or with semi-automated tools. Possible sources of conflict have been studied and classified in earlier parts of this chapter. The emerging consensus is that the manual approach is long and error-prone while tools are not powerful enough yet to act on large and complex models (military models for example). To streamline the process, mapping can be decomposed into three distinct sub-processes:

- *Conceptual Mapping:* This is the inevitable human-in-the-loop aspect of mapping. Experts in both domains (source and target) must agree on concept similarity. At this level, it is important to know if the models have something in common (intersect) and to extract ontologies if possible. Two models intersect if any of their subsets are identical or any of their elements can be derived from one another directly or through a transformation. Two concepts are deemed identical if they represent the same real world view. Spaccapietra et al [8] assert that "if a correspondence can be defined such that it holds for every element in an identifiable set (e.g., the population of a type), the correspondence is stated at the schema level. This intentional definition of a correspondence is called an inter-database correspondence assertion (ICA)." Conceptual mapping is the listing of all existing ICA.
- *Attribute mapping:* The next logical step is to identify similar atributes. At this level special attention has to be paid to synonyms, homonyms and the inherent context of attributes. Two attributes are said to be equal is they describe the same real world property. It is possible to say for example that an attribute "amount" in a cash register model is the same as the attribute "quantity" in another model if they both refer to "the total number of a product sold to a customer". As this example shows, attribute mapping cannot be done out of context. If "amount" was referring to "the amount of money given to the cashier" then the correspondence no longer holds.
- *Content Mapping:* Most mapping efforts tend to conglomerate content mapping with attribute mapping. Two values are said to be identical if they can be derived from one another. For example "<Total Price>" equals "<total purchase> + (<state tax>*<total purchase>)). This example does not say anything about the relationship between the attribute "total price" on one side and the other three attributes on the other side. At the attribute level, equivalence between real world properties is established while the content level deals with how attribute values are derived from one another.

The complexity of any mapping effort is directly related to the complexity of these individual components. The amount of effort can be measured by the size of the area of intersection, the similarity in concepts and to a lesser extent the disparity in attributes and the derivability of content.

6.2.6 Applying Data Management to the Example

Data Management is the next logical step in the integration effort discussed in the example. Modelers must focus on identifying and mapping concepts, attributes and content. Let us apply the mapping steps identified earlier.

Conceptual Mapping
In this effort, it seems obvious that the concepts of car and parts are identical in both models. However, modelers must decide whether the concept of Dealership is similar to that of Customer. It might be that the manufacturer distinguishes between individual costumers that might order online or have specific demands (the information about individual customers might be captured in an "Ind_Orders" object or table for example) and dealerships which are licensed vendors of their products (the "Dealership" object or table). This decision greatly affects the outcome of the integration effort because of the domino effect it has on Attribute and Content Mapping. In this case the decision is that the concepts of Dealership and Customer are related and therefore identical (see figure 6.5). It turns out that this is the closest possible match because the manufacturer does not take orders directly from individuals. All transactions are done through a dealership. We will see later how this affects the outcome. The concept of Manufacturer is represented as an attribute (Rental Company) in one model and as an object in the other; however, it is clear from the schemas that these are conceptually identical.

SOURCE	TARGET
CAR	CAR
CUSTOMER	CUSTOMER
PARTS	PARTS

Fig. 6.5. Conceptual Mapping

There is only one figure because of the relative simplicity of the conceptual mapping in this case. Simply put the figure shows that these concepts are equivalent wherever they exist within the three models.

Attribute Mapping
At this level, similar attributes must be identified. Through a good Data Administration process, a close examination of the schemas yields the results presented in figures 6.6, 6.7 and 6.8. The mapping process has to be performed for each interface.

Figure 6 shows that some attributes in the source have an unknown correspondence in the target. We will see how this issue is resolved during Data Alignment. Additionally the figure does not identify how these attributes are related; this is done during content mapping.

Figure 6.7 shows that attributes of one object can have a mapping in another object. In this case <P-Num>, an attribute of Customer maps to <Serial_Number> an attribute of Parts.

EZ RENTAL	A&C DEALERSHIP
Car_ID	VIN_Number
Type	Unknown
Mileage	Unknown
Manufacturer	Make
Manufacturer	Model

Fig. 6.6. Car Attribute Mapping from EZ Rental to A&C

A&C DEALERSHIP	CheapCar Manufacturer
Name	Name
Location	Location
Customer_Type	Unknown
Policy_Number	Unknown
VIN_Number	Car.VIN
P_Num	Parts.Serial_Number

Fig. 6.7. Customer Attribute Mapping from A&C to CheapCar

A&C DEALERSHIP	CheapCar Manufacturer
Name	Name
P_Num	Serial_Number
Part_Type	Type
Make	Car_Model
Model	Car_Model

Fig. 6.8. Parts Attribute Mapping from A&C to CheapCar

Having identified the attributes and their images, the team can now focus on deriving attributes values from the source to the target.

Content Mapping

The process of content mapping corresponds to generating functions that map the values of attributes to one another. Figure 6.9 shows the content mapping between the attributes of a car in the rental model and its counterpart in the dealership model.

The functions show that for example the contents of the attribute <Manufacturer> must be decomposed into a make component and a model component and then mapped to <Make> and <Model> respectively. Modelers have to build similar tables for each set of attributes. Figure 6.10 shows an implementation of the mapping of the car object from EZ rental to A&C dealership using a standard tool.

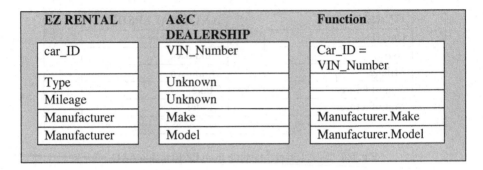

EZ RENTAL	A&C DEALERSHIP	Function
car_ID	VIN_Number	Car_ID = VIN_Number
Type	Unknown	
Mileage	Unknown	
Manufacturer	Make	Manufacturer.Make
Manufacturer	Model	Manufacturer.Model

Fig. 6.9. Content Mapping of the Car Attribute

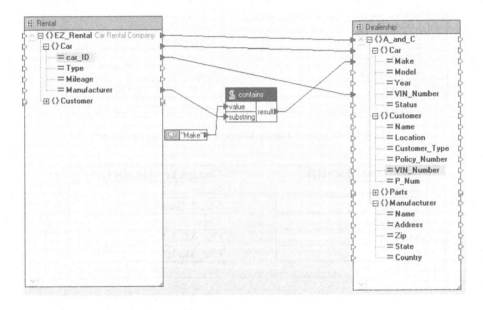

Fig. 6.10. Car Mapping Implementation

6.2.7 Data Alignment

The goal of data alignment is to identify gaps between the source and the target. The focus at this level is to map the non-intersecting (see figure 6.6) areas of the two models by either merging them or introducing a reference model that intersects with the complement. A complete Data Alignment process ensures completeness in mapping and protects the integrity of information exchange.

Figure 6.11 shows the relative complement of set A in B. In terms of Data Alignment if A is the source and B is the target, this means that B has to be extended to contain the elements of A that are missing or find a larger set C such that $A \bigcup B \in C$.

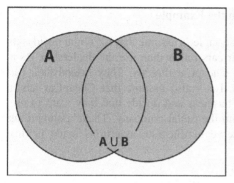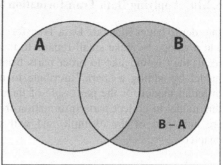

Fig. 6.11. Complement of A in B and the Union of A and B

6.2.8 Applying Data Alignment to the Example

Data Alignment addresses the holes represented by the "Unknown" fields in Figures 6.6, 6.7 and 6.9. The recommended approach here is to either extend the target model or simply leave these attributes out because they are not important to the target model. In the car example, the modelers mapping the car concept from EZ rental to A&C recognize that "type" is an attribute proper to the rental business and therefore decide not to include it in the exchange. The Mileage attribute on the other hand is very important to the dealership because it is a trigger in their decision making process. As a result they agree to extend their model by adding a "mileage" attribute to the car object. Figure 6.12 shows the resulting mapping.

EZ RENTAL	A&C DEALERSHIP	Function
car_ID	VIN_Number	Car_ID = VIN_Number
Mileage	Mileage (extended)	Mileage = Mileage
Manufacturer	Make	Manufacturer.Make
Manufacturer	Model	Manufacturer.Model

Fig. 6.12. Mapping of the Car Attribute after Data Alignment

6.2.9 Data Transformation

The goal of Data Transformation is to align models in terms of their level of resolution. The assumption that models are expressed at the same level of detail is not true. Furthermore, information that is deemed vital in one model might not hold the same value in another due to disparities in focus, goal and approach between the two. As a result objects need to be aggregated or disaggregated during the mapping process in order to establish correspondences.

6.2.10 Applying Data Transformation to the Example

In order to better illustrate Data Transformation, let's assume that EZ rental decides to add a garage to make small repairs to their cars rather than use the dealership. As a result they would like to order parts from CheapCar directly. They extend their car model by adding a "parts" attribute to it. Let's also assume that CheapCar has a different model for the parts side of their business and decide that they want to use this model to collect parts information from the rental company. These assumptions are now part of the example and will serve to illustrate assertions made in this chapter.

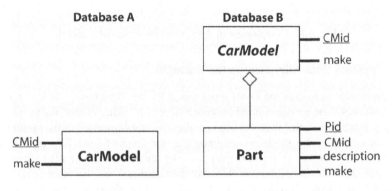

Fig. 6.13. Illustration of Aggregation

Figure 6.13 illustrates an example discussed in [8] that demonstrates a situation similar to this scenario. It shows model A representing a car as a whole entity "CarModel" with an id and a make, and model B representing parts of a car. When mapping these two models, parts have to be aggregated yielding the superset of "CarModel" with additional "id" and "make" attributes and then matched with the car entity in A. A systematic application of the Data Engineering process results in stable and well documented solutions.

6.2.11 Model-Based Data Engineering

Data Engineering presents a series of processes for modelers to use during integration projects in static environments. However, in rapidly changing environments, modelers need the capability to add emerging models without having to start over from scratch. The authors advocate the introduction of a Common Reference Model (CRM) as an integral part of Data engineering, which leads to Model Based Data Engineering (MBDE). This chapter will discuss how MBDE improves the integration process.

The car example will serve as an illustration in this case too. In most integration projects, it is rare to have to completely map the models. Instead, the goal of the integration dictates the information exchange needs in the form of Business Objects and Business Rules. The authors will show how making Business Objects and Business Rules part of Data Engineering simplifies the process and reduces the integration effort to a set of manageable semantic data entities. Here again, the car example will serve as a practical application.

Defining a Data Engineering process is a good step towards the goal of interoperability. However, solutions derived from a systematic application of the Data Engineering process often lack flexibility and reusability. The traditional approach when integrating heterogeneous models is to create proprietary connecting interfaces. This results in P2P connections that are satisfactory in static environments. However, in this era of rapid communication and globalization, the need to add new models can arise at any moment. In order to add a new model, the Data Engineering process must be reapplied and the end result is another P2P solution. Mathematically speaking, federating N models using P2P connections will result in $N*(N-1) \div 2$ interfaces (see figure 7.18). For example, federating ten heterogeneous systems would require forty-five unique interfaces, as shown in figure 7.14.

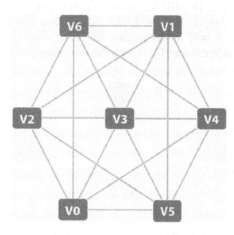

Fig. 7.14. Federation of seven models using P2P Connections

In past decades, flexibility and reuse have been neglected and more effort has been rightly directed at identifying and resolving mapping issues. In order to avoid those same pitfalls, the Data Engineering process must include rules and guidelines addressing these issues. For the solution to be flexible and reusable, Tolk and Diallo [6] argue that implementing Data Engineering must include the implementation of a CRM. A valid CRM must represent:

- Property values: For any model M containing a list of independent enumerated values V_1, V_2...V_n susceptible of being exchanged, there must be an exhaustive set S of unique enumerated values in the reference model such that $S_V = \{V_1, V_2...V_n\}$. S_V can be extended as more models are added to the federation.
- Properties: For any model M containing a list of independent attributes A_1, A_2,..., A_n susceptible of being exchanged, there must be an exhaustive set S of attributes in the reference model such that $S_A = \{A_1, A_2...A_n\}$. S_A can be extended as more models are added to the federation.
- Propertied Concepts: For real life objects $O1$, O_2...O_n susceptible of being exchanged, there must be a set S of independent objects in the reference model

such that $S_O = \{O_1, O_2...O_n\}$. Objects can be added as new models join the federation

- Associated Concepts: For any set of Objects O, linked through a relationship R describing real world concepts C1,C2...C_n susceptible of being exchanged there must be an exhaustive set S of concepts in the reference model such that $S_C = \{ C_1, C_2...C_n \}$.

This extended framework advocating the creation of a CRM during the implementation of Data Engineering is known as MBDE. The main advantage of MBDE is the creation of a series of information exchange requirements with a specific input/output data set and format to which all participating models have to abide. It becomes in fact the common language spoken and understood by all members of a federation. In MBDE, models interoperate through the CRM. Each model understands the language of the CRM and can therefore exchange information with any other model. A new model joining the federation must only interface with the CRM and it automatically interfaces with the rest of the federation.

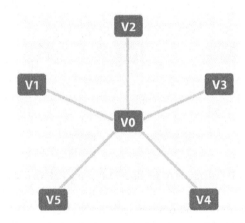

Fig. 6.15. Federation of six models with V0 as CRM

A CRM reduces the number of total interfaces in a federation of N models (including the CRM) from $N*(N-1) \div 2$ interfaces (P2P) to N-1 interfaces (see figure 6.19). It is worth noting that an existing model can be a member of the federation and act as the CRM provided it fulfills the requirements specified earlier.

6.2.12 Model-Based Data Engineering for Web Services

Having described Data Engineering and discussed how MBDE improves the process this chapter addresses the more practical issue of implementation. The authors will show that Service Oriented Architectures in general and web services in particular are a natural fit for implementing MBDE. The main idea behind SOA in general and web services in particular is to offer software applications as services following standards defined by the World Wide Web Consortium (W3C) [31]. Conformant to those standards, web services must be describable, discoverable and possess a communication protocol. Most web services are described by the Web Service Description Language (WSDL), discovered through a Universal Description, Discovery and Integration (UDDI) directory and use the Simple Object Access Protocol (SOAP) as the means of communication (see figure 6.16).

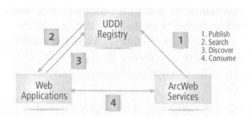

Fig. 6.16. Web Services Architecture

The architecture in figure 6.16 shows that services are published via a broker commonly referred to as a UDDI (1). Applications search the UDDI for services (2, 3) and consume them if they are available (4). A SOAP layer between these three nodes acts as the medium of communication.

SOA have many inherent advantages namely

- Web Services are asynchronous: The state of the client does not affect the state of the service. The client requests information in a clearly defined format through an interface, and the service provides the information in a fashion unknown to the client.
- Web Services are loosely coupled: The disappearance of a client does not affect the overall system (which is not true for traditional architectures). Other clients continue to be served as needed. This means that only the service is aware of all of the participants thus making it the only mediator between the participating systems.
- Web Services can be offered as a family of independent services: Data mapping and data mining can be offered as distinctive components of the same service. This allows a substantial reduction of overhead data during the information exchange since static or setup data will only be communicated once.

The SOA approach reduces the role of the client (subscribing systems) and puts the onus on the service to internally perform all the tasks and only return pertinent results. XML is the standard language for web services description, discovery, and communication. The purpose of XML is to describe, identify, and even qualify the data contained in a document. XML is used for traditional data processing, document-driven programming, archiving and binding. In addition, XML has the advantage of being hierarchical, linkable, easily reusable, and easily processed. It makes perfect sense and becomes common practice to use XML as a mean to describe database contents [10].

Through these standards, web services inherently possess what traditional point-to-point solutions lack. They are flexible and reusable as well as portable and stable because the status of a service is independent of that of their clients. Additionally web services inherently define a data format (XML) and a communication protocol (SOAP). Data modelers and experts on both sides can now focus on the important task of meaningful information exchange which has to be distinguished from mere data exchange. It is a common error to equate the exchange of bits and bytes (which is a technical issue relatively simple to solve) with that of information exchange. While SOA is a very good architecture in terms of data exchange it is still lacking when it comes to information exchange. In order for information exchange to occur, services have to be orchestrated in such a way that the atomic aspects of the model can be combined to produce meaningful information. As stated earlier in this chapter, web services must be combined with a reference model in order to obtain a truly flexible and reusable architecture. Three essential components are needed to facilitate systems integration via web services:

- A reference model is defined as any data model containing constructs that covers the entire scope of the simulation space. Establishing a common language allows virtually any system within the domain to participate via a translator.
- A translator or data mediator is a program or script capable of converting a system's native dialect to that of the reference model and vice versa. Therefore a translator is the only component of the web service architecture that needs to be system specific. The data engineering process is used to define what information is exchanged.
- A family of Web Services: As described earlier, web services are self-described, discoverable and have a clearly defined communication protocol. Services and methods can be developed to monitor and manage data flow (update, insert, delete) in the reference model [4].

Furthermore the translator component can be offered as a service of services or a composition of services. The family of services can be separated into two distinct categories:

- Atomic Services: *An Atomic Service is used to exchange SIE by using the defining properties of the related propertied concept as parameters.* The propertied concepts of the CRM can be related to each other using associations (building the associated concepts that can become as complex as necessary for an application). If two or more propertied concepts are associated with each other in the CRM, they can be used to construct a composite information element, thus leading to the following definition:
- Composite Services: *A Composite Service is used to exchange semantically related SIE by exposing the defining properties of all related propertied concepts that belong to the associated concept representing the composite on the CRM side as parameters.* There are three classes of operations possible for composites (and in principle also to atomic information, but in practice this is observed for higher information elements only): *Filtering, Transformation, and Aggregation.* Furthermore we have to distinguish between *weak composites* which can only be used to retrieve information form the CRM and *strong composites* which can insert and or retrieve information from the CRM

Data can be represented either as individual elements or as a collective whole. Web Services reduce data into the individual items that they understand and are able to communicate. A human in the loop has to assemble and interpret data in order to turn it into information. In order for web services to provide information and not just individual data items, services have to be orchestrated. The end result is that services are combined into super services that ultimately fulfill the user's need. Beyond just providing information rather than data to users, orchestrating services reduces the amount of traffic on the network. Service orchestration is based on two main principles:

- *Concept filtering:* This consists of dividing the reference model into well-defined concepts (organizations, resources, locations, reporting objects). Most models are deliberately structured in this manner, making this process relatively simple.
- *Access flexibility:* In order to support service orchestration it is necessary to have access to data on multiple levels. The first level is defined as atomic access. This is a set of services that provide access to individual data items within any object. The second level is the composite access. These are services that provide access to a combined set of objects describing a given concept. Access flexibility allows data to be recombined according to the user's needs without having to redefine a new set of services.

It should be pointed out that a CRM is not necessarily a data model implemented as a database. The CRM captures the information structure and exchange constraints to ensure complete and consistent information exchange. That such a complete and consistent information exchange is captured in the form of valid transactions in a database is one of many implementations of the CRM. Valid Web service calls, XML schema, and other techniques are possible as well. In other words: *The CRM captures the content, structure, and underlying business rules for the information exchange requests valid in the supported domain; the family of Web services ensures the consistent and unambiguous information exchange based on these principles.*

The ideas of MBDE for web services were first published in [6]. In the meantime, they have been successfully implemented in support of the Joint Rapid Scenario Generation capability of the US Joint Forces. Within this effort, several authoritative databases made their data available via web services in support of generating scenarios using authoritative data. These data had to be modified to the needs of several simulation systems. Data providers as well as data users had different data models. The application of MBDE let to the definition of more than 1,000 web services that where implemented on a server provided by industry partners enabling the exchange of these heterogeneous data via composable web service without modification of the participating system. Perme et al. [11] give an overview of this system. Another application for Homeland Security applications is given by Tolk et al. in [12].

6.3 Leading Towards a New Paradigm for Information Exchange

While the ideas and processes presented thus far have already been proven feasible and applicable, the combination of SOA and MDBE point towards a paradigm shift in information exchange. Before we delve in the paradigm shift in more details it is important to recognize that there are several levels of interoperability and our ultimate goal is to have meaningful information exchange between systems with as little human-in-the-loop as possible.

6.3.1 The Levels of Conceptual Interoperability Model

The Levels of Conceptual Interoperability Model (LCIM) has been developed to provide both a metric of the degree of conceptual representation that exists between interoperating systems and also as a guide showing what is necessary to accommodate a targeted degree of conceptual representation between systems. The model was originally developed to support the interoperability of simulation systems, but has been shown to be useful for other domains, as well. It was first published by Tolk and Muguira [13] and later enhanced in response to ongoing research, in particular reflecting ideas of Page et al. [14] and Hofmann [15]. The following figure shows the current LCIM:

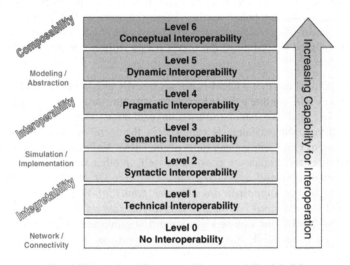

Fig. 6.17. Levels of Conceptual Interoperability Model

The current version of the LCIM distinguishes seven layers, starting with stand-alone systems. The underlying approach is best understood from the by bottom-up.

- Level 0: Stand-alone systems which need *No Interoperability*. No connection exists between any systems.
- Level 1: On the level of *Technical Interoperability*, a communication protocol exists for exchanging data between participating systems. At this level, a communication infrastructure is established allowing exchanging bits and bytes; the underlying networks and communication protocols are unambiguously defined. This is the minimal form of connectivity required for interoperation of systems: being technically connected.
- Level 2: The *Syntactic Interoperability* level introduces a common structure to exchange information; a common data format is applied where the format of the information exchange is unambiguously defined. Examples for such common formats are the use of XML or the use of the Object Model Templates of the

IEEE1516 Standard High Level Architecture (HLA). Although a common format is used, nothing ensures that the receiver understands what the sender wants to communicate.

- Level 3: If a common information exchange reference model is used, the level of *Semantic Interoperability* is reached. On this level, the meaning of the data is shared; the content of the information exchange requests are unambiguously defined. Examples are the Protocol Data Units (PDU) of the IEEE1278 standard Distributed Interactive Simulation (DIS). Also, if the semantic meaning of the data is captured in the form of a controlled vocabulary with an associated dictionary, this level is supported.

- Level 4: *Pragmatic Interoperability* is reached when the interoperating systems are aware of the methods and procedures that each other are employing. In other words, the use of the data – or the context of its application – is understood by the participating systems; the context in which the information is exchanged is unambiguously defined. In particular in systems in which the necessary information can be submitted in several successive communication instances (like sending several HLA messages or object that in summary comprise the required information, or if more than one PDU is needed to cover the information request), the business objects associated with this sort of work flow between the systems must be known. Another way to think about pragmatic interoperability is that individual meaning of data elements is placed into the context of how the data is used within the functionality of the resulting system.

- Level 5: As system functions and services operate on data over time, the state of that system will change, and this includes the assumptions and constraints that affect its data interchange. If systems have attained *Dynamic Interoperability*, then they are able to comprehend the state changes that occur in the assumptions and constraints that each other is making over time, and are able to take advantage of those changes; the effect of the information exchange within the participating systems is unambiguously defined. As a minimum, the input data is connected to output data including temporal aspects of the systems (white box with behavior). A complete open solution with everything revealed to the user (such as in open source including the system specification of the platform on which the service will be executed) marks the high end of this level.

- Level 6: Finally, if the conceptual models – i.e. the assumptions and constraints of the purposeful abstraction of reality – are aligned, the highest level of interoperability is reached: *Conceptual Interoperability*. On this level, the assumptions and constraints describing the purposeful abstraction of reality are unambiguously defined. This requires that conceptual models will be documented based on engineering methods enabling their interpretation and evaluation by other engineers. In other words, on this we need a "fully specified but implementation independent model" as requested in Davis and Anderson [16] and not just a text describing the conceptual idea. As pointed out by Robinson [17], this is one of the major challenges.

Page at al. [14] introduced the three categories used as well in figure 6.17: *Integratability* manages the physical and technical realms and challenges of connections between systems, which include hardware and firmware, and protocols. This is the domain of networks and other physical connections between systems. *Interoperability* deals with the software and implementation details of interoperation, including exchange of data elements based on a common data interpretation, which can be mapped to the levels of syntactic and semantic interoperability. Here we are on the simulation side of M&S and how the models are actually implemented and executed. *Composability* addresses the alignment of issues at the model level. The underlying models are meaningful abstractions of reality used for the conceptualization being implemented by the resulting simulation systems.

The LCIM has been successfully applied in various domains, such as the Department of Defense and the Department of Energy. It has been evaluated using rigor mathematical approaches [18] and was used as a reference model for alternative layered approaches by leading research institutes [19, 20]. It can be regarded as a mature model in support of interoperation evaluation.

3.2 Self-organizing Information Exchange

The idea of self-organizing information exchange is popular in several semantic web applications. Su et al. [21] describe prototypical experiments on automatable transformations between XML documents. Most of these applications are not very complex and focus on quasi-static data, such as "data describing addresses" or "data describing references to journal papers." Ambiguities and different levels of resolution are not dealt with. Therefore, the existing proposed solutions are not applicable without extensions and modifications.

The current IEEE1278 (Distributed Interactive Simulation Protocol – DIS) and IEEE1516 (High Level Architecture – HLA) standards support simulation interoperability. They mandate a set of Protocol Data Units in DIS or a Federated Object Model in HLA to exchange information. How the system specific information objects are mapped to the data used for information exchange is not documented. It is not of general concern if the participating systems can provide or use the information at all. Although it starts to become a best practice to capture the information in the federation agreements, this is not part of the standard and is not done in a machine understandable way.

The alternative proposed below relies on metadata describing system characteristics on all levels of the LCIM using MDBE. As a result machines or software applications should be able to:

- parse and compute the information (which requires the use of a formal language, such as XML or OWL)
- select systems for a task based on their described capabilities (using OWL-S for web services)
- choreograph the execution, and
- orchestrate the process in order to optimize the process support.

The ideas and experiments in this direction are further captured in Yilmaz and Paspuletti [22] and Yilmaz and Tolk [23].

The resulting information exchange paradigm shift replaces the prescriptive use of a common information exchange model with a more flexible one that allows systems to specify their information exchange capabilities and discover – guided by the principles of data engineering and supported by intelligent agents – what information is available. System can also match their view of the world with an appropriate selection and orchestration of weak composites.

By applying the methods of MBDE, we suggest a model view in alignment with the ideas of Model Theory [24] for each system interface. This is also a knowledge representation of concepts and associations as described in various implementations by Sowa [25]. Each description comprises:

- information elements to be exchanged in the form of sets of properties that are used to describe the concepts in the application,
- associations between these propertied concepts reflecting represented relations between concepts in the application,
- definitions of valid property values for each property,
- property value constraints based on application specific rules (such as: if value of property a = A then value of property B cannot be A as well),
- propertied concept constraints based on application specific rules (such as: propertied concept C1 cannot be instantiated if propertied concept C2 is nonexistent),
- association constraints based on application specific rules (such as: the context CT requires transmitting the associated propertied concepts PC1 and PC2 in one transaction).

On the implementation level each of these system specific descriptions must be consistent in itself and represents a structure M. Pillay [24] defines this structure as: *"simply a set X, say, equipped with a distinguished family of functions from X_n to X (various n) and a distinguished family of subsets of X_n (various n). Here X_n is the Cartesian product X x ... x X, n times."*

The metadata sets captured during the MDBE process are necessary and sufficient to describe these structures. Furthermore, two structures *M1* and *M2* are equivalent if and only if each structure element of *M1* is mapped to exactly one structure element of *M2* under all *n*.

As shown by Pillay [24] and Sowa [25], such structures describe various implementations for information technology specific solutions, such as databases, artificial languages (i.e., computer understandable languages, such as specified by grammars), ontological representations (i.e., in computer understandable form, such as captured in OWL), and other examples. In other words, it is possible to algorithmically evaluate if two structures represent equivalent views.

MBDE introduces for each system specific description a system independent description (on the second-tier level, which is the conceptual representation), which in itself is a structure. In the current approach Data Engineers ensure the mapping between implementing entities and representing concepts is equivalent to that of the implementing structure with the conceptual structure.

If this is consistently captured in metadata using MDBE, this work can be executed, verified, and validated by intelligent software agents as envisioned by Yilmaz and Paspuletti [22]. However, in order to support such a vision, the assumptions on constraints of simulation systems identified in efforts described by Davis and Anderson [16], Hofmann [15] or Yilmaz [26] and Yilmaz and Tolk [23] need to be captured in a standardized way and made accessible by intelligent agents.

Such standards are targeting the conceptual level, which means the modeling part of M&S and clearly contributes to the research agenda proposed by Robinson [17].

6.4 Summary

Developers can use the MBDE framework and common reference data models in related application fields that require services composition to enable overarching pattern recognition (as in homeland security applications). These same concepts might also support a cascading Web services framework connecting various reference data models, e.g., when connecting independently developed models of different communities of interest. While agencies' data engineering processes remain independent, the resulting data mediation services can be agency-oriented or enable highly efficient peer-to-peer results in special cases. In emergencies, for example, first responders need to share information. Police, the National Guard, and local health organizations can maintain common information-exchange data models for their own systems, but in emergencies, the unambiguous police tag set must map to the National Guard and health organization tag sets, and vice versa. This can occur through a mediation service that translates between the information spaces of all first-responder organizations. The result would be, for example, that police could use their own information system to coordinate their work with the National Guard, as well as request beds in the local hospital. With the right metadata, these services can even be composed by intelligent agents on the mid term, and community of interest specific data models can be generated from participating information exchange specifications by intelligent software on the long term as well. We are approaching the next step towards self-aware and self-configuring software.

Until then, the applicable methods we present here are technically mature enough to be applied to support communities of interests in service oriented architectures. Initial prototypes have demonstrated their feasibility and efficiency. What is currently missing is the community-wide will to agree to such a common way to do business; cultural gaps, rather than technical gaps, are the main obstacles. In our case, commercial industry partners are increasing their support for our methods, which bodes well for the future of a common path in our domain.

References

[1] Spaccapietra, S., Parent, C., Dupont, Y.: Model Independent Assertions for Integration of Heterogeneous Schemas. Very Large Database (VLDB) Journal 1(1), 81–126 (1992)

[2] Microsoft Inc (last accessed in March 2006), http://www.microsoft.com/technet/prodtechnol/sql/2005/intro2is.mspx

[3] Peter Wood (last accessed in March 2006), http://www.dcs.bbk.ac.uk/~ptw/teaching/ssd/toc.html

[4] Website Indexing 2nd Edn. (last accessed in March 2006), http://members.optusnet.com.au/~webindexing/Webbook2Ed/glossary.htm

[5] Tolk, A.: Common Data Administration, Data Management, and Data Alignment as a Necessary Requirement for Coupling C4ISR Systems and M&S Systems. Information & Security: An International Journal 12(2), 164–174 (2003)

[6] Tolk, A., Diallo, S.: Model-Based Data Engineering for Web Services. IEEE Internet Computing 9(4), 65–70 (2005)

[7] http://mathworld.wolfram.com/Bijection.html (last accessed September 12, 2007)

[8] Parent, C., Spaccapietra, S.: Database Integration: The Key to Data Interoperability Advances in Object-Oriented Data Modeling 2000, pp. 221–253 (2000)

[9] World Wide Web Consortium (last accessed September 12, 2007), http://www.w3.org/2002/ws

[10] Tolk, A.: Moving towards a Lingua Franca for M&S and C3I – Developments concerning the C2IEDM. In: ACM Proceedings of the European Simulation Interoperability Workshop (2004)

[11] Perme, D., et al.: Joint Event Data Initialization Services (JEDIS) – Implementing a Service Oriented Architecture for Initialization. In: IEEE Proceedings of the Spring Simulation Interoperability Workshop (2007)

[12] Tolk, A., Diallo, S.Y., Turnitsa, C.D.: Model-Based Alignment and Orchestration of Heterogeneous Homeland Security Applications enabling Composition of System of Systems. In: IEEE Winter Simulation Conference (2007)

[13] Tolk, A., Muguira, J.A.: The Levels of Conceptual Interoperability Model (LCIM). In: IEEE Proceedings of the Fall Simulation Interoperability Workshop (2003)

[14] Page, E.H., Briggs, R., Tufarolo, J.A.: Toward a Family of Maturity Models for the Simulation Interconnection Problem. In: IEEE Proceedings of the Spring Simulation Interoperability Workshop (2004)

[15] Hofmann, M.: Challenges of Model Interoperation in Military Simulations. SIMULATION 80, 659–667 (2004)

[16] Davis, P.K., Anderson, R.H.: Improving the Composability of Department of Defense Models and Simulations. RAND Corporation (2003)

[17] Robinson, S.: Issues in Conceptual Modeling for Simulation: Setting a Research Agenda. In: 2006 OR Society 3rd Simulation Workshop, Ashorne, UK (2006)

[18] Lei, Y., Wang, W., Li, Q., Wang, W.: Concepts and Evaluation of Simulation Model Reusability. In: IEEE Proceedings of the Fall Simulation Interoperability Workshop (2007)

[19] Morris, E., Levine, L., Meyers, C., Place, P., Plakosh, D.: System of Systems Interoperability (SOSI): Final Report. Carnegie Mellon Software Engineering Institute, Technical Report CMU/SEI-2004-TR-004 ESC-TR-2004-004 (2004)

[20] Zeigler, B.P., Hammonds, P.E.: Modeling and Simulation-based Data Engineering. Academic Press, London (2007)

[21] Su, H., Kuno, H.: Rundensteiner E.: Automating the transformation of XML documents. In: 3rd International Workshop on Web Information and Data Management, pp. 68–75. ACM Press, New York (2001)

[22] Yilmaz, L., Paspuletti, S.: Toward a Meta-Level Framework for Agent-supported Interoperation of Defense Simulations. Journal of Defense Modeling and Simulation 2(3), 161–175 (2005)

[23] Yilmaz, L., Tolk, A.: Engineering ab initio dynamic interoperability and composability via agent-mediated introspective simulation. In: IEEE Winter Simulation Conference, pp. 1075–1182. IEEE CS Press, Los Alamitos (2006)

[24] Pillay, A.: Model Theory. Notices of AMS 47(11), 1373–1381 (2000)

[25] Sowa, J.F.: Knowledge Representation: Logical, Philosophical, and Computational Foundations. Brooks Cole Publishing Co. (2000)

[26] Yilmaz, L.: On the Need for Contextualized Introspective Simulation Models to Improve Reuse and Composability of Defense Simulations. Journal of Defense Modeling and Simulation 1(3), 135–145 (2004)

SAM: Semantic Advanced Matchmaker

Erdem Savas Ilhan and Ayse Basar Bener

Bogazici University Department of Computer Engineering
P.K. 2 TR-34342 Bebek, Istanbul, Turkey

Abstract. As the number of available Web services increase finding appropriate Web services to fulfill a given request becomes an important task. Most of the current solutions and approaches in Web service discovery are limited in the sense that they are strictly defined, and they do not use the full power of semantic and ontological representation. Service matchmaking, which deals with similarity between service definitions, is highly important for an effective discovery. Studies have shown that use of semantic Web technologies improves the efficiency and accuracy of matchmaking process. In this research we focus on one of the most challenging tasks in service discovery and composition: Service matchmaking. We make use of current semantic Web technologies like OWL and OWL-S to describe services and define ontologies. We introduce an efficient matchmaking algorithm based on bipartite graphs. We have seen that bipartite matchmaking has advantages over other approaches in the literature for parameter pairing problem, which deals with finding the semantically matching parameters in a service pair. Our proposed algorithm ranks the services in a candidate set according to their semantic similarity to a certain request. Our matchmaker performs the semantic similarity assignment implementing the following approaches: Subsumption-based similarity, property-level similarity, similarity distance annotations and WordNet-based similarity. Our results show that the proposed matchmaker enhances the captured semantic similarity, providing a fine-grained approach in semantic matchmaking.

7.1 Introduction

7.1.1 Motivation

In recent years, Web services became the dominant technology in providing the interoperability among different systems throughout the Web. If Web service is used in limited business domain or with strict rules with known business partners everything will be fine. The problem of finding the right and most suitable Web services for user needs emerges when open e-commerce systems are widely used and user requirements dynamically change over time.

Although there are currently proposed technologies for discovery of Web services, such as UDDI [5], they do not satisfy the full discovery requirements. This discovery process is based on syntactical matching and keyword search that does not allow the automatic processing of Web services. To solve the problem of automatic discovery and processing of Web services, the Semantic Web [6] vision is proposed. Semantic Web is an effort by the W3C consortium [7], and one of its main purposes is to facilitate the discovery of Web resources.

There are different efforts and frameworks for semantic annotation and discovery of Web services [10, 11, 12]. For Web service discovery the researchers also propose

R. Nayak et al. (Eds.): Evolution of the Web in Artificial Intel. Environ., SCI 130, pp. 163–190, 2008.
springerlink.com

some techniques and algorithms. However, they mostly classify the discovered Web services as set-based. They do not focus on rating the Web services using semantic distance information [13].

The evolution of Web services, from conventional services to semantic services, caused service descriptions contain extra information about functional or non-functional properties of Web services. The semantic information included in the service descriptions enables the development of advanced matchmaking schemes that are capable of assigning degrees of match to the discovered services. Semantic discovery of Web services means semantic reasoning over a knowledge base, where a goal describes the required Web service capability as input. Semantic discovery adds accuracy to the search results in comparison to traditional Web service discovery techniques, which are based on syntactical searches over keywords contained in the Web service descriptions [3].

Improvement in matching process could be gained by the use of ontological information in a useful form. With the use of this information, it would be possible to rate the services found in discovery process. As in real life, users/ agents should be able to define how they see the relation of ontological concepts from their own perspective. Similarity measures have been widely used in information systems [14, 15, 16], software engineering [68, 69] and AI [17, 18, 19]. So integration of knowledge from these techniques can improve the matching process.

By using semantic distance definition information, we aim to get a rated and ordered set of Web services as the general result of the discovery process. We believe that this would be better than set-based classification of discovered services. In this research, we propose a new scheme of matchmaking that aims to improve retrieval effectiveness of semantic matchmaking process. Our main argument is that conventional evaluation schemes do not fully capture the added value of service semantics and they do not consider the assigned degrees of match, which are supported by the majority of discovery engines. The existing approach to service matchmaking contains subsumption values regarding the concept that the service supports. In our proposed approach, we add semantic relatedness values onto existing subsumption-based procedures. Our matchmaker performs the semantic similarity assignment implementing the following value-added approaches: Subsumption-based similarity, property-level similarity, similarity distance annotations and WordNet-based similarity.

7.1.2 Outline

In Section 2, we refer to related work in this field and point out the contributions they make, emphasizing how they can be extended in a way to support better results. In section 3 we state the problem that we focus on in this research. We list the problems in semantic service matchmaking that we are seeking answers for.

Section 4 presents our approach to service matchmaking. We give details of our proposed approach and introduce the techniques and methods we apply in formalism. Section 5 presents the evaluation results of the test runs which we performed on a prototype system that we implemented. We present numerical results on test runs and discuss these results, clearly demonstrating the advantages of our proposed approach with a comparison to a conventional service matchmaker. Finally, Section 6 concludes our research with the future directions for extending the capabilities of our service matchmaker.

7.2 Related Work

Semantic Web services aim to realize the vision of the Semantic Web, i.e. turning the Internet from an information repository for human consumption into a worldwide system for distributed Web computing [6]. The system is a machine-understandable media where all the data is combined with semantic metadata. The domain level formalizations of concepts form up the main element within this system, which is called ontology [11]. Ontology represents concepts and relations between the concepts; these can be hierarchical relations, whole-part relations, or any other meaningful type of linkage between the concepts [4].

The semantic matchmaking process is based on ontology formalizations over domains. In the upcoming section we present some of the selective research on the matchmaking process considering the concepts that we build our research on. Matchmaking of Web services considers the relationship between two services. The first one is called the advertisement and the other is called the request [2]. Advertisement denotes the services description of the existing services while the request indicates the picture of service requirements [19].

Traditional approaches to modeling semantic similarity between Web Services compute subsume relationship for function parameters in service profiles within a single ontology. In [20] a graph theoretic framework based on bipartite graph matching for finding the best correspondences among function parameters belonging to advertisement and request is introduced. On computing semantic similarity between a pair of function parameters, a similarity function is introduced, determining similar entity, which relaxes the requirement of a single ontology and accounts for the different ontology specifications. The function presented for semantic similarity across different ontologies provides an approach to detect similar parameters. The method is based on a matching process over weighted sum of synonym sets, semantic neighborhood, and distinguishing features. They make use of WordNet to capture the similarity between parameter names and also consider properties of concepts in matchmaking as distinguishing features. However, the method lacks use of contextual information or user preferences in matchmaking process.

In [22], a semantic ranking MSC is designed to rank the results of advertisements matchmaking. MSC stands for the initials of three factors' second words: Semantic Matching Degree (to capture the semantic aspects of attributes), Semantic Support (to describe the interestingness or potential usefulness of an attribute) and Relational Confidence (to capture the association relationships among attributes). Three categories of attributes are defined in advertisements matchmaking: *Generalizable Nominal Attribute (GNA)* whose values can form a concept hierarchy; *Numeric Attribute (NUA)*, called quantitative attribute, whose values are numeral; Nominal Attribute (NOA) whose values are neither numeral nor can form a concept hierarchy. However, in this study, the presented factors excluding Semantic Matching Degree, are not directly dealing with assessing service similarity to a certain request.

In [65], the authors introduce a step-by-step matchmaking process. They consider profile, input-output and non-functional attributes matching in the process. They also provide ranking of services according to their similarity to a certain request. However, they consider discrete scales of similarity measures: Exact, plug-in, subsume, intersection and fail. Although they provide ranking, they apply subsumption based reasoning and do not consider properties of concepts and contextual information in matchmaking. In [66], the authors describe capability matchmaking as matching inputs, outputs, preconditions and effects of services. They focus on the input-output matching. However, only a discrete scale classification of matching is provided. Besides, contextual information and properties of OWL concepts are not taken into account.

In [55], authors approach the problem of service similarity in a category-based approach. They define the following categories: Lexical similarity, which considers the textual similarity of service names; attribute similarity, which evaluates semantic similarity between Web service attributes and interface similarity, which considers the input-output parameter type and name similarity. They define a conceptual model of Web services, where a Web service is identified with its attributes and operations. Their study mostly focuses on the attribute similarity and focus on QoS attributes. The input-output parameter type similarity is only performed by a parameter type mapping table. The properties for those types defined in ontology are not considered. Besides, contextual information or user preferences are not considered in the matching process.

7.3 Problem Statement

The current industry standard for service discovery is UDDI [53]. However, UDDI has some shortcomings in that it returns coarse results for a keyword-based search and more importantly it lack semantics. It is basically a framework that supports category-based search.

Semantic service discovery tries to solve the above problem by utilizing capability-based search mechanism [25]. Capability based search enables reasoning on service input, output parameters, preconditions and effects. Service discovery tries to find all services that satisfy a certain request whereas service matchmaking deals with the relationship between two services [26]. Service discovery needs support of service matchmaking process.

The problem that we focus on in this research is to find suitable Web services that satisfy a certain request by applying a matchmaking process. A conventional matchmaker as described by Paolucci et al. [27] would distinguish four different relations between two concepts: Exact, subsume, plug-in and fail. This matchmaker would also determine the matching degree of a service to the request by the lowest degree of match. Considering the parameters, if one of them fails to match with request parameters the service would be considered as a fail to match. However, this scale of discrimination may not be appropriate in certain situations. The experimental research so far has shown that simple subsumption based matchmaking is not sufficient to capture semantic similarity [1, 22, 54, 55].

A discrete scale matchmaking methodology may result in false negatives. Some of the services in the candidate set might be eliminated due to not fitting those discrete scales. However, semantic relation has many dimensions other than subsumption. Next section describes those dimensions that we take into account, namely the properties of concepts and explicit annotations on concept similarity. A complete system should also evaluate all the candidate services and rank them according to their semantic similarity to a given request. We believe that considering semantic relations in depth provides a better matchmaking process.

Contextual information is also important to a matchmaking process. A matchmaking approach considering only raw OWL ontology definitions would operate without considering context. Matchmaking should also consider user preferences to be able to fully capture semantic similarity upon a user request.

Service composition also benefits from semantic Web to support automatic composition. Compositions can be generated dynamically by utilizing semantic descriptions of services to organize them in a workflow. In most of the composition approaches services are added to the composition one by one [56, 57]. As each service is added a matchmaking step is needed to make sure that the service supports the IOPE (Input, output, precondition and effect) constraints of the workflow node. Besides, the selected service's IOPE affects the matchmaking process for the next workflow node. Thus, a rich matchmaking framework is needed to obtain successful compositions. Considering alternative services in a composition is crucial for evaluation. So, a feature-rich and ranking based matchmaker is crucial for a composition engine.

In this research, we aim to provide an efficient and accurate matchmaking algorithm using scoring and ranking based on similarity distance information, extended subsumption and property level similarity assessment in a general semantic Web service discovery framework.

7.4 Proposed Solution

In this research, we propose a method that enhances simple subsumption based matchmaking approaches. Our main motivation is to capture semantic similarity between services in a more efficient way and eliminate false negatives. We consider service input and output parameters and perform the matchmaking considering the I/O (Input/Output) interface of services. Our proposed solution uses decision modules that can be plugged in and out. We have implemented some of these modules to add semantic relatedness values onto existing subsumption-based procedures. Our proposed matchmaker agent, SAM – Semantic Advanced Matchmaker, mainly provides ranking and scoring based on concept similarity. We also introduce the use of similarity distance annotations in an OWL document. Similarity distance supports explicitly annotating concept similarity in a numerical fashion. These annotations might refer to user's view of ontology and the similarity degree of concepts according to the user. Similarity distance is actually a method to represent context information in matchmaking process as described in Section 7.3.

Figure 7.1 overviews the matchmaking process in SAM. We assume that a conventional service discovery is performed on a request and as a result we have a relevant service set to apply matchmaking on. SAM gets the relevant service set and the request as inputs. The output of SAM is a ranked list of relevant service set. The focus of this study is the methods and procedures that take place in Matchmaking box described in Figure 7.1.

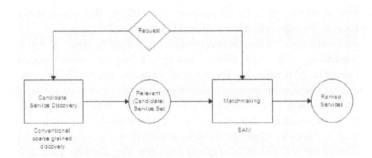

Fig. 7.1. SAM matchmaking process

The components of the proposed architecture are shown in Figure 7.2. Request service definition and the corresponding relevant services set, which are discovered through conventional discovery mechanisms, are presented as input to the system.

Matchmaker component organizes the framework for comparing each of the services in the relevant service set with the request. It communicates with the Scoring Module to obtain similarity score for each parameter pair for a certain request-service pair. Finally, matchmaker component generates a bipartite graph for each request-service pair cooperating with Bipartite Graph Module. Each bipartite graph represents the service and request parameters with the similarity score assigned on every edge. As we are interested in retrieving best matching services compared to a request, bipartite graph matching algorithm will find the maximum weight match in each bipartite graph. This step ends up with parameter pairings in bipartite graphs representing maximum weight match in each service and request pair. At the end of the matchmaking process, the services in the relevant service set are ranked according to their scores in bipartite-graphs, which of course represent their semantic similarity to the request. SAM Bipartite Matcher module uses the Hungarian algorithm of Kuhn (1955), as improved and presented by Lawler (1976). The bipartite matcher engine finds maximum-weight matching in a complete bipartite graph.

Scoring Module is the part of the system where similarity between concepts is scored according to several criteria. We propose a plug-in architecture here, so that additional scoring modules can be plugged in or out. Currently, we implemented three scoring modules: Subsumption based scoring, similarity distance scoring and WordNet similarity scoring. Pellet reasoner is used in association with the Scoring Module for inference in ontology. Details on how these scoring modules work are described in the following sections.

The main software components of our proposed matchmaking agent are shown in Figure 7.3. The top layer represents our matchmaker Semantic Advanced Matchmaker (SAM). OWL-S API models the service, profile, process and grounding ontologies of OWL-S in an easy to use manner. It is a widely used API in semantic applications. OWL-S API also presents interfaces for reasoning operations and utilizes Jena constructs at the back-end. At the bottom of the hierarchy we have Pellet reasoner for OWL reasoning operations.

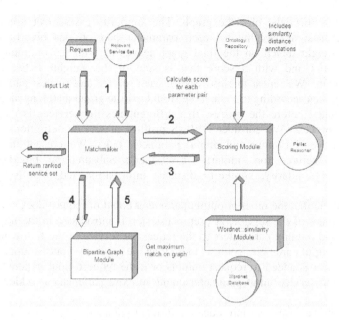

Fig. 7.2. Matchmaking agent components

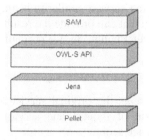

Fig. 7.3. Software components of matchmaking agent

We believe that a discrete scale (exact, plug-in, subsume, and fail) of service classification is not sufficient for a matchmaking process. On the other hand, semantic ranking of services can capture a set of services that are lost in a discrete scale match. Semantic similarity assessment is a crucial step for the ranking process. In our proposed architecture, we present value-added similarity assessment approaches between service and request parameter pairs, which are described below.

7.4.2 Matching Algorithm

Previous research has shown that bipartite graph matching algorithm is a good fit for finding matching parameters in a service and request pair [9]. Bipartite graph matching provides us a solution for parameter pairing problem. We consider the inputs and outputs as separate cases and partition the service parameters and request parameters to form the bipartite graph. The similarity assessment process of our matchmaker assigns weights for each parameter pair on this bipartite graph. A maximum weight match on the final graph leaves us with the optimum matching parameter pairs and with a score that is sum of the weights between matched parameter pairs. We repeat this process for each service and request pair and finally rank the services according to their score from bipartite graph matching algorithm.

As we stated before the process that differentiates the services is the similarity assessment process. We consider OWL-S profiles of service definitions and assign similarity scores for input and output parameter pairs. We present the following value-added features for similarity assessment: Subsumption based similarity, property-level similarity, similarity distance information and WordNet similarity assessment.

It is possible that the input or output parameter count of request does not match the parameter count of the service. A preprocess step is introduced in the matchmaking process to add dummy parameters to the request or service in the following cases: If the service output parameter count is more than the request output parameter count an extra parameter is added to request outputs or if the request input parameter count is more than the service input parameter count an extra parameter is added to service inputs.

For the above cases, we introduced additional parameters to service or request to make the parameter count equal and to be able to support perfect bipartite matching. We ignored the additional output parameter after the matchmaking process. However, for the inputs we normalize the service similarity score. The idea is to apply a fair matchmaking so that a service that can provide the results with less input does not get a lower score due to its parameter count. So, we normalize the score of such a service using the following equation:

$$S_{new} = |P_r|/|P_s| * S_{old} \tag{7.1}$$

In equation 7.1, S_{old} represents the original service similarity score, $|P_r|$ and $|P_s|$ represent the parameter count of request and service, respectively. Finally, S_{new} represents the normalized service score.

An exception case is defined for the output parameters. If the service has less output parameters than the request, we do not directly eliminate the service with a zero score. We believe that any piece of provided information is valuable considering the outputs. So we classify the service as partially satisfying the request. The score for the matched output parameters are considered in the final evaluation. However, this is not the case for the input parameters. As a request with insufficient parameter set will not be able to invoke a certain service.

Another preprocess step in our matchmaking agent is to decompose any parameter type if it can be represented as a owl:unionOf element. This step ensures that all the parameters involved in matchmaking are atomic and reasoning on parameter count can be performed safely. As an example, let us suppose the concept "address" is expressed as a union of concepts "street" and "home". If the request requires the parameters street and home as output and the service provides the parameter address, we may not capture the exact match between parameter pairs without such decomposition. The matchmaker could even return a failure to match as the service provides less output then required by the request.

After the decomposition of request and service parameters, the matchmaker returns a failure to match for the following cases: If the service has less output parameters then required by the request or if the request has less input parameters then required by the service.

After all the services in the candidate set are evaluated against the request, SAM stores the following information for each service and request pair: Match type (exact, subsume, plug-in, fail, property-level and ontology distance) for each matching parameter pair, similarity score (range between 0 and 1) for each matching parameter pair and total similarity score of service and request matching.

We provide ranking of services according to the above scores and in terms of their input and output parameters. However, considering both the input and output parameters we prioritize output matching as the outputs of a service are more important for client of a matchmaker. The following equation is used for this weighted calculation:

$$S_{final} = w_{input}*S_{inputs} + w_{output}*S_{outputs} \qquad (7.2)$$

In the above equation, S_{inputs} and $S_{outputs}$ represent the similarity score considering only the input parameters and output parameters respectively. w_{input} and w_{output} represents the weights for the input and output similarity scores, where they are fixed to 0.4 and 0.6 after several runs of matchmaker considering the expected outcome. Finally, S_{final} represents the final score of similarity considering both the input and output parameters.

$$S_{x,y} = w_{sub}*Subsumption_{x,y} + w_{word}*WordNet_{x,y} \qquad (7.3)$$

$S_{x,y}$, in the above equation, represents final similarity score between concepts x and y. $Subsumption_{x,y}$ represents semantic score obtained through subsumption, property level matching and semantic distance. $WordNet_{x,y}$ represents the WordNet score for concept names. The coefficients for subsumption and WordNet are fixed at 0.9 and 0.1 after making experimental runs. We plan to apply a neural network training approach to determine values for coefficients utilizing a large training data in future.

As we have indicated before, a feature-rich matchmaking process is crucial for capturing semantic similarity between parameter pairs. Following sections describe the value-added approaches that our matchmaking agent applies in order to evaluate semantic similarity and rank the services according to the final similarity score.

7.4.1.1 Subsumption Based Similarity Assessment

We make use of OWL-DL constructs *subClassOf, disjointWith, complementOf, unionOf and intersectionOf* to assess concept similarity based on subsumption. If two concepts are explicitly stated to be complement or disjoint, a zero score is directly assigned. Otherwise, we check for subconcept relation and also assess according to property level assessment procedure described below.

We wanted to capture similarity values in bipartite graph since it is important to decompose concepts that include the characteristic of "a union of". Following this approach, we always pair and assess score for atomic concepts in matchmaking process.

Considering the input and output parameter matching, the following cases are favored in subsumption: Request input parameter subsumed by the service input parameter or service output parameter subsumed by the request output parameter.

The following equation explains the above scoring differentiation for subsumption relation in parameters:

$$S_{final} = w * S_{x,y} \tag{7.4}$$

$S_{x,y}$ represents the semantic similarity score between parameters x and y, where w represents the weight that is adjusted according to the subsumption relation. For the above described favored subsumption cases we assign 0.6 to w and 0.4 otherwise. S_{final} represents the final adjusted similarity score between the concepts.

7.4.1.2 Property-Level Similarity Assessment

We believe that in matchmaking it is also important to have properties and their associated range in measuring the degree of match. Such as, if two concepts have similar properties (properties having subclass relation) and their range classes are similar, then this improves their level of similarity.

Our proposed matchmaker checks for property level similarity in a recursive fashion. A complete similarity assessment is performed for range classes of each matching property. This means that if the range classes of similar properties have also common properties, property level matching is again applied to capture similarity at that level.

As the properties of a parent concept are inherited by its child concepts in a semantic network, property-level matching is crucial to capture semantic similarity between sibling concepts. Such concepts are not in a subsumption relation. However, since they have a common parent, it is highly likely that they contain common properties with similar range types.

The following equation describes how property-level matching is used in scoring:

$$Sp_{x,y} = w_p * \sum S_{m,n} \tag{7.5}$$

In the above equation x and y represents the concepts that have common properties. The property-level matching score for the concepts x and y is expressed as Spx,y. Let us suppose m and n represent associated range classes for each matching property of x and y. Sm,n represents the similarity score between concepts m and n. wp represents the weight for property-level matching affecting the total similarity score, where we defined it to be equal to 0.1 after several runs. Then, Spx,y equals to the adjusted sum of similarity scores for each m and n. Note that SAM considers object type properties for subsumption. For data type properties similarity will be simply equal to 0 or 1.

Using property level similarity assessment ranks a service that would normally be eliminated by a conventional matchmaker. For example, a user request may favour a particular author for a novel. A service, which returns articles that are written by that particular author, will have a high score even though the concept of "an article" does not compare to the concept of "a novel". Therefore our proposed architecture returns positive results for concepts that have similar properties as well as the similar concepts.

7.4.1.3 Similarity Distance Based Assessment

Similarity distance represents the semantic similarity between two concepts in terms of a value in the range as [0,1]. It is an explicit annotation on the ontology, where contextual information is represented. Similarity distance annotations can be considered as a user's point of view on semantic relations of concepts or similarity relations of concepts in a certain context.

Similarity distance annotations can be introduced into ontology by the ontology creators or by the user's of a system dynamically. A system can provide the user with an interface to assess similarity relations between concepts in a user-friendly way. The user can classify concepts as being equal, very similar, similar or not related so that the system can map these relation classes to the range [0,1]. Our implementation does not focus on the semantic distance annotation step and assumes that the ontology with semantic distance annotation is given.

In [24], authors make use of semantic distance information by representing annotations in a custom XML file accompanying the ontology document. We further improve this approach and apply a standards based approach. We defined a similarity distance ontology which represents how similarity values between concepts are defined. Importing similarity distance ontology and annotating the concepts with similarity distance concept will enable any ontology to represent similarity relation between its concepts. This way we provide a standard representation and enable an ontology which has similarity distance annotations to be processed in any OWL processor.

To represent similarity distance information we applied N-ary relation pattern in OWL, which is used to represent additional attributes on a property [58]. The additional attribute in our case is the similarity distance value. Figure 7.4 shows how this pattern is organized:

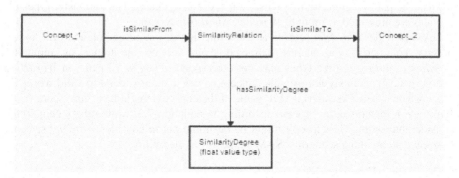

Fig. 7.4. N-ary relation pattern in OWL, representing similarity distance information

SimilarityRelation concept is introduced as a class with this pattern and the similarity distance value is represented as the range of *hasSimilarityDegree* property of this concept. The similar classes are represented as *Concept_1* and *Concept_2* in Figure 7.4. The relations *isSimilarFrom* and *isSimilarTo* attach the concepts to the similarity relation and indeed they are symmetric.

The similarity distance formulation is defined as follows:

$$Sd_{x,y} = Sd_{x,t} * Sd_{t,k} * ... * Sd_{m,y} \tag{7.6}$$

In the above equation, $Sd_{a,b} \in [0,1]$ for any a and b pair. $Sd_{x,y}$ represents similarity distance between concepts x and y. The product of similarity distance values on the path from x and y gives the value for $Sd_{x,y}$. If the concepts are not subclasses of each other then we take the path including their first common parent in the hierarchy. If there are more than one path between two concepts (occurs if a concept has more than one parent) we take the path with higher score.

Similarity distance assignment approach is not strictly defined. Indeed, we encourage similarity assignment to be performed in a consistent way, considering the ontology as a whole. In addition, if similarity distance annotation is not found between two concepts, then a default distance value is assigned according to the following equation:

$$Sd_{x,y} = 1/|subClassOf(x)_{direct}| \tag{7.7}$$

In the above equation $Sd_{x,y}$ represents similarity distance between concepts x and y and $|subClassOf(x)_{direct}|$ represents the number of elements in set of direct subclasses of concept x.

7.4.1.4 WordNet Based Similarity Assessment

WordNet organizes concepts in synonym sets and provide information on semantic relations between synsets by making use of pointers. In our architecture we take *WordNet* as a secondary source of information in addition to the ontology repository. We aimed at reasoning with these highly structured information sources in order to get more reliable result sets.

We make use of the *Wordnet::Similarity* project to assess similarity score among words. The path length criterion is used for score assignment. The parameter types of services are presented as input to *Wordnet::Similarity* module.

SAM provides *WordNet* scoring as a module that can be turned on and off. As we present parameter type names extracted from service ontologies as input to *Wordnet::Similarity*, some types may not correspond to any word name in *WordNet*. Ontology designers may not provide concept names that correspond to valid words in *WordNet* database. Considering this, some of the concepts might have zero score from *WordNet* in matchmaking process according to equation 7.3, while others gain some considerable score. Therefore, it might be desirable not to consider *WordNet* scoring in some matchmaking scenarios. SAM supports this flexibility.

7.5 Experimental Results

7.5.1 Test Ontology

The ontology and services we used are retrieved from "OWL-S Service Retrieval Test Collection version 2.1". The services in the collection are mostly extracted from public UDDI registries, providing 582 Web services described in OWL-S and from seven different domains. The OWL-S Test collection version 2.1 contains 29 queries, each of which associated with a set of 10 to 15 services [8].

In order to evaluate the performance of our proposed matchmaking agent we extended the book ontology in OWL-S Service Retrieval Test Collection (OWL-S TC) and also modified related request and service definitions accordingly [8]. As shown in Figure 7, we added subclasses of *Magazine*, namely *Foreign-Magazine* and *Local-Magazine* classes. We introduced subclasses for *Publisher* concept, *Author* and *Newspaper* concepts. We also annotated the book ontology with similarity distance information, making use of similarity distance ontology that we have imported.

The ontology contains information on printed material classification and related concepts such as authors, publishers, publishing intervals in terms of time and date and several other concepts. Figures 7.5 to 7.7 are sections from the book ontology.

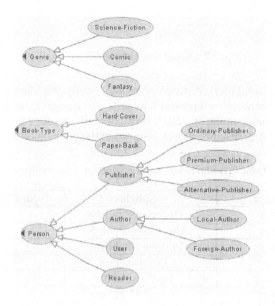

Fig. 7.5. A section of book ontology

As shown above we introduced subclasses for Publisher and Author concepts. This will provide further differentiation in matchmaking process when considering the above concepts in parameter types.

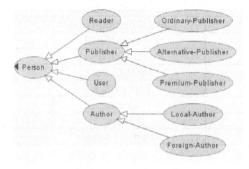

Fig. 7.6. Person ontology section

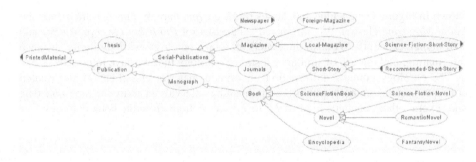

Fig. 7.7. Printed Material ontology section

We imported similarity distance ontology and introduced subclasses of SimilarityRelation concept to represent similarity distance annotations. This way, we can represent semantic similarity values between book ontology concepts. Following table lists the similarity distance values assigned in the book ontology:

Table 7.1. Similarity distance values in book ontology

Superclass	Subclass	Similarity Distance
Publisher	Ordinary-Publisher	0.2
Publisher	Alternative-Publisher	0.5
Publisher	Premium-Publisher	0.3
Author	Local-Author	0.3
Author	Foreign-Author	0.7
Magazine	Local-Magazine	0.3
Magazine	Foreign-Magazine	0.7
Newspaper	Local-Newspaper	0.3
Newspaper	Foreign-Newspaper	0.7
Book	Short-Story	0.7
Book	Science-Fiction-Book	0.4
Book	Novel	0.3
Book	Encyclopedia	0.1
Novel	Science-Fiction-Novel	0.6
Novel	Fantasy-Novel	0.2
Novel	Romantic-Novel	0.2
Book-Type	Hard-Cover	0.7
Book-Type	Paper-Back	0.3
Genre	Comic	0.3
Genre	Fantasy	0.2
Genre	Science-Fiction	0.5

The above similarity distance values are annotated in the ontology as described in Table 7.2 below. The similarity distance value in super concept *SimilarityRelation* is overridden by the annotation. *Owl:hasValue* construct is used to represent the

Table 7.2. Similarity distance annotation

```
<owl:Class rdf:ID="Similarity_Author_LocalAuthor">
<rdfs:subClassOf
 rdf:resource="http://127.0.0.1/ontology/similarityOntology.owl#SimilarityRelation"/>
<rdfs:subClassOf>
  <owl:Restriction>
    <owl:hasValue rdf:datatype="http://www.w3.org/2001/XMLSchema#float"> 0.3
    </owl:hasValue>
    <owl:onProperty rdf:resource=
    "http://127.0.0.1/ontology/similarityOntology.owl#hasSimilarityDistance"/>
  </owl:Restriction>
</rdfs:subClassOf>
<rdfs:subClassOf>
  <owl:Restriction>
    <owl:onProperty rdf:resource=
    "http://127.0.0.1/ontology/similarityOntology.owl#isSimilarFrom"/>
    <owl:allValuesFrom rdf:resource="#Author"/>
  </owl:Restriction>
</rdfs:subClassOf>
<rdfs:subClassOf>
  <owl:Restriction>
    <owl:allValuesFrom>
     <owl:Class rdf:about="#Local-Author"/>
    </owl:allValuesFrom>
    <owl:onProperty rdf:resource=
    "http://127.0.0.1/ontology/similarityOntology.owl#isSimilarTo"/>
  </owl:Restriction>
</rdfs:subClassOf>
</owl:Class>
```

similarity distance. The concepts that are described as similar are represented with *isSimilarFrom* and *isSimilarTo* properties using *owl:Restriction* construct.

7.5.2 Test Environment

SAM is developed in Java with Eclipse IDE. Java is a good choice considering its power in open source community and semantic Web projects that are developed in Java [29, 30, 31, 32]. As described in section 7.4, SAM makes use of OWL-S API for processing OWL-S documents and Jena for reasoning on OWL. Apache Web server is used to publish OWL ontology documents and services described with OWL-S. We also used Protégé ontology editor to edit and visualize OWL documents.

7.5.3 Test Results

In order to demonstrate the value-added features of our proposed matchmaker, we present a test case for request-service matchmaking. The service and request pairs are described in Table 7.3.

Results for input parameter matching for test case are listed in Table 7.4.

Table 7.3. Test case description

Service Name	Inputs	Outputs
Request	Ordinary-Publisher, Novel, Paper-Back	Local-Author, Genre
Service 1	Publisher, ScienceFictionBook	Author, Price
Service 2	Book, Alternative-Publisher, Book-Type	Publisher, Price, Date
Service 3	FantasyNovel, Author	Price, Comic
Service 4	Newspaper, Book-Type, Person	Review, Fantasy
Service 5	Publication, Book-Type, Reader	Genre, Publisher

Table 7.4. Input parameter matching for test case

Service Name	Parameter Pairing	Service Score
Service 1	*Novel to Science-Fiction-Book (Property Level) : 0.13176 *Ordinary-Publisher to Publisher (Request Subsumed) : 0.108 *Novel to Book (Request Subsumed + WordNet) : 0.266	0.35964
Service 2	*Paper-Back to Book-Type (Request Subsumed) : 0.162 *Ordinary-Publisher to Alternative-Publisher (Only Distance) : 0.0108	0.4388
Service 3	*Novel to Fantansy-Novel (Service Subsumed) : 0.12 *Ordinary-Publisher to Author (Only Distance) : 0.00018	0.18026

Table 7.4. (*continued*)

Service 4	*Novel to Newspaper (Property Level + WordNet) : 0.07225 *Paper-Back to Book-Type (Request Subsumed) : 0.162 *Ordinary-Publisher to Person (Request Subsumed) : 0.00211	0.23636
Service 5	*Novel to Publication (Request Subsumed + WordNet) : 0.155 *Paper-Back to Book-Type (Request Subsumed) : 0.162 *Ordinary-Publisher to Reader (Only Distance) : 0.00018	0.31718

Service 2 has the highest score considering input parameters only. The request input parameter *Novel* is a direct subclass of service parameter *Book*. For the input parameter matching, service parameter subsuming the request parameter is favored considering equation 4. An extra score of 0.5 is also received from WordNet similarity for *Novel* and *Book* parameters. We might not have considered WordNet score, considering that concept names such as *Paper-Back* or *Book-Type* are not included in WordNet database. Even if that was the case Service 2 still has the highest score. The same favorable case also holds for *Paper-Back* and *Book-Type* concepts as they have direct subclass relationship. The last parameter pair, *Ordinary-Publisher* and *Alternative-Publisher*, has no property in common and they are sibling concepts. Because of that we just consider their distance in ontology calculating with the similarity distance annotations according to equation 7.6.

Service 1 has the next highest score although it operates with only 2 input parameters. We normalized Service 1's score proportional to 3/2, according to equation 7.1. *Novel* and *Science-Fiction-Book* concepts are sibling nodes in ontology and they have the following common properties: *publishedBy*, *datePublished*, *timePublished*, *writtenBy*, *hasType* and *isTitled*. We apply property level matching and compare the semantic similarity for the range types of these common properties. The range types are exact matches in terms of similarity except *publishedBy* property. *Novel* has type *Publisher* and *Science-Fiction-Book* has type *Alternative-Publisher* respectively. So, a property-level similarity score for this subsumption relation is added to the parameter pair scores, considering equation 7.5. The other parameter pair, *Ordinary-Publisher* and *Publisher* has also subsumption relation in the favorable case, where the service parameter subsumes the request.

Service 5 also demonstrates parameter pairs having WordNet similarity, *Novel* and *Publication*. *Ordinary-Publisher* and *Reader* concepts both have the *Person* concept as parent. But they do not have a subsumption relation and we calculate their similarity score by considering the path in ontology between them and the similarity distance annotations on this path.

Novel and *Newspaper* concepts in Service 4, has *Publication* concept as common parent but they are on different paths in ontology. However, they have common properties and we consider those property ranges in terms of similarity and add to the final score.

Service 3 is also an example of score normalization in input parameter matching case. It has the weakest match among all services: A subsumption relation with an unfavored case, where request parameter subsumes the request and similarity distance score from *Ordinary-Publisher* to *Author*, where they do not have any properties in common.

The following table lists the results for output parameter matching for our test case.

Table 7.5. Output parameter matching for test case

Service Name	Parameter Pairing	Service Score
Service 1	*Genre to Price (No Match + WordNet) : 0.01429 *Local-Author to Author (Request Subsumed) : 0.108	0.12229
Service 2	*Genre to Comic (Service Subsumed + WordNet) : 0.17033	0.17033
Service 3	*Genre to Price (No Match + WordNet) : 0.01429 *Local-Author to Publisher (Only Distance) : 0.00018	0.01447
Service 4	*Genre to Genre (Exact + WordNet) : 1.0 *Local-Author to Publisher (Only Distance) : 0.00018	1.00018
Service 5	*Genre to Fantasy (Service Subsumed + WordNet) : 0.12229	0.12229

Service 4 has a great score advantage considering the exact match for *Genre* parameter in outputs. In contrast to input parameter matching, request parameter subsuming the service parameter is favorable in subsumption relation for outputs. We have this property is Service 2 and Service 5.

We can see that Service 1 has no semantic similarity for *Genre* and *Price* considering the book ontology. However, we have a score of 0.01429 coming from WordNet database. For Service 1 and Service 5 we have the same output score and we do rank randomly for such cases.

Table 7.6. Final service rank for test case

Service Name	Service Score
Service 1	0.21723
Service 2	0.27771
Service 3	0.08078
Service 4	0.69465
Service 5	0.20024

Considering both the input and output matching with the equation 2, we have the above ranking in table 7.6. Service 4 has the advantage of exact match in output parameters. Service 2 has favorable similarity relations in input parameters and it is also the second in ranking for output parameters. Finally, Service 3 has the weakest match for both inputs and outputs, which makes it the last in ranking.

To better demonstrate the advantages of using SAM as compared to a conventional matchmaker, we have also implemented a conventional matchmaker. This matchmaker does not have property-level similarity matchmaking or similarity distance annotation capabilities. So, if the concepts do not have a subsumption relation they will have a zero matchmaking score. If there is a subsumption relation then the following equation will apply in addition to equation 7.4:

$$S = 1 / D_{x,y} \tag{7.8}$$

In equation 7.8, S represents the subsumption similarity score, where D is the path length incremented by 1, between the concepts x and y in ontology. Finally, according to equation 7.4 this similarity score is multiplied by an adjustment factor for subsume or plug-in cases.

As shown in table 7.7, we have a loss of fine-grained discrimination between services. Although Service 1 had property-level matching in SAM, now it loses the advantage and ranked below Service 4 and Service 5. Service 4 and Service 5 are also equal in ranking due to similarity distance and property-level matching non-existence.

Table 7.8 demonstrates the output parameter matching with a conventional matchmaker. Although the difference is not as obvious as the case in inputs, we can observe that Service 1 has lost the advantage of WordNet score, which is a second source of information in SAM. Service 3 also now gets a zero score as similarity distance and WordNet is not considered, which means it is totally left out in this match.

Table 7.7. Conventional matchmaker - Input parameter matching for test case

Service Name	Parameter Pairing	Service Score
Service 1	*Ordinary-Publisher to Publisher (Request Subsumed) : 0.54	0.81
Service 2	*Novel to Book (Request Subsumed) : 0.54 *Paper-Back to Book-Type (Request Subsumed) : 0.54	1.08
Service 3	*Novel to FantansyNovel (Service Subsumed) : 0.36	0.54
Service 4	*Paper-Back to Book-Type (Request Subsumed) : 0.54 *Ordinary-Publisher to Person (Request Subsumed) : 0.54	1.08
Service 5	*Novel to Publication (Request Subsumed) : 0.54 *Paper-Back to Book-Type (Request Subsumed) : 0.54	1.08

Table 7.8. Conventional matchmaker - Output parameter matching for test case

Service Name	Parameter Pairing	Service Score
Service 1	*Local-Author to Author (Request Subsumed) : 0.36	0.36
Service 2	*Genre to Comic (Service Subsumed) : 0.54	0.54
Service 3	No Match	0
Service 4	*Genre to Genre (Exact) : 0.9	0.9
Service 5	*Genre to Fantasy (Service Subsumed) : 0.54	0.54

Table 7.9. Conventional matchmaker - Final service rank for test case

Service Name	Service Score
Service 1	0.54
Service 2	0.756
Service 3	0.216
Service 4	0.972
Service 5	0.756

According to table 7.9 the final ranking changes for Service 5 and Service 1. However, as we have more test scenarios and more services in the relevant set, the difference will be more obvious and a fine-grained scoring as in SAM will be needed for a clear differentiation.

The above test runs demonstrate that matchmaking in SAM provides fine-grained differentiation for services, taking into account context information with similarity distance annotations. Besides, SAM can capture semantic similarity observing common properties in concepts. The advantages of SAM as compared to a conventional matchmaker will be far greater, when we have more complex service sets with a great number of services having parameters referring to several ontologies.

7.5.4   Threats to Validity

SAM addresses several challenges in the Web service matchmaking process. In this section we describe these challenges, describing how SAM overcomes them.

Considering input or output parameter matching, number of parameters in the request or service interface may not be equal. As mentioned in section 7.4, SAM applies normalization to similarity score if the input parameter count of a service is less than the request parameters.

The above case indeed raises an issue of imperfect matching. Although the service might satisfy the request with less input parameters, the interface of the service and request does not match perfectly. However, SAM does not make any classification of match type like perfect or imperfect. It just evaluates the amount of similarity score and provides a relative ranking between candidate services. But considering the parameter pairing SAM provides a match type with the similarity score. As mentioned before, the interpretation of scores and similarity distance information by the user is not considered in SAM. We assume that an automated system can provide the user with an interface to assign similarity distance values in a user-friendly way.

WordNet similarity score is considered as a second source of information. However, some parameter types defined in ontologies may not correspond to valid WordNet entries. In order to provide a fair matchmaking process WordNet scoring is enabled or disabled in SAM.

Matching and ranking involves the consideration of priorities. The similarity equations that are described by SAM make use of coefficients that represent priorities. Such as the output similarity score being prioritized over input similarity scores, as the outcome of a service is much more important for a request. Another example is the WordNet score having a less priority than the ontology based similarity distance score.

SAM is tested and evaluated on the book ontology provided with the OWL-S TC library. We presented two test scenarios with 9 services and two requests. Testing SAM with different ontologies and services may better demonstrate the value-added features of it. The coefficients in presented equations will converge to certain values as SAM is tested with different ontologies. However, the relative ratio of the coefficients will be very similar to current values. Current coefficient values are adjusted considering the book ontology and changing these values might affect the ranking of services. As stated, further testing will make SAM much more reliable.

7.5.5 Comparing SAM with Existing Matchmakers

As described in section 7.2, web services matchmaking has been an active research area. Several matchmakers have been introduced with prototype implementations. Each of these studies focus on certain aspects of matchmaking. In this section we will discuss the advantages and disadvantages of SAM in comparison to other matchmaker proposals.

In [55], authors Jian Wu and Zhaohui Wu, introduce the study "Similarity-based Web Service Matchmaking". They present four similarity assessment methods as part of their matchmaker. As in SAM, they make use of WordNet but for a different purpose. SAM benefits from WordNet as a second source of information base for service parameter type similarity assessment. In their proposed matchmaker, WordNet is used to assess the similarity of service profile categories and parameter names instead of parameter types. Their approach deals with lexical similarity more than semantics. They consider the parameter type similarity through a data type similarity lookup table. SAM makes use of ontologies for parameter type similarity, which we believe is a better method considering semantics and standards. One advantage of their study is the consideration of QoS attribute similarity. However, SAM has the advantage of subsumption similarity assessment with a level down to properties of concepts and makes use of similarity distance information considering as the context in matchmaking.

In [73], one of the best known systems *Larks* is presented. The authors define an agent capability description language named *Larks* and deal with the problem of matchmaking. Their matchmaking algorithm makes use of this specialized language as opposed to SAM, where standard semantics technologies like OWL and OWL-S are employed. Besides, Larks has a discrete scale of similarity classification. On the other hand SAM assigns similarity scores and ranks the services. Larks ignore a service if one of the attributes is in a fail state. SAM just considers the total score a service retrieves, not eliminating and leaving the choice to the consumer. Besides, SAM makes use of WordNet, property-level matching and similarity distance information that Larks does not include.

In [74], Lei Li and Ian Harrocks introduce a service matchmaker prototype which works on DAML-S definitions. Their matchmaking algorithm uses a discrete scale of match as in Larks: Exact, plug-in, subsume, intersection and fail. One advantage over SAM is that their matchmaker provides service profile category match, where SAM assumes that the category match is performed and a candidate service set is presented. However, SAM has many advantages over Li and Harrocks's matchmaker such as property-level matching, WordNet scoring and use of similarity distance information to consider user preferences in matching.

In [54] authors introduce a matchmaking approach based on a custom capability description (OSDL) and query language (OSQL). However, different from other matchmakers they also consider property-level matching. They categorize the similarity of two concepts as equal, inherited (subsumption), property relation (one concept is a property of another) and mixed relation (a transitive relation where one concept is a property of another which is subsumed by a third concept). In their definition similarity of concept X to concept Y is not always equal to the reverse case. They argue that the similarity of subsuming concepts depends on the number of properties they possess. In SAM the similarity relation is symmetric. Besides, SAM does not consider concepts being properties of each other in matchmaking. We suspect that the property of a concept should contain similarities to the concept itself. SAM considers concepts having similar properties. Both SAM and their matchmaker focus on interface similarity but SAM introduces value-added approaches like similarity distance and WordNet scoring in addition.

Considering the above comparisons we can conclude that SAM contributes value-added approaches in matchmaking. Similarity distance information is used in a standard context improving MS-Matchmaker [24]. Property-level matching and WordNet scoring are all combined in a service ranking matchmaker. The following table summarizes this comparison.

Table 7.10. SAM in comparison to other matchmakers

Feature / Matchmaker	SAM	MS-Matchmaker	Larks	Li & Harrocks	Wu & Wu	OSDL
Subsumption Service	+	+	+	+	+	+
Interface Matching	+	+	+	+	+	+
Constraints	-	-	+	-	-	-
Property Level Matching	+	-	-	-	-	+
Similarity Distance	+	+	-	-	-	-
WordNet	+	-	-	-	+ (lexical)	-
Nonfunctional Matching Service	-	-	-	-	+	+
Category Matching	-	+	+	+	+	-
OWL-S	+	-	-	-	-	-

7.6 Conclusions and Future Work

We proposed a novel advanced matchmaker architecture, which introduces new value-added approaches like semantic distance based similarity assessment, property level assessment and WordNet similarity scoring. Instead of classifying candidate Web services in a discrete scale, our matchmaking agent applies a scoring scheme to rank candidate Web services according to their relevancy to the request.

The ranking property enables to include some of the relevant Web services in the final result set whereas they would have been discarded in a discrete scale classification. Additionally, our proposed matchmaking agent improves subsumption- based matchmaking by utilizing OWL constructs efficiently and by considering down to a level of concept properties in the process. An improvement at this point can be to consider similarity between properties in addition to similarity of property range objects.

We also introduced semantic distance annotation in ontology to represent relevancy of concepts to the user in a numerical way. Semantic distance annotations improve the relevancy of returned Web service set as they actually represent user's view of ontology. WordNet similarity measurement is also presented as a value-added feature, which acts as a secondary source of information, strengthening the power of reasoning.

Considering the value-added approaches that SAM introduces in matchmaking, we can conclude that in a service discovery architecture SAM will improve the precision and accuracy of the discovery process. As a result businesses can find partners in a distributed environment easily and with alternatives considered. This will lead to better collaboration among partners and improvement of business processes.

We think that preconditions and results of a service can also be considered for a complete matchmaking process. At that point, use of SWRL (Semantic Web Rule Language) in both service advertisements and request description will enhance the capabilities of our matchmaking agent. Current architecture of SAM supports such a rule based extension.

Non-functional attributes of Web services can also be taken into account in matchmaking process. QoS attributes of a Web service can be of great concern to a consumer. So, extending our matchmaking algorithm to include QoS attributes can be another improvement.

SAM currently presents a relative ordering of services as the output with similarity scores compared to a certain request. In order to eliminate some services with ignorable similarity scores a threshold value can be introduced. We leave this to the client agent that makes use of SAM. However, such a threshold can also be implemented in SAM as a cut-off score in order to classify only a subset of candidate service set as relevant.

Another improvement will be to add context aware decision-making capabilities, enabling our matchmaking agent to reason based on user profiles, preferences, past actions etc. The architecture that we have presented can be considered as a basis for the development of context-aware agent.

References

1. Wang, H., Li, Z., Fan, L.: An Unabridged Method Concerning Capability Matchmaking of Web Services. In: Proceedings of the 2006 IEEE/WIC/ACM International Conference on Web Intelligence (2006)
2. Klusch, M., Fries, B., Khalid, M., Sycara, K.: OWLS-MX: Hybrid Semantic Web Service Retrieval. In: 1st Intl. AAAI Fall Symposium on Agents and the Semantic Web. AAAI Press, Arlington VA (2005)

3. Keller, U., Lara, R., Polleres, A., Toma, I., Kifer, M., Fensel, D.: WSMO Web Service Discovery, http://www.wsmo.org/2004/d5/d5.1/v0.1/20041112/

4. El-Ghalayini, H., Odeh, M., McClatchey, R., Solomonides, T.: Reverse Engineering Ontology to Conceptual Data Models. In: Proceeding (454) Databases and Applications (2005)

5. Universal Discovery Description and Integration Protocol (2006), http://www.uddi.org

6. Semantic Web, W3C (2006), http://www.w3.org/2001/sw/

7. W3C, World Wide Web Consortium (2006), http://www.w3.org/

8. Khalid, M.A., Fries, B., Kapahnke, P.: OWL-S Service Retrieval Test Collection Version 2.1, Deutsches Forschungszentrum für Künstliche Intelligenz GmbH Saarbrücken, Germany (2006)

9. Saip, H.A.B., Lucchesi, C.L.: Matching Algorithms for Bipartite Graphs, Relatorio Tecnico DCC-03/93

10. Motta, E., Domingue, J.B., Cabral, L.S., Gaspari, M.: IRS-II: A Framework and Infrastructure for Semantic Web Services. In: Fensel, D., Sycara, K.P., Mylopoulos, J. (eds.) ISWC 2003. LNCS, vol. 2870, pp. 306–318. Springer, Heidelberg (2003)

11. OWL-S Submission (2004), http://www.w3.org/Submission/OWL-S

12. Fensel, D., Bussler, C.: The Web Service Modeling Framework: WSMF. Electronic Commerce: Research and Applications 1(2), 113–137 (2002)

13. Klein, M., Konig-Ries, B., Muussig, M.: What is needed for semantic service descriptions? A proposal for suitable language constructs. Proceedings of Inernational Journal of Web and Grid Services 1(3/4), 328–364 (2005)

14. Voorhees, E.: Using WordNet for Text Retrieval. In: Fellbaum, C. (ed.) WordNet: An Electronic Lexical Database, pp. 285–303. The MIT Press, Cambridge (1998)

15. Ginsberg, A.: A Unified Approach to Automatic Indexing and Information Retrieval. IEEE Expert 8(5), 46–56 (1993)

16. Lee, J., Kim, M., Lee, Y.: Information Retrieval Based on Conceptual Distance in IS-A Hierarchies. Journal of Documentation 49(2), 188–207 (1993)

17. Agirre, E., Rigau, G.: Word Sense Disambiguation Using Conceptual Density. In: Proceedings of the 16th Conference on ComputationalLlinguistics, vol. 1, pp. 16–22 (1996)

18. Hovy, E.: Combining and Standardizing Large-scale, Practical Ontologies for Machine Translation and Other Uses. In: Proceedings of the 1st International Conference on Language Resources and Evaluation (LREC), Granada, Spain (1998)

19. Wang, Y., Stroulia, E.: Semantic Structure Matching for Assessing Web-Service Similarity. In: Proceedings of the 1st International Conference on Service Oriented Computing, Trento, Italy (2003)

20. Guo, R., Chen, D., Le, J.: Matching Semantic Web Services accross Heterogeneous Ontologies. In: Proceedings of the 2005 The Fifth International Conference on Computer and Information Technology (CIT 2005) (2005)

21. Paolucci, M., Kawamura, T., Payne, T., Sycara, K.: Semantic matching of Web services capabilities. In: Horrocks, I., Hendler, J. (eds.) ISWC 2002. LNCS, vol. 2342, pp. 333–347. Springer, Heidelberg (2002)

22. Shen, X., Jin Bie, X., Sun, Y.: MSC: A Semantic Ranking for Hitting Results of Matchmaking of Services. In: Proceedings of the 30th Annual International Computer Software and Applications Conference (COMPSAC 2006) (2006)

23. Osman, T., Thakker, D., Al-Dabass, D.: Semantic-Driven Matchmaking of Web Services Using Case-Based Reasoning. In: IEEE International Conference on Web Services (2006)

24. Şenvar, M., Bener, A.: Matchmaking of Semantic Web Services Using Semantic- Distance Information. In: Yakhno, T., Neuhold, E.J. (eds.) ADVIS 2006. LNCS, vol. 4243. Springer, Heidelberg (2006)

25. Srinivasan, N., Paolucci, M., Sycara, K.: Semantic Web Service Discovery in the OWL-SIDE. In: Proceedings of the 39th Hawaii International Conference on System Sciences (2006)

26. Wang, H., Li, Z.: A Semantic Matchmaking Method of Web Services Based On SHOIN+(D). In: Proceedings of the 2006 IEEE Asia-Pacific Conference on Services Computing (APSCC 2006) (2006)

27. Paolucci, M., Kawamura, T., Payne, T.R., Sycara, K.: Semantic Matching of Web Services Capabilities. In: International Semantic Web Services Conference (2002)

28. Bunke, H.: Graph Matching: Theoretical Foundations, Algorithms and Applications. In: Proc. Vision Interface 2000, Montreal, pp. 82–88 (2000)

29. Jena – A Semantic Web Framework for Java, http://jena.sourceforge.net/

30. McBride, B.: Jena: A Semantic Web Toolkit, IEEE Internet Computing (2002)

31. CMU OWL-S API, http://www.daml.ri.cmu.edu/owlsapi

32. OWL-S API, http://www.mindswap.org/2004/owl-s/api/doc/

33. Resource Description Framework (RDF) (2004), http://www.w3.org/RDF/

34. Castillo, J.G., Trastour, D., Bartolini, C.: Description Logics for Matchmaking of Services, Trusted E-Services Laboratory HP Laboratories Bristol (2001)

35. Daconta, M.C., Obrst, L.J., Smith, K.T.: The Semantic Web: A Guide to the Future of XML, Web Services, and Knowledge Management. John Wiley & Sons, Chichester (2003)

36. Semantic Web Vision, http://www.hpl.hp.com/semweb/sw-vision.htm

37. Shadbolt, N., Hall, W., Lee, T.B.: The Semantic Web Revisited. IEEE Intelligent Systems 21(3), 96–101 (2006)

38. Decker, S., Mitra, P., Melnik, S.: Framework for the Semantic Web: An RDF Tutorial. IEEE Internet Computing (2000)

39. OWL Web Ontology Language Overview (2004), http://www.w3.org/TR/owl-features/

40. OWL Web Ontology Language Guide (2004), http://www.w3.org/TR/owl-guide/

41. OWL-S 1.1 Release, http://www.daml.org/services/owl-s/1.1/

42. OWL-S: Semantic Markup for Web Services, http://www.daml.org/services/owl-s/1.1/overview/

43. Balzer, S., Liebig, T., Wagner, M.: Pitfalls of OWL-S – A Practical Semantic Web Use Case. In: ICSOC (2004)

44. Jørstad, I., Dustdar, S., Thanh, D.V.: A Service Oriented Architecture Framework for Collaborative Services. In: Proceedings of the 14th IEEE International Workshops on Enabling Technologies: Infrastructure for Collaborative Enterprise (2005)

45. Hashimi, S.: Service-Oriented Architecture Explained (2003), http://www.ondotnet.com/pub/a/dotnet/2003/08/18/soa_explained.html

46. Pasley, J.: How BPEL and SOA are Changing Web Services Development. IEEE Internet Computing (2005)

47. Service Oriented Architecture, http://msdn2.microsoft.com/en-us/architecture/aa948857.aspx

48. Web Services Description Language (WSDL) 1.1 (2001), http://www.w3.org/TR/wsdl

49. SOAP Version 1.2 Part 0: Primer (2003), http://www.w3.org/TR/2003/REC-soap12-part0-20030624/

50. SOAP Version 1.2 Part 1: Messaging Framework (Second Edition) (2007), http://www.w3.org/TR/soap12-part1/

51. Extensible Markup Language (XML), http://www.w3.org/XML/

52. Chester, T.M.: Cross-Platform Integration with XML and SOAP, IT PRO (2001)

53. Introduction to UDDI: Important Features and Functional Concepts, OASIS (2004), http://uddi.xml.org/

54. Kuang, L., Wu, J., Deng, S., Li, Y., Shi, W., Wu, Z.: Exploring Semantic Technologies in Service Matchmaking. In: Proceedings of the Third European Conference on Web Services (2005)
55. Wu, J., Wu, Z.: Similarity-based Web Service Matchmaking. In: Proceedings of the 2005 IEEE International Conference on Services Computing (2005)
56. Sirin, E., Parsia, B., Hendler, J.: Composition-driven Filtering and Selection of Semantic Web Services. In: AAAI Spring Symposium on Semantic Web Services (March 2004)
57. Akkiraju, R., Srivastava, B., Ivan, A., Goodwin, R., Mahmood, T.S.: Semantic Matching to Achieve Web Service Discovery and Composition. In: Proceedings of the 8th IEEE International Conference on E-Commerce Technology and the 3rd IEEE International Conference on Enterprise Computing, E-Commerce, and E-Services (2006)
58. Defining N-ary Relations on the Semantic Web (2006), http://www.w3.org/TR/swbp-n-aryRelations/
59. OASIS-UDDI, http://www.uddi.org/
60. Business Process Execution Language for Web Services version 1.1 (2007), http://www-106.ibm.com/developerworks/webservices/library/ws-bpel/
61. Schmitz, D., Lakemeyer, G., Gans, G., Jarke, M.: Using BPEL Process Descriptions for Building up Strategic Models for Inter-Organizational Networks. In: International Workshop on Modeling Inter-Organizational Systems (MIOS), Larnaca, Cyprus, October 2004. LNCS. Springer, Heidelberg (2004)
62. Miller, G.A.: WordNet: A Lexical Database for English. Communicating of the ACM (1995)
63. Petersen, T., Patwardhan, S., Michelizzi, J.: WordNet:Similarity - Measuring the Relatedness of Concepts. In: Proceedings of the Nineteenth National Conference on Artificial Intelligence (AAAI 2004) (2004)
64. Liu, P.Y., Zhao, T.J., Yu, X.F.: Application-Oriented Comparison and Evaluation of Six Semantic Similarity Measures Based on WordNet. In: Proceedings of the Fifth International Conference on Machine Learning and Cybernetics, Dalian (2006)
65. Yao, Y., Su, S., Yang, F.: Service Matching based on Semantic Descriptions. In: Proceedings of the Advanced International Conference on Telecommunications and International Conference on Internet and Web Applications and Services (2006)
66. Guo1, R., Le, J., Xia, X.L.: Capability Matching of Web Services Based on OWL-S. In: Proceedings of the 16th International Workshop on Database and Expert Systems Applications (2005)
67. Colucci, S., Noia, T.D., Sciascio, E.D., Donini, F.M., Mongiello, M.: Description Logics Approach to Semantic Matching of Web Services. In: 25th Int. Conf. Information Technology Interfaces (2003)
68. Lim, J.E., Choi, O.H., Na, H.S., Baik, D.K.: A Methodology for Semantic Similarity Measurement among Metadata based Information System. In: Proceedings of the Fourth International Conference on Software Engineering Research (2006)
69. Marcus, A., Maletic, J.I.: Identification of High-Level Concept Clones in Source Code. In: Proceedings Automated Software Engineering (ASE 2001), San Diego, CA, November 26-29, 2001, pp. 107–114 (2001)
70. OASIS, http://www.oasis-open.org
71. Ma, K.J.: Web Services: What's Real and What's Not? IEEE IT PRO (2005)
72. Munkres' Assignment Algorithm, http://www.public.iastate.edu/~ddoty/HungarianAlgorithm.html
73. Sycara, K., Klusch, M., Widoff, S., Lu, J.: Dynamic Service Matchmaking Among Agents in Open Information Environments. SIGMOD Record 28(1) (March 1999)

74. Li, L., Harrocks, I.: A Software Framework For Matchmaking Based on Semantic Web Technology. In: Proceedings of the Twelfth International World Wide Web Conference (WWW 2003) (2003)
75. Ilhan, E.S., Akkus, G.B., Bener, A.B.: SAM: Semantic Advanced Matchmaker. In: The Nineteenth International Conference on Software Engineering and Knowledge Engineering (SEKE 2007), Boston, USA (July 2007)
76. Ilhan, E.S., Akkus, G.B., Bener, A.B.: Improved Service Ranking and Scoring: Semantic Advanced Matchmaker (SAM). In: 2nd International Working Conference on Evaluation of Novel Approaches to Software Engineering, Barcelona, Spain (July 2007)
77. Özyilmaz, S., Akkus, G.B., Bener, A.: Matchmaking in semantically enhanced Web services: Inductive ranking methodology. In: ICSSEA 2007, Paris, France, December 4-6 (2007)

8

Clarifying the Meta

Vladan Devedžić[1], Dragan Gašević[2], and Dragan Djurić[1]

[1] FON – School of Business Administration, University of Belgrade, POB 52,
11000 Belgrade, Serbia
devedzic@fon.bg.ac.yu, dragandj@gmail.com
http://fon.fon.bg.ac.yu/~devedzic/
[2] School of Computing and Information Systems, Athabasca University,
Athabasca, AB T9S 3A3, Canada

Abstract. The word and the concept of meta are used in virtually all disciplines of computing. Professionals often use it intuitively and in specific contexts in which the meaning of a word containing the meta- prefix is well understood, at least by specialists. For example, all experts in artificial intelligence know what metarules are and where they are useful. However, the meta can imply much more than understanding some concepts in a specific domain. This paper surveys some of the most interesting cases of using the concept of meta in computing in order to help build the big picture.

8.1 Introduction

Metacomputing, metaprogramming, metaclasses, metaatributes, metadata, metasyntax, metalanguage, metainformation, metamodeling, metalevel, metareasoning, metarules, metasearch, meta-CASE tools, MetaCrystal,... Had enough? No? OK, here's more – metaphysics, metalogic, metamathematics, metaphilosophy, metatheory, metasystem, metamemory, metaconcept, metacognition, metameaning, metaheuristics, metauncertainty, metaontologies, metamorphosis,...

The word and the concept of *meta* are used in virtually all disciplines of computing. Professionals often use it intuitively and in specific contexts in which the meaning of a word containing the *meta-* prefix is well understood, at least by specialists. For example, all experts in artificial intelligence know what metarules are and where they are useful. However, the *meta* can imply much more than understanding some concepts in a specific domain. Surveying some of the most interesting cases of using the concept of *meta* in computing can help build the big picture.

8.2 The Meaning of *Meta*

In ancient Greek, the word *meta* has several meanings. One of them is preposition taking the genitive case, that translates as the English *with* (Wikipedia, The Free Encyclopedia (http://en.wikipedia.org/wiki/Main_Page)). But there are other meanings as well – *changed in position*, *beyond*, *on a higher level*, *transcending*, etc. All these meanings are referring to the body of knowledge about a body of knowledge or about a field

R. Nayak et al. (Eds.): Evolution of the Web in Artificial Intel. Environ., SCI 130, pp. 191–200, 2008.
springerlink.com

of study (Web Dictionary of Cybernetics and Systems, http://pespmc1.vub.ac.be/ASC/indexASC.html), such as in epistemology, where the prefix *meta-* is explicitly used to mean *about*. *Meta* is also often used for connoting purposes, as in *metaporphosis* where it connotes *change*.

FOLDOC dictionary of computing (http://foldoc.doc.ic.ac.uk/foldoc/index.html) defines *meta* even more precisely as "a prefix meaning one level of description higher. If X is some concept then meta-X is data about, or processes operating on, X." For example, if a computer stores and processes some data, then metadata is data about the data – things like who has produced the data, when, and what format the data is in. If we are studying a language, then metalanguage is a language used to talk about the language. The syntax of the language can be specified by means of a meta-syntax.

8.3   Meta and Meta-meta

Implications of the phrase "one level of description higher" are layering and hierarchy. Whenever we describe things at a certain level of abstraction, we use terms and concepts from a higher level of abstraction. For example, to describe a software system that models a process or processes from a real world, we can use UML concepts such as objects, classes, and relations. These are metalevel concepts used to describe real-world ones. The metalevel concepts are, in turn, defined at yet another higher level of abstraction, i.e. at a meta-metalevel. And so on.

A naturally arising question is: how far can we go on that way? In theory – very far. Consider some XML-based Web resources that end users are interested in. They are data. Consider also some other data derived directly from the first data, such as XML metadata and indexes of Web sites. These are metadata about the first data, but are also data in their own right. Deriving such metadata from data is performed using a metaoperation, such as different Web mining and resource discovery functions. In theory, another (or even the same) metaoperation can be used on the metadata to derive meta-metadata. This is inherently recursive [5].

However, the recursion is never too deep if it is to make sense in practice. If the Web resources the end users are interested in are e-learning resources such as courses, lessons, lesson plans, curricula, teaching strategies and media (audio and video), then indexes of these resources are of interest as well (e-learning metadata repositories, digital "library cards", and educational Web portals). Metadata about these indexes can be of interest as well, and so can be some other kinds of metadata (such as how semantically similar are two or more resources). But it is currently difficult to envision many systems of practical value that would go much further than that in the meta-meta direction.

Although this *meta-meta depth factor* is never too high, different fields of computing implement the meta and meta-meta philosophy differently.

8.4   Metacomputing

Today, computing resources are often distributed among different sites. It implies different software environments, no file-system sharing (even on machines running the same operating systems), and different security policies. These shortcomings may be mitigated by using single very large computers instead of distributed resources, but it is an expensive solution.

Metacomputing is based on the idea of connecting many modest systems with a total power far outweighing the few supercomputers available. The whole can be greater than the sum of the parts. If a single application can run on two different parallel computers with 100 CPUs each, with some additional effort it can be made to run as a single 200-CPU application on a single logical *metacomputer* that physically comprises both machines. Likewise, there may exist a need to use two applications that are each suited to a different architecture of machine and combine them in a closely-coupled manner. The applications can communicate with each other on a single metacomputer from the two machines using standard parallel programming techniques. Hence metacomputing is essentially distributed computing, and is developed as a natural progress of supercomputing, parallel computing, cluster computing, and network computing.

A geographically separated collection of people, computers and storage connected through a fast network and supported by specific software is called a *metasystem* (http://legion.virginia.edu/presentations/metacomputing/). The software makes the entire collection of machines as easy to use as a single desktop computer. It also makes collaboration of the people involved very easy. High performance, security, fault tolerance, and transparency of such a metasystem is achieved in spite of its distributed nature.

A metasystem can run a *metaapplication* – a multi-component program, pieces of which are separate programs, optimized for running on different hardware platforms and computer architectures. Such a metaaplication can use big datasets that live on geographically-remote sites, and can read/write files transparently regardless of execution location.

In other words, metacomputing allows multiple computers to tackle shared tasks. The metacomputer is a collection of computers held together by state-of-the-art technology and "balanced" so that, to the individual user, it looks and acts like a single computer (http://www.epcc.ed.ac.uk/DIRECT/grid/node10.html).

Large-scale metacomputing involving massive computational tasks using high-speed networks is called *grid computing* [1], [4]. It is supposed to eventually transform computing from an individual and corporate activity into a general utility. Proposed software implementations for grid computing are layered ones (e.g., see http://www.gridforum/org/iga.html), including:

- low-level support for high-speed network i/o;
- different grid services (such as authentification, authorisation, resource location, event services, etc.);
- application toolkits for data analysis and visualisation, distributed computing, collaborations, problem solving environments;
- grid-aware metaapplications implemented in terms of grid services and application toolkit components as building blocks.

Meta-meta depth factor here is usually 1 – typical perceived beneficiaries of metacomputing and grid technologies include governments (e.g., tasks, processes, and activities related to disaster response and national defense) and institutions (e.g., hospitals can use relatively low-cost and smaller-scale private grids with central management). However, the factor of 2 is envisioned as well in the form of "virtual grid" (multi-institution collaboration) and "public grid" (enormous community of service, resource, and network providers and consumers) [4].

8.5 Meta in Programming

Metaprogramming is the writing of programs that write or manipulate other programs (or themselves) as their data or that do part of the work that is otherwise done at run-time during compile time (http://en.wikipedia.org/wiki/Metaprogramming_ (programming)). The most common metaprogramming tool is a compiler. It allows programmers to write code using a high-level programming language and transform it into an equivalent assembly language code or into another equivalent code suitable for interpretation (such as Java bytecodes). Another characteristic example of metap-rogramming is writing a script that generates code, allowing programmers to produce a larger amount of code and get more done in the same amount of time as they would take to write all the code manually. C++ templates and Common Lisp macros support metaprogramming by defining the contents of (part of) a program by writing code that generates it. Finally, reflection is also a valuable language feature for facilitating metaprogramming.

Recent developments suggest using refactoring as metaprogramming [7]. Originally, refactoring is the process of improving an existing program by transforming it into a new, better version. It is performed either manually, or using modern IDEs like Eclipse (http://www.eclipse.org), or other sophisticated tools that support navigation, querying, and visualization of static and dynamic program structure and behavior (e.g., Jquery, http://www.cs.ubc.ca/labs/spl/projects/jquery/). Metaprogramming by refactoring would also require a specific *refactoring language* to allow the developer to express to express current refactorings as well complex queries and program trans-formations, such as splitting and combining program fragments.

8.6 Meta in Software Design

One of the best practices in software design is that of using design patterns as general-ized, common solutions to recurring design problems. Some researchers and develop-ers call design patterns meta-level design tools. Others believe it is more appropriate to talk about patterns as generalized solutions, and pattern instances as domain- or application-specific instantiations of such generalized solutions.

Metapatterns are patterns describing other patterns [8]. They are useful when de-signing a large framework, such as a GUI framework – a set of classes designed to work together in creating GUIs. Each framework contains one or more generic parts that are further configured and adapted by the user when applying the framework. Such parts are often called hot spots. Metapatterns are used to describe generalized design of such hot spots.

Technically, metapatterns are often represented using the notion of abstract (tem-plate) classes and methods. For example, the metapattern called Unification describes how to specify only the skeleton of an algorithm, leaving some details open – the trick is to use a template method that relies on one or more abstract ("hook") methods. Thus some steps of the algorithm are deferred to implementation of the abstract methods in subclasses. The user is expected to use subclasses to override the hook methods to achieve the desired functionality.

Depending on whether design patterns are considered meta-level design tools or not, the meta-meta depth factor here is 1 or 2.

8.7 Meta Object Facility

Model Driven Architecture (MDA) is an increasingly popular trend and standard-in-the-making in software engineering [6]. It defines three viewpoints (levels of abstraction) from which a certain system can be analyzed. Starting from a specific viewpoint, we can define the system representation (viewpoint model). The representations/models/viewpoints are *Computation Independent Model* (CIM), *Platform Independent Model* (PIM) and *Platform Specific Model* (PSM).

MDA also defines a four-layer system-modeling architecture, Fig. 8.1. At the M0 layer are things from the real world. The models of the real world are at the M1 layer (e.g., system models specified using UML notation). They are represented using the concepts defined in the corresponding metamodel at the M2 layer (e.g. UML metamodel). The best part here is the topmost, M3 layer – the meta-metamodel based on the *Meta Object Facility (MOF)* standard, that defines both the metamodels at the M2 layer and itself! Hence the meta-meta depth factor here is strictly 2. In fact, MOF defines an abstract language and a framework for specifying, constructing and managing technology-neutral metamodels. It is the foundation for defining any modeling

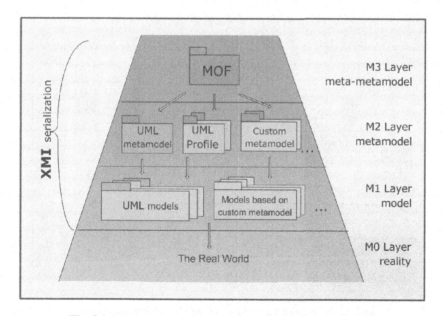

Fig. 8.1. MDA four-layer MOF-based metamodeling architecture

language, such as UML. MOF also defines a framework for implementing repositories that store metadata (e.g. models) described by metamodels. The purpose of the four layers with common meta-metamodel is to support multiple metamodels and models and their scaling – to enable their extensibility, integration and generic model and metamodel management.

8.8   Meta on the Semantic Web

The Semantic Web (http://www.semanticweb.org) was originally conceived as a layer of machine-understandable metadata on top of the existing Web. The idea is that by annotating existing documents and information on the Web with some metadata, agents will be able to understand what the information is and what it is about. It is possible only if along with additional descriptions in the form of metadata, Web pages and documents also contain links to a web of ontologies [2] that provide vocabularies and meanings of terms and concepts used in metadata. In spite of the fact that a number of useful ontologies are still lacking, the Semantic Web is a trendy topic in computing.

Both syntactically and semantically, there is a lot of meta involved with the Semantic Web. That fact is best understood having in mind the Semantic Web "layer cake", Fig. 8.2. Note that higher-level languages use the syntax and semantics of the lower levels. All Semantic Web languages use XML syntax; in fact, XML is a metalanguage for representing other Semantic Web languages. For example, XML Schema defines a class of XML documents using the XML syntax. RDF provides a framework for representing metadata about Web resources, and is XML-based as well. RDF Schema, OWL, and other ontology languages also use the XML syntax.

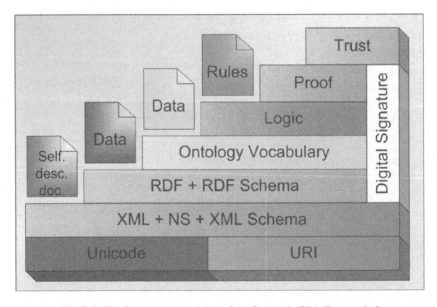

Fig. 8.2. Tim Berners-Lee's vision of the Semantic Web "layer cake"

Semantically, a crucial concept related to the ontologies today is *Ontology Definition Metamodel* [3]. ODM specifies all essential concepts used to define ontologies (e.g., Resource, Class, Property, etc.). Thus ODM is essentially a meta-ontology that includes multiple ontology languages, and currently grows quickly to include mappings to different ontology languages like OWL, RDFS, etc. A similar MOF-compatible ontology metamodel is implemented in Protégé, a popular Semantic Web tool that supports import, export, and conversion from/to different ontology languages through plug-ins. Protégé can be even adapted to support a new ontology language by inserting new *metaclasses* and *metaslots* into a Protégé ontology. Hence Protégé is a meta-tool for ontology development and knowledge acquisition.

The meta-meta depth factor here is 1 both in terms of syntax (e.g., there is a concept of *meta-schema* – a schema that defines the validity of XML Schema definition documents) and in terms of semantics (ODM).

8.9 Software (Meta)Modeling and Modeling Spaces

In its essence, software engineering is all about creating models. Some hard-core programmers would say: "Hey, that is not true – we use UML diagrams rarely, and even when we do, it is only because our managers forced us to." But, what else is our precious Java, C++ or even assembly code if not a model? A model is a simplified abstraction of reality. We build software to make a simplified snapshot of reality that we are interested in.

How do we define software models? By using metamodels, which define concepts that are the building blocks of models. These metamodels are defined using some other concepts that belong to meta-metamodel. We could continue this meta-meta layering but it is not necessary – meta-metamodel can be self-defined. Model Driven Architecture is a standard that clarifies these meta-meta relations using a well-structured, layered architecture.

MDA is primarily targeted at MOF-based and UML-related design approaches. However, it can be generalized to encompass not only object-oriented systems but also earlier, well-known modeling approaches. Using this generalization, we could identify various *modeling spaces* depending on the meta-metamodel used in describing metamodels – MOF modeling space, EBNF modeling space, etc. Even models from everyday modeling spaces can fit in – art, literature, etc.

Fig. 8.3 shows some common examples of models put in a layered architecture conceptually inspired by MDA – a famous painting (*Mona Lisa*, also known as *La Gioconda*, painted by Leonardo da Vinci in the 16th century), and a part of a written music score (of the song "Smoke on the Water"). A noble Renaissance woman and a rock song are things from the real-world, at the M0 layer. A painting and a music sheet are obviously abstractions of real-world things. Therefore, they are models and can be put at the M1 (model) layer. Metamodels are used to define these models. In the case of a music score, which is a written interpretation of a song, the metamodel (M2) is a set of concepts – stave, note, etc. – and rules that define what each note means and how we can arrange them on five staves. In this context, the meta-metamodel (M3) includes self-defined concepts that then define concepts: stave, note etc. Although this architecture is imprecise and definitely not perfect from the perspective of music theory, at least it captures a formal interpretation of music.

Things get harder in the case of painting. Is it possible to specify a metamodel that can define such a complex and ambiguous model as a masterpiece painting? A simplistic view, based on physical appearance only, may lead to the definition of this metamodel in terms of concepts like line, circle, color, etc. The meta-metamodel would then be a set of concepts used to define line, circle, color and their meanings. However, *Mona Lisa*, like any other artwork, is much more than just lines and colors. It has much to do with human psychology and experience and can be interpreted in many ways. It is much more difficult, if not impossible, to define a formal metamodel or meta-metamodel in this case. We may anticipate that they exist, but they are extremely complex and implicit.

Another important issue is that, although *Mona Lisa* or written notes are models, they are also things from the real world. We can hold them in our hands (if the guards in Louvre let us do this, in the case of *Mona Lisa*!) and they can be items entered in the information system that stores information about art.

The previous analysis shows us that something can be taken as a model if it is an abstraction of things from the real world, but it is simultaneously a thing from the real world. Whether we take it as a model or as a real-world thing depends on the context, i.e. on the point of view. Also, models can be defined using metamodeling concepts formally or implicitly.

Any software is primarily a model of the real world, but in some phases of its development it is also treated as the thing from the real world. That often leads to confusion with UML and programming language code – both are models, but at same stages of development UML models are used to represent code as a thing from the real world.

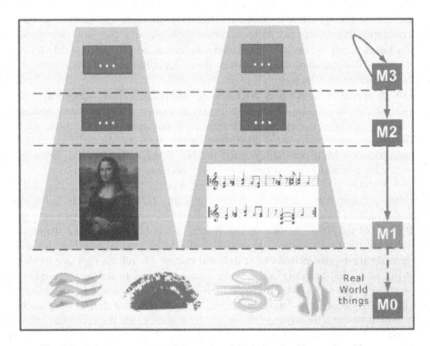

Fig. 8.3. A few common models put in a MDA-inspired layered architecture

8.10 Meta to Meta…

An XSLT stylesheet is an XML document defining a transformation from one class of XML documents into another, thus specifying a meta-to-meta transformation at the syntactic level. A much more comprehensive approach is enabling transformation from one modeling space to another. Software engineers and developers often need to analyze and describe their artifacts and the domains their software models from different perspectives. Fig. 8.4 illustrates this idea using the MDA four-layer system-modeling architecture. It is possible to model the same set of real-world things in two different modeling spaces, but in another way. In this case, the relation between these modeling spaces is oriented towards pure transformation, bridging from one space to another. Examples of such parallel modeling spaces are MOF and ECore (ECore is a meta-metamodel very similar to but also different from MOF; although ECore is a little simpler than MOF, they are both based on similar object-oriented concepts). In such cases, things like XSLT and Java programs can be used to perform the actual transformations.

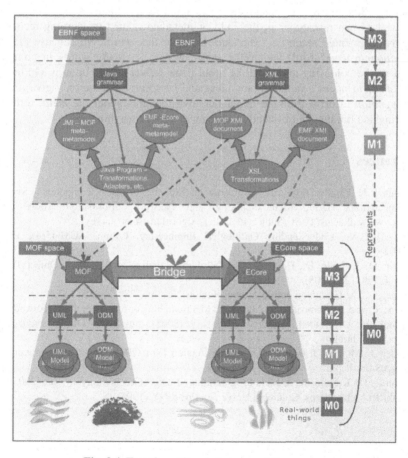

Fig. 8.4. Transformations between modeling spaces

It is also possible to take models from one modeling space as real-world things to be modeled and represented in another space. Both MOF and ECore spaces are represented in other, more concrete modeling spaces. For example, they can be implemented using repositories, serialized into XMI etc. Figure 3 shows them as Java program codes and XML documents in the EBNF space. Models from the MOF space are modeled in Java code according to JMI standard, and in XML according to the MOF XMI.

8.11 ....and Beyond

There are numerous other advanced domains of computing in which meta-level computational tasks are well developed and largely applied. For example, *metaheuristics* is a top-level general strategy which guides other heuristics to search for feasible solutions in domains where the task is hard (e.g., Tabu Search, simulated annealing, and genetic algorithms). In dialogue modeling and analysis, *meta-statements* are turns that are about the discussion rather than part of the discussion pertaining to the topic – "What does John mean?" is a meta-statement about John's statement.

A well-known researcher in the field of metadata related to learning technology systems commented recently: "Metadata... I hate that word. We can use the word *description* instead." Still, the prefix *meta-* used in so many words in computing should not be confusing and repelling at all if we reflect a little upon it. On the contrary, the *meta* opens a range of new ideas and perspectives if we only give it a little *metacognition* effort – thinking about our own thoughts about the *meta*, reflecting, exploring and linking, and basically understanding our own thoughts.

References

1. Allan, R.J.: Parallel Application Software on High Performance Computers - Survey of Computational Grid, Meta-computing and Network Information Tools (2000), http://www.dl.ac.uk/TCSC/Subjects/Parallel_Algorithms/steer_survey/
2. Devedžić, V.: Understanding Ontological Engineering. Comm. ACM 45(4), 136–144 (2002)
3. Đurić, D., Gašević, D., Devedžić, V.: Ontology Modeling and MDA. J. Object Technology 4, 109–128 (2005)
4. Foster, I.: The Grid: Computing without Bounds. Sci. Am. 288, 78–85 (2003)
5. Ip, A., Currie, M., Morrison, I., Mason, J.: Metasearching or Megasearching: Toward a Data Model for Distributed Resource Discovery (2005), http://www3.dls.au.com/metadata/DataModel.html
6. Miller, J., Mukerji, J. (eds.): MDA Guide Version 1.0.1 (2005), http://www.omg.org/docs/omg/03-06-01.pdf
7. Thomas, D.: Refactoring as Meta Programming? J. Object Technology 4, 7–11 (2005)
8. Volk, T.: Metapatterns. Columbia University Press, NY (1995)

9

Emergent Semantics in Distributed Knowledge Management

Carola Aiello[2], Tiziana Catarci[2], Paolo Ceravolo[1], Ernesto Damiani[1],
Monica Scannapieco[2], and Marco Viviani[1]

[1] Dipartimento di Tecnologie dell'Informazione- Università di Milano
via Bramante, 65 - 26013, Crema (CR), Italy
{damiani,ceravolo,viviani}@dti.UNIMI.it
[2] Dipartimento di Informatica e Sistemistica - Sapienza Università di Roma,
Via Ariosto 25, - 00185, Roma, Italy
{carola.aiello,tiziana.catarci,monica.scannapieco}@dis.uniroma1.it

Abstract. Organizations and enterprises have developed complex data and information exchange systems that are now vital for their daily operations. Currently available
systems, however, face a major challenge. On todays global information infrastructure,
data semantics is more and more context- and time-dependent, and cannot be fixed
once and for all at design time. Identifying emerging relationships among previously
unrelated information items (e.g., during data interchange) may dramatically increase
their business value. This chapter introduce and discuss the notion of Emergent Semantics (ES), where both the representation of semantics and the discovery of the proper
interpretation of symbols are seen as the result of a selforganizing process performed
by distributed agents, exchanging symbols and adaptively developing the proper interpretation via multi-party cooperation and conflict resolution. Emergent data semantics
is dynamically dependent on the collective behaviour of large communities of agents,
which may have different and even conflicting interests and agendas. This is a research
paradigm interpreting semantics from a pragmatic prospective. The chapter introduce
this notion providing a discussion on the principles, research area and current state of
the art.

9.1 Introduction

On todays global information infrastructure, access to information involves interaction with distributed sources. Moreover network agents want to access information efficiently, protecting their sensible information and preserving their
autonomy. One of the most important ways of improving the effectiveness of
information retrieval and service sharing is by explicitly describing information
services semantics. Ontology serves this purpose: it consists of explicit, partial
definitions of the intended meaning of symbols for a domain of discourse. Unfor-
tunately, building shared ontology is a complex process and top-down ontology
design, even when done collaboratively, is known not to scale well. Also domains
are rapidly evolving and the semantics of data cannot be fixed once and for all

R. Nayak et al. (Eds.): Evolution of the Web in Artificial Intel. Environ., SCI 130, pp. 201–220, 2008.
springerlink.com

at design time. This requires to make semantics more and more context- and time-dependent.

Emergent Semantics has been proposed as a solution to the semantic interoperability problem. The general idea is that if we renounce to a centralized control over semantic description, we can improve them by exploiting implicit information emerging during data exchange. For instance repeated downloads are a confirmation of data quality, while the frequency of interactions can define a degree of correlation. This paradigm can be applied in many environments such as human-computer interaction, language games for robot-robot and robot-human communication, scientific databases, e.g. biological data, where data is captured through experiments and subsequently analyzed. In this work, we limited our analysis to distributed knowledge management. This is an environment characterized by the multiple information agents interacting with different level of cooperations. In this context data exchange can be taken as a source of additional information to be used in a process of enrichment of semantic representation. The chapter is organized as follow. In Section 9.2 we introduce the notion of Emergent Semantics. In Section 9.3 we discuss the main principles of this research paradigm. In Section 9.4 we summarize the principal research areas, such as P2P data integration, service discovery or trust and reputation management. Finally, in Section 9.5 we provide some conclusive remarks.

9.2   The Notion of Emergent Semantics

The notion of semantics has various definitions in different domains. In the domain of Programming Languages, semantics basically refer to rules which relate inputs and outputs [74]. In logic, semantics is the Kripke's theory of truth, usually expressed by a a set-theoretic interpretation [62]. On the web, semantics is often intended as the metadata used for annotating resources, in this case the term "equivalent semantics" designates for equivalent annotation imposed on different resources, using a common vocabulary, as for instance in [20]. For some applications in the field of image retrieval the semantic interpretation of the image content is available as the result of an automatic or semi-automatic image analysis process, applied to images belonging to specific domains and described in advance [63]. Here the semantics of a resource is a typical pattern related to that resource. As a least common denominator, we can characterize semantics as a relationship or a mapping established between an information items syntactic structure and some external domain.

The Emergent Semantics approach consolidates the local semantics held by autonomous information agents into a global, population-wide semantics that results from the continuous interaction of the agents among themselves. The large-scale structures emerging from these continuous interactions dynamically provide meaning to the local symbols. Semantics constructed incrementally in this way is called Emergent Semantics (ES). This is a research paradigm interpreting semantics from a pragmatic prospective. More complete presentations of this paradigm can be found in [22], [2], [71].

9.3 Principles and Main Features of the Emergent Semantics Paradigm

Emergent semantics is the form of semantic interoperability viewed as an emergent phenomenon [2] constructed incrementally among data sources in a bottom-up, semi-automatic manner without relying on pre-existing, global semantic models. In such a dinamic scenario, global information is highly evolutionary: documents of already existing sources may be updated, added or deleted; new sources and services may appear and some may disappear (definitively or not). At any given point in time, the state of the semantic interoperability depends on the frequency, the quality and the efficiency with which negotiations can be conducted to reach agreements on common interpretations within the context of a given task. The main principles and features of the emergent semantics paradigm can be summarized as follows:

- *Semantic agreement.* Two or more peers need to establish a semantic agreement, that is to share the meaning of a model, like a conceptual model, or other relevant meta-data information to the task at hand.
- *Agreement negotiation.* Agreements are negotiated between peers, and are dynamically built and maintaned. Such agreements constitute the basis for communication between peers which is realized in terms of message exchanges. Negotiations of semantic agreements are performed on local basis, that is a peer directly contacts all the peers it wants to communicate with. In this way the number of interactions is greater than the one characterizing environments that involve third parties with the role of mediating the communication. However, though the agreements are built on a local basis, a global agreement is nevertheless obtained as a result of their propagation. This is, in a sense, the real essence of the emergent semantics paradigm: it is not necessary to contact all the network, in fact the communication can be incrementally realized by exploiting semantic agreements. Agreements are established on the basis of the peer's specific goals; hence for each distinct goal a peer could establish distinct agreements.
- *Self organization.* The emergent semantics paradigm relies, at the end, on a self-organization process. Self-organization processes have been studied in many scientific disciplines such as physics or biology for a long time and are recently being investigated in computer science as well. Under this perspective, emergent semantics is another application of this successful principle in the computer science field.

9.4 Principal Research Areas

Important efforts in many research areas are needed to achieve semantic interoperability by the emergent semantics paradigm. In this section we detail the principal involved research areas : *(i)* P2P Data Integration, *(ii)* Service Discovery and *(iii)* Trust and Reputation.

9.4.1 P2P Data Integration

Information systems have been characterized by a multitude of autonomous, heterogeneous information repositories. The problem of how to provide transparent access to heterogeneous information sources while maintaining their autonomy is a time-honored one. Information integration systems typically provide a uniform query interface to a collection of distributed and heterogeneous information sources, giving users or other agents the illusion of querying a centralized and homogeneous information system. As such, they are considered as mediation systems between users and multiple data sources which can be syntactically or semantically heterogeneous while being related to the same domain. Existing mediator-based information systems can be distinguished according to: (1) the type of mappings between the mediated schema (on global schema) and the schemas of the sources: there exist basically two approaches for such mappings, the Global As View (GAV) and the Local As View (LAV). The global-as-view approach describes the global schema as a view over all local schemas, whereas the local-as-view approach describes each local schema as a view over the global schema; (2) the languages used for modelling the global schema and the source descriptions and (3) the expressivity of the global schema. Starting from this main thread, a recent research line aims to integrate data in a peer-to-peer (P2P) environment and is identified as peer data management or peer-to-peer data integration. A P2P system is characterized by a structure constituted by various autonomous nodes that hold information and that are linked to other nodes by means of mappings. No global ontology a peer can refer to is actually available. The goal of P2P data integration is to provide unified access to this set of heterogeneous data sources. The lack of any agreed-upon global schema or ontology makes it very difficult for the participating parties in the system to reach a global consensus on semantic data.

In the following we describe the principal research issues related to P2P data integration: *(i)* Formal Semantics, *(ii)* Ontology Matching, *(iii)* Query Processing and *(iv) Data Quality.*

Formal Semantics. Initial approaches rely on some pre-defined corpus of terms serving as an initial context for defining new concepts [41] or make use of gossiping and local translation mappings to incrementally foster interoperability in the large [1]. However, there is still a fundamental lack of understanding behind the basic issues of data integration in P2P systems. Indeed, since no single actor is in charge of the whole system, it is unrealistic to assume restrictions on the overall topology of the P2P mappings [42, 31]. Hence, one has to take into account the fact that the mappings may have an arbitrary structure, possibly involving cycles among various nodes. This needs to be addressed both from the point of view of modeling the system and characterizing its semantics (see, e.g., [17, 42] for a first order logic characterization and [13, 14] for an alternative semantics proposal) and from the point of view of computing answers to queries posed to a peer. Query answering has difficulties that arise from the necessity of distributing the overall computation to the single nodes, exploiting their local processing capabilities and the underlying technological framework. A recent proposal highlighting the peculiarities of P2P data integration systems is [37].

Ontology Matching. Dynamic ontology matching techniques can be used as the basis for driving the negotiation of agreements in order to discover the mappings between concepts of different peer ontologies and maintain them.

The general goal of ontology matching techniques is to compare different ontological descriptions for finding concepts that have a semantic affinity. A survey of ontology matching techniques is provided in [70], where formal and heuristic approaches are classified. The former are founded on model theoretic semantics and automated matching techniques [52, 10, 38], while the latter are based on the idea of guessing mappings that may hold between similar graph structures through a combination of analysis, matching, and learning techniques [27]. In [16], heuristic techniques for performing dynamic ontology matching in open, peer-based networked contexts are proposed. Peculiar features of these techniques are the capability of performing the matching process in a flexible and adaptable way, by dynamically configuring the matching algorithm with the most appropriate matching model for the specific matching case. This is done by taking into account the level of richness of ontological knowledge description as well as the requested degree of mix between the importance of linguistic and contextual features of concepts in the evaluation of their matching degree.

Query Processing. A P2P data integration system consists of a set of (physical) peers. Each peer has an associated schema that represents its domain of interest. Some peers store actual data, and describe which data they store relative to their schema; the stored data does not necessarily cover all aspects of the peers' schema. Peers are linked through peer mappings. A mapping is the the medium for exchanging data and reformulating queries among different schemas; in particular, the mapping defines the overlapping parts of acquaintances' schemas. Peer mappings describe the semantic relationship between the schemas of pairs (or small sets of) peers. Given a query over a peer P_i the system will use the peer mappings to reformulate the query over the schemas of the neighboring peers. Typically, when a peer joins a P2P data integration system, it will supply mappings, as is most convenient, to a small number of peers. Semi-automatic schema mapping techniques have been investigated in [64, 25, 26].

The key step of query processing in a P2P data integration system is reformulating a peer's query over other peers on the available semantic paths. Broadly speaking, the P2P data integration system starts from the querying peer and reformulates the query over its immediate neighbors, then over their immediate neighbors, and so on. Whenever the reformulation reaches a peer that stores data, the appropriate query is posed on that peer, and additional answers may be found. Since peers typically do not contain complete information about a domain, any relevant peer may add new answers.

Among the principal approaches to query processing in P2P systems we cite Piazza [42], Hyperion [50] and PeerDB [61].

In Piazza two types of mappings, *peer* and *definitional* mappings are defined, and used for performing query answering. Peer mappings describe data within the stored relations (generally with respect to one or more peer relations); definitional mappings are instead between the schemas of the peers. This approach is alternative to the one used by the Hyperion project [3]. Hyperion mappings rely on the usage of mapping tables that store the correspondence between values. As outlined in [50, 51], mapping tables are often the result of expert knowledge and are manually created by domain specialists. However, mechanisms to partially support the task of mapping discovery can be used: new mappings can be inferred from already existing mappings present in the mapping table.

PeerDB [61] is another P2P-based system for distributed sharing of relational data. Similar to Piazza, PeerDB does not require a global schema, but it doesn't use schema mappings for mediating between peers. Instead, PeerDB employs an information retrieval based approach for query reformulation. In their approach, a peer relation (and each of its columns) is associated with a set of keywords. Given a query over a peer schema, PeerDB reformulates the query into other peer schemas by matching the keywords associated with the two schemas. Therefore, PeerDB does not have to follow semantic paths to reach a distant peer. The resulting reformulated queries in PeerDB may not be semantically meaningful, and ultimately the system requires user input to decide which queries are to be executed.

Besides mappings, also routing indices can be used for propagating queries in the system [21, 43], that provide aggregated information about records that are retrieved in the query processing phase. Hence, such indexes can be used for query optimization purposes.

A further approach for indexing data in peer-to-peer systems is based on communities. A community is built on the basis of the similarity of the schema (or the schema mappings) hold by peers. Query processing benefits from the fact that if a query is posed on a certain peer than members of its community will also have similar data. An example of such an approach is provided in ESTEEM [30].

Data Quality. In peer-to-peer environments, where information is exchanged between heterogeneous data sources, the quality of data exchanged and provided by different peers is extremely important. A lack of attention to data quality can imply data of low quality to spread all over the system.

In [56], the problem of the quality of web-available information has been faced in order to select data with high quality coming from distinct sources: every source has to evaluate some pre-defined data quality parameters, and to make their values available through the exposition of metadata.

When considering the issue of exchanging data and the associated quality, a model to export both data and quality data needs to be defined. Some conceptual models to associate quality information to data have been proposed that include an extension of the Entity-Relationship model [73], and a data warehouse conceptual model with quality features described through the Description Logic formalism [44]. Both models are thought for a specific purpose: the former to introduce quality elements in relational database design; the latter to introduce quality elements in the data warehouse design. In [69], a model for associating quality metadata to data exported in the context of a cooperative and distributed system is described.

An important step for quality-aware data integration is the assessment of the quality of the data owned by each peer. For this task, some of the results already achieved for traditional systems can be borrowed, such as record linkage techniques or data cleaning tools [5].

Data integration must take into account data sources heterogeneity. As described in [75], when performing data integration two different types of conflicts may arise: semantic conflicts, due to heterogeneous source models, and instance-level conflicts, due to what happens when sources record inconsistent values on the same objects. The Data Quality Broker described in [30] is a system solving instance-level conflicts. Other notable examples of data integration systems within the same category are AURORA [75] and the system described in [68]. AURORA supports conflict tolerant queries, i.e. it provides a dynamic mechanism to resolve conflicts by means of defined conflict resolution functions. The system described in [68] describes how to solve both semantic and instance-level conflicts. The proposed solution is based on a multidatabase query language, called FraQL, which is an extension of SQL with conflict resolution mechanisms. Similarly to both such systems, the Data Quality Broker supports dynamic conflict resolution, but differently from them the Data Quality Broker relies onto quality metadata for solving instance-level conflicts. A system that also takes into account metadata for instance-level conflict resolution is described in [33]. Such a system adopts the ideas of the Context Interchange framework [11]; therefore, context dependent and independent conflicts are distinguished and accordingly to this very specific direction, conversion rules are discovered on pairs of systems.

The ESTEEM architecture [30] is an example of an emergent semantics system with data quality support. In the Esteem architecture the data quality profile module involves the computation of data quality metrics on the peer data that are available to other peers. More specifically, each peer has the possibility of associating quality metadata to the exported data (at value level). Such metadata represent data quality measures corresponding to some specific quality dimensions. Metrics for the most common quality dimensions (column completeness, format consistency, accuracy and internal consistency (see [6] for the definition of such metrics) are already implemented and the model is ready to be extended to other dimensions. Besides the data quality profile module, a quality manager module is invoked during query processing in order to exploit quality metadata and to take data inconsistencies into account. More specifically, it is assumed that data can exhibit key-level conflicts [7]. This implies that a record linkage step must be performed in order to provide answers to user queries.

9.4.2 Service Discovery

The emergence of semantics is a key issue to enforce timely discovery and dynamic composition of distributed services. Recently, many organizations have heavily invested in Web Service technologies and, as a consequence, a growing number of services is being made available. Service proliferation over the Web has been facilitated by the development of several standards, like WSDL for service description, UDDI for service registry, SOAP for message exchange and BPEL4WS for service orchestration.

The discovery of services is the most important functionality in distributed and service-oriented environments. Standards like UDDI or WSDL support description of services and discovery functionalities from a syntactic perspective. But the major problem remains: the semantics of the service description. Often, the same services are described by users and service providers in different ways. As services and their descriptions are evolving quickly responding to market changes, it is in general impossible to keep up with all requirements in time. A decentralized discovery service exploiting emergent semantics approaches to extend the standards in a controlled way and distribute the changes among the peers appears as a concrete possible solution. With such an approach, a peer could "learn" about new descriptions and mappings incrementally departing from existing standards used for bootstrapping the process. Modern approaches for service discovery have to address the treatment of dynamical aspects both with respect to the continuous addition and removal of services in a highly variable environment and with respect to different contexts in which a service could be invoked [15]. Advanced techniques and tools for enabling semantic service discovery are therefore highly desired and required. In particular, it is necessary that services are described in a formal way and service semantics is well captured. In the literature, ontology-based approaches are being developed to exploit the benefits of the ontology technology, such as inferencing, in the context of service discovery. In the Semantic Web, the ontology description languages OWL [24] and OWL-S [19] have been recently proposed. Service description is composed by a service profile (what the service does), a service model (how the service works) and a service grounding (how to invoke the service).

The Semantic Web Services Initiative (SWSI, see www.swsi.org) relaxes the constraint of using a description logic formalism for defining service workflow, and uses a first-order logic based language. In the Unified Problem Solving Method Development Language (UPML) framework [34] logical expressions defined in goals, mediators, ontologies and Web Services are expressed using frame logic. UPML distinguishes between domain models, task models, problem solving methods and bridges, and is also the basis of the Internet Reasoning Service (IRS) [57], a knowledge-based approach to Semantic Web Services. Domain models are effectively the domain ontology, while task models provide a generic description of tasks to be solved. Problem solving methods provide implementation-independent descriptions of tasks, while the bridges map between the various components.

Service matchmaking has been addressed by several approaches in literature: given a request R and a set of advertisements S, the matching procedure must return the set of advertised services that match better with R, possibly ranked with respect to their level of matching (if it can be evaluated). In most approaches the starting point is the UDDI registry, where service descriptions are published; UDDI registry offers searching functionalities that use traditional

keyword-based techniques, featured by low precision and recall. To provide semantic matching between service descriptions, some approaches consider concept
definitions within ontologies (concept-based techniques). In [58] a framework for
semi-automatically marking up Web service descriptions with ontologies is proposed with algorithms to match and annotate WSDL files with relevant ontologies; domain ontologies are used to categorize Web services into domains. The
use of ontologies enables service matchmaking in the discovery process. In fact,
the elements used for service capability description refer to concepts that can be
properly defined and semantically related in domain ontologies. Semantic relationships between concepts are then exploited to establish the type of matching
between advertisements and requests. Dynamic discovery of distributed services
is based on semantic interoperability. In [76] a service ontology specifies domain
concepts with a set of synonyms to allow a flexible search and a set of service
classes to define the properties of services, its attributes and operations. In [12] a
new technique for Web service discovery which features a flexible matchmaking
by exploiting DAML-S ontologies is proposed. In [49] a Web Service Modeling
Ontology (WSMO) is expressed by using the formal F-Logic language to describe
various aspects related to Semantic Web Services. They start from the Web Service Modeling Framework (WSMF)[35], that consists of four elements: ontologies
that provide terminology used by other elements (concepts, axioms, relations and
instances), goals of Web Services (by means of pre- and post-conditions), Web
Service description (non functional properties, choreography and orchestration
aspects) and mediators which bypass interoperability problems.

9.4.3 Trust and Reputation

As outlined in previous sections, today's Web infrastructure is increasingly used
for semantics-driven access to services and resources. This problem is twofold.
The first aspect is related to information retrieval, and can be adressed by intelligent search and selection techniques. The second is deciding which among
many sources is most reliable and it is usually presented by the notions of *trust*
or *reputation*.

In human society, trust and reputation are social knowledge allowing to evaluate which agents can be considered as a reliable sources of information or services.
In computer science Trust is not a new research topic in itself; however even, if
there is a rapidly growing literature on the theory and applications of trust in
different kind of settings, there is a considerable confusion around the terminology used to describe them. In fact, depending on the area where the concept of
trust is used – security and access control in computer networks, reliability in
distributed systems, game theory and agent systems, and policies for decision
making under uncertainty – it varies in these different communities in how it is
represented, computed, and used.

The concept of trust is often connected to the mechanism that verify the
identity of a remote source of information; in this context, it is investigated it
is in association with signatures and encryption mechanisms, whose purpose is
to provide protection against malicious parties. The policies used to manage

authorizations, allowing to distinguish trusted and untrusted parties, are defined a-priori by a central authority. In a community, however, trust in strictly connected to the notion of relationship among parties. In distributed community-oriented scenarios, for evaluating the reliability of resources we need to deal with the notion of relationship between a *trustor*, the subject that trusts a target entity, and a *trustee*, the entity that istrusted. The formalization of this notion can significantly improve the quality of the retrieved resources. As stated in [53] and [47], trust can be an important factor in decision-making, because it forms the basis for allowing a trustee to use or manipulate resources owned by a trustor or may influence a trustor's decision to use a service provided by a trustee. According to the Emergent Semantics approach the subjectivity of knowledge is seen as a potential source of value and local misalignment is used as a way for improving and evolving semantic mappings.

Defining Trust. *Trust* is a complex subject relating to belief in the honesty, truthfulness, competence, reliability, etc., of the trusted person or service. There is no consensus in the literature on what trust is and on what constitutes trust management, because the term trust is being used with a variety of meanings [55]. Many researchers have recognized the value of modeling and reasoning about trust computationally; however, here is no entierly shared notion of trust nature as many authors in the field have noted, the meaning of trust as used by each researcher differs across the span of existing work.

In [46] two definitions of trust are introduced: *reliability trust* and *decision trust* respectively. The first one is introduced by means of the Gambetta definition of trust [36] as "the subjective probability by which an individual, A, expects that another individual, B, performs a given action on which its welfare depends". This definition includes the concept of *dependence* on the trusted party, and the *reliability* (probability) of the trusted party as seen by the trusting party.

However having high (reliability) trust in a person in general is not necessarily enough to decide to enter into a situation of dependence on that person [32]. Jøsang et al. introduces the definition inspired by McKnight & Chervany [55] where "decision trust" is "the extent to which one party is willing to depend on something or somebody in a given situation with a feeling of relative security, even though negative consequences are possible".

An alternative definition given in Mui et al. [59], which refers to past encounters, and may be thought as *reputation-based* trust, described as "a subjective expectation an agent has about another's future behavior based on the history of their encounters".

Another interesting definition affirms that trust is "the firm belief in the competence of an entity to act dependably, securely, and reliably within a specified context" (assuming dependability covers reliability and timeliness) [39].

The close relationship between trust and belief is emphasized by the definition by Olmedilla et al. [60], which refers to actions and not competence like the previous one: "Trust of a party A to a party B for a service X is the measurable belief of A in that B behaves dependably for a specified period within a specified context (in relation to service X)".

Depending on the environment where the notion of trust must be applied the suitable definition can spotlight different features. In a distributed environment one of the main features to be considered is the "dynamic nature" of trust, as discussed in [72]. As time passes, the trust one entity has in another might not stay the same. In particular trust can change depending on the *experience* that a trustor has about a trustee, and this experience is always related to a *context*. Another important aspect is the difference between trust and reputation. While the concept of reputation refers to a perception that is generally said or believed about an agent; trust ultimately is an individual phenomenon that is based on various factors or evidence, some of which carry more weight than others. The difference between trust and reputation can be illustrated by the following statements [46]: (*a*) "I trust you because of your good reputation" and (*b*) "I trust you despite your bad reputation".

A distributed system relaying on the notion of trust must support two important tasks: learning reputation and reasoning on trust. An agent can learn reputations interacting with other agents and aggregating trust evaluations of other agents. While the action of reasoning on trust describes the process in which an agent integrates the reputations from other agents, with a trust model (public or private) and its own beliefs, to update its local trust model.

Modeling Trust. In the previous Section we have introduced the definition of trust. Now we focus our discussion on how realizing this notion can be supported by a model. Models for computing trust can belong to two categories: (*i*) *policies-based* models or (*ii*) *reputation-based* models [8]. The definition of "hard evidence" used in policies opposed to the subjective estimation of trust used in reputation systems, as appear in [4], reflect the difference between the term *hard security* used for traditional mechanisms like authentication and access control, and *soft security* for social control mechanisms in general, of which trust and reputations systems are examples. The difference between these two approaches was first described by Rasmusson & Jansson in 1996 [66].

Policies-based models describe the conditions necessary to obtain trust, and can also prescribe actions and outcomes if certain conditions are met. Policies frequently involve the exchange or verification of credentials, which are information issued (and sometimes endorsed using a digital signature) by one entity, and may describe qualities or features of another entity. In this field the terms authorization and authentication are often connected to trust. *Authorization* can be seen as the outcome of the refinement of a more abstract trust relationship. We define authorization as a policy decision assigning access control rights for a subject to perform specific actions on a specific target with defined constraints. *Authentication* is the verification of an identity of an entity, which may be performed by means of a password, a trusted authentication service, or using certificates. There is then an issue of the degree of trust in the entity that issued the certificate. Note that authorization may not be necessarily specified in terms of an identity. Anonymous authorization can be implemented using capabilities or certificates.

Reputation-based systems model an assessment on the history of interactions with or observations of an entity, either directly with the evaluator (personal experience) or as reported by others (recommendations or third party verification). How these histories are combined can vary, and recursive problems of trust can occur when using information from others (i.e., can I trust an entity's recommendation about another entity?).

At a basic level, both credentials and reputation involve the transfer of trust from one entity to another, but each approach has its own unique problems which have motivated much of the existing work in trust.

Trust Research Classification. Due to the growing interest about trust and the resulting growing corpus of literature on it, there is no shared taxonomy of trust research. However some survey literature does exist approaches have been divided in different areas in literature. In particular [4] organizes trust research in four major areas:

1. *Policy-based trust.* Trust is established simply by obtaining a sufficient amount of credentials pertaining to a specific party, and applying the policies to grant that party certain access rights. The recursive problem of trusting the credentials is frequently solved by using a trusted third party to serve as an authority for issuing and verifying credentials.
2. *Reputation-based trust.* History of an entity's actions/behavior is used to compute trust, and may use referral-based trust (information from others) in the absence of (or in addition to) first-hand knowledge.
3. *General models of trust.* Trust models are useful for analyzing human and agentized trust decisions and for computable models of trust operational. Work in trust modeling describes values or factors that play a role in computing trust, and leans more on work in psychology and sociology for a decomposition of what trust comprises. Modeling research ranges from simple access control polices (which specify who can be trusted when accessing data or resources) to analyses of competence, beliefs, risk, importance, utility, etc.
4. *Trust in information resources.* Trust is an an increasingly common theme in Web related research regarding whether Web resources and Web sites are reliable. With the advent of the Semantic Web, new work in trust is harnessing both the potential gained from machine understanding, and addressing the problems of reliance on the content available in the web so that agents in the Semantic Web can ultimately make trust decisions autonomously. Provenance of information is key to support trust decisions, as is automated detection of opinions as distinct from objective information.

In his short survey Griffiths [40] divides trust researches in three areas:

1. *Security-oriented trust.* A mechanism for ensuring security, encompassing issues of authentication, authorization, access control, privacy.
2. *Service-oriented trust.* A mechanism for achieving, maintaining, and reasoning about quality of services and interactions.

3. *Socially-oriented trust.* A social notion for modeling and reasoning about the relationships between agents, influenced by social science, psychology or philosophy.

In addition to the taxonomies given by [4] and [40], in [65] trust research is categorized according to the definitions of *individual-* or *system-level* trust; in the former, individual agents model and reason about others, while in the latter agents are forced to be trustworthy by externally imposed regulatory protocols and mechanisms (this category includes the already described area of policy-based trust). Finally, depending on the fact that agents either trust others directly based on their experiences, or base their trust on the recommendations of others, models can divided in *direct-trust-based* or *recommendation-based.*

Trust and Emergent Semantics. As seen in Section 9.3, in an ES setting, the notion of uncertain knowledge is crucial. In [29] authors analyze Knowledge Management Systems describing the shared knowledge distinguish to two dimensions: (i) the number of domains involved and (ii) the number of conceptualization used in order to describe these domains. While in a centralized system domains and conceptualizations are usually in a 1 : 1 relationship, and if multiple domains are taken into account a single conceptualization is used ($n : 1$); distributed systems have $1 : n$ or $n : m$ relationship. This simple remark clearly shows how the mapping among the representations of different peers are subject to uncertain knowledge. Trust can be one of the ways for modeling some aspects related to uncertain knowledge. Reconsidering the discussion proposed in Section 9.3, we can see that trust can have an impact on many typical actions of peer-to-peer environments. Trust related P2P actors include:

- *Grouping:* a group of peers that agree to be considered as a single entity, at least with respect to a given task, by other peers. To become a member of a peer must provide some credential that can vary from proving to manage some specific type of resource, to providing a token.
- *Discovery:* is a mechanism that allows the user to discover resources in the peer to peer network.
- *Query Propagation:* when a provider receives a query, it might propagate that query to another provider it considers expert on what it believes is the meaning of the request. In order to decide where to propagate a query a peer has two possibilities: (i) a proximity criteria, i.e. the query will be sent to known peers (i. e. by using the discovery functionality) and selection will be done according to a quantitative criteria (number of peers, number of possible re-routings hops-, etc.); this way peers or providers that are not directly reachable by the seeker or that have just joined the system, can advertise their presence and contribute to the resolution of the query; (ii) a semantic criteria: if the provider computes some matching between a query and concepts in its own context, the query resolution mechanism might look for addresses of other peers that have been associated to the matching concept. Here propagation is done on the base of explicit trust since the provider defines other peers as "experts" on the query topic.

Projects and Applications. In this section we discuss the state of the art of projects and applications relying on ES for managing Trust in Distributed Knowledge Management. Several projects adopt an approach that can be associated to the ES principles. For example in [9] authors state that Knowledge Management systems should be designed in order to support the interplay between two qualitatively different processes: the autonomous management of knowledge of individual groups and the coordination required in order to exchange knowledge among them. The authors introduces the notion of a *proximity* criteria, that is related to semantic closure and trustworthiness among peers. Another project directly focusing on Distributed Knowledge Management is SWAP, [28]. Here the notion of Trust is adopted in order to rate the resources shared in a semantic-based peer-to-peer system. However this work does not propose a process for managing the resource rating. A first work discussing architectures and functions form managing aggregation and distribution of reputations on semantic is [45]. In [67] the same problem is addressed using path algebra. In [18] the authors provide a algorithm for aggregating rating on trustworthiness. No mature system implemented on these algorithms exist right now.

Enlarging our attention to the notion of imprecise knowledge another group of researches is available. Here again we note the lack of implemented solutions. In [48] authors provide a preliminary discussion on the role of belief and trust in managing emergence in the context of information and service semantics. A taxonomy of belief level is proposed and a measure of data quality based on belief and trust is scratched. In [1] the issue of managing semantic interoperability among data sources in a bottom-up, semi-automatic manner without relying on pre-existing, global semantic models is addressed. A solution is proposed based on an heuristic focusing on the cyclic exchange of local mappings. This work strongly underlines the need for a theoretical progress on the uncertainty reduction problem in a distribute environment. A contribution on this line is given in [54] where authors discus how fault tolerance can be managed in a system where trustworthiness on data is constructed by computing a reputation chain. Also in [23] a solution is proposed based on the use of decentralized probabilistic models to reduce uncertainty on schema mappings.

9.5   Conclusions and Open Issues

In a distributed environment of information agents such as in the Semantic Web or in peer-to-peer systems, where information is spread over heterogeneous sources and no global ontology nor centralized control are available, the aim is to enable agents to interoperate irrespective of the source of their initial/local semantics. Semantic interoperability is a crucial element for making distributed information systems usable.

In this chapter we have presented the emergent semantics paradigm, its features and the principal involved research areas. Emergent semantics consolidates the local semantics of autonomous information agents into a global, population-wide semantics that results from the continuous interaction of the agents among

themselves, for this, emergent semantics has been proposed as a solution for the
semantic interoperability problem.

Of course, as discussed in previous sections, a lot af open issues need to be addressed by future research. Major challenges to be addressed in future researches
are:

- A robust theoretical foundation to advanced methodologies for evolvable semantics representation.
- Developing proof-of-concept knowledge management systems for information-
  bound organisations and communities, capable of extracting actionable
  meaning from social interaction patterns.

At first glance, the second challenge might seem narrower than the first one,
but it has become clear that emergent semantics applications are legion, so that
extraction and representation must be somewhat tunable to domain-specific requirements. In particular focusing on the theoretical foundation researches have
to explores the notion of *Incomplete Knowledge*. In a large-scale distributed environment of autonomous agents, information and information needs can no longer
be expressed concisely, as expected by database and semantic web technologies,
but have to deal with numerous sources of knowledge that can be inconsistent
or can express uncertain information. Also inconsistency can insist on global
level but not on local level, and this two dimensions must be managed by a
composition not effacing disalignment of local sources.

In distributed environments, several qualitatively different sources of uncertainty have to be dealt with as well. Besides uncertainty about users information
needs, a comprehensive approach must deal with diverse uncertainty types such
as uncertainty on knowledge conceptualizations, uncertainty on metadata assertions and uncertainty on trustworthiness of the information providers. Current
reasoning techniques for handling uncertainty have been developed for isolated
problems. It is a well-known fact that complete, probabilistic reasoning is as computationally intractable as reasoning in full first order logic is. A number of other
formalisms are available for reasoning under uncertainty, such as Bayesian networks or possibilistic and fuzzy logic. Trust is another important notion related
with uncertainty and formalized according to different approaches. Also there
are various application domains where no definite semantics can be attached to
facts captured in the real world. This is the case for scientific databases, e.g.
biological data, where data is captured through experiments and subsequently
analyzed to guess what is the phenomenon that may actually be materialized
into that data. Another example is forensic science, where investigators may
try to understand what is the situation behind the factual evidences that have
been collected here and there. Potential semantics are associated by formulating hypotheses, which are then tested to see if they can be supported by the
collection of available data. If yes the running hypothesis can be confirmed. If
not another hypothesis will be formulated and tested, till a plausible semantics
is found. Semantics emerges gradually, thanks to a reformulation and test cycle
rather than agreement between different agents. In general it is important to
stress that ES approach needs a common abstraction framework for reasoning

under uncertainty, handling complex conditional relationships between various sources of uncertainty and their models.

Another way to face the problem is to focus on the notion of data quality. Data stored electronically is usually affected by a number of quality problems, ranging from poor accuracy at a syntactic level (data is wrong, for example as a consequence of errors during data collection or manipulation), to various forms or inconsistency, both in a single source or across multiple sources, to problems related to their currency, and so on. This is true also in classical data integration settings, but the risk of poor data quality becomes even more critical in emergent semantics settings. Due to the open nature of these networks, no restriction at all can be enforced on the quality of the data shared by the peers. Thus, it is essential that data quality control be enforced both through the trust and privacy mechanisms, and with proper measures at query processing time that explicitly take quality into account.

Acknowledgments

This work was partly funded by the Italian Ministry of Research under FIRB contract RBNE05FKZ2_004 (project TEKNE).

References

1. Aberer, K., Cudré-Mauroux, P., Hauswirth, M.: The Chatty Web: Emergent Semantics Through Gossiping. In: Proceedings of the 12th International World Wide Web Conference (2003)
2. Aberer, K., Mauroux, P.C., Ouksel, A.M., Catarci, T., Hacid, M.S., Illarramendi, A., Kashyap, V., Mecella, M., Mena, E., Neuhold, E.J.: Emergent semantics principles and issues. LNCS, pp. 25–38. Springer, Heidelberg (March 2004)
3. Arenas, M., Kantere, V., Kementsietsidis, A., Kiringa, I., Miller, R.J., Mylopoulos, J.: The hyperion project: From data integration to data coordination. SIGMOD Record 32(3), 53–58 (2003)
4. Artz, D., Gil, Y.: A survey of trust in computer science and the semantic web. Journal ofWeb Semantics (to appear, 2007)
5. Batini, C., Scannapieco, M. (eds.): Data Quality: Concepts, Methodologies, and Techniques. Springer, Heidelberg (2006)
6. Batini, C., Scannapieco, M. (eds.): Data Quality: Concepts, Methodologies, and Techniques ch. 2. Springer, Heidelberg (2006)
7. Batini, C., Scannapieco, M. (eds.): Data Quality: Concepts, Methodologies, and Techniques ch. 5. Springer, Heidelberg (2006)
8. Bonatti, P., Olmedilla, D.: Driving and monitoring provisional trust negotiation with metapolicies. In: POLICY 2005: Proceedings of the Sixth IEEE International Workshop on Policies for Distributed Systems and Networks (POLICY 2005), pp. 14–23. IEEE Computer Society, Washington (2005)
9. Bonifacio, M., Cuel, R., Mameli, G., Nori, M.: A peer-to-peer architecture for distributed knowledge management. In: Proceedings of the 3rd International Symposium on Multi-Agent Systems, Large Complex Systems, and E-Businesses (2002)

10. Bouquet, P., Magnini, B., Serafini, L., Zanobini, S.: A sat-based algorithm for context matching. In: CONTEXT, pp. 66–79 (2003)
11. Bressan, S., Goh, C.H., Fynn, K., Jakobisiak, M., Hussein, K., Kon, H., Lee, T., Madnick, S., Pena, T., Qu, J., Shum, A., Siegel, M.: The Context Interchange Mediator Prototype. In: Proc. SIGMOD 1997, Tucson, AZ (1997)
12. Brogi, A., Corfini, S., Popescu, R.: Flexible Matchmaking of Web Services Using DAML-S Ontologies. In: Proc. Forum of the Second Int. Conference on Service Oriented Computing (ICSOC 2004), New York City, NY, USA (2004)
13. Calvanese, D., Damaggio, E., De Giacomo, G., Lenzerini, M., Rosati, R.: Semantic data integration in p2p systems. In: DBISP2P, pp. 77–90 (2003)
14. Calvanese, D., De Giacomo, G., Lenzerini, M., Rosati, R.: Logical foundations of peer-to-peer data integration. In: PODS, pp. 241–251 (2004)
15. Casati, F., Shan, M.-C., Georakopoulos, D.: The VLDB Journal: Special Issue on E-Services 10(1) (2001)
16. Castano, S., Ferrara, A., Montanelli, S.: H-match: an algorithm for dynamically matching ontologies in peer-based systems. In: SWDB, pp. 231–250 (2003)
17. Catarci, T., Lenzerini, M.: Representing and Using Interschema Knowledge in Cooperative Information Systems. Journal of Intelligent and Cooperative Information Systems 2(4) (1993)
18. Ceravolo, P., Damiani, E., Viviani, M.: Bottom-up extraction and trust-based refinement of ontology metadata. IEEE Transactions on Knowledge and Data Engineering 19(2), 149–163 (2007)
19. The OWL Service Coalition. OWL-S 1.1 release (2004),
    http://www.daml.org/services/owl-s/1.1/
20. Corcho, Ó.: Ontology based document annotation: trends and open research problems. IJMSO 1(1), 47–57 (2006)
21. Crespo, A., Garcia-Molina, H.: Routing indices for peer-to-peer systems. In: ICDCS, pp. 23 (2002)
22. Cudré-Mauroux, P., Aberer, K., Abdelmoty, A.I., Catarci, T., Damiani, E., Illaramendi, A., Jarrar, M., Meersman, R., Neuhold, E.J., Parent, C., Sattler, K.-U., Scannapieco, M., Spaccapietra, S., Spyns, P., Tré, G.D.: Viewpoints on Emergent Semantics. In: Spaccapietra, S., Aberer, K., Cudré-Mauroux, P. (eds.) Journal on Data Semantics VI. LNCS, vol. 4090, pp. 1–27. Springer, Heidelberg (2006)
23. Cudre-Mauroux, P., Aberer, K., Feher, A.: Probabilistic message passing in peer data management systems. In: ICDE 2006: Proceedings of the 22nd International Conference on Data Engineering (ICDE 2006), p. 41. IEEE Computer Society, Washington (2006)
24. Dean, M., Schreiber, G., Bechhofer, S., van Harmelen, F., Hendler, J., Horrocks, I., McGuinness, D.L., Patel-Schneider, P.F., Stein, L.A.: OWL Web Ontology Language W3C Recommendation (2004),
    http://www.w3.org/TR/2004/REC-owl-ref-20040210/
25. Do, H.H., Rahm, E.: Coma - a system for flexible combination of schema matching approaches. In: VLDB, pp. 610–621 (2002)
26. Doan, A., Domingos, P., Halevy, A.Y.: Reconciling schemas of disparate data sources: A machine-learning approach. In: SIGMOD Conference, pp. 509–520 (2001)
27. Doan, A., Madhavan, J., Dhamankar, R., Domingos, P., Halevy, A.Y.: Learning to match ontologies on the semantic web. VLDB J 12(4), 303–319 (2003)
28. Ehrig, M., Haase, P., van Harmelen, F., Siebes, R., Staab, S., Stuckenschmidt, H., Studer, R., Tempich, C.: The swap data and metadata model for semantics-based peerto- peer systems. In: Schillo, M., Klusch, M., Müller, J., Tianfield, H. (eds.) MATES 2003. LNCS (LNAI), vol. 2831, pp. 144–155. Springer, Heidelberg (2003)

29. Ehrig, M., Schmitz, C., Staab, S., Tane, J., Tempich, C.: Towards evaluation of peer-to-peer-based distributed information management systems. In: van Elst, L., Dignum, V., Abecker, A. (eds.) AMKM 2003. LNCS (LNAI), vol. 2926, Springer, Heidelberg (2004)
30. Esteem Team. Emergent Semantics and Cooperation in MultiKnowledge Environments: the ESTEEM Architecture. In: Proc. of the VLDB Int. Workshop on Semantic Data and Service Integration (SDSI 2007). Vienna, Austria (to appear, 2007)
31. Fagin, R., Kolaitis, P.G., Miller, R.J., Popa, L.: Data exchange: Semantics and query answering. In: ICDT, pp. 207–224 (2003)
32. Falcone, R., Castelfranchi, C.: Social Trust: A Cognitive Approach. In: Trust and deception in virtual societies, pp. 55–90. Kluwer Academic Publishers, Norwell (2001)
33. Fan, W., Lu, H., Madnick, S., Cheungd, D.: Discovering and Reconciling Value Conflicts for Numerical Data Integration. Information Systems 26(8) (2001)
34. Fensel, D., Benjamins, V.R., Motta, E., Wielinga, B.: UPML: A Framework for Knowledge System Reuse. In: Proceedings of the Sixteenth International Joint Conference on Artificial Intelligence (IJCAI 1999), Stockholm, Sweden, pp. 16–23 (1999)
35. Fensel, D., Bussler, C., Ding, Y., Omelayenko, B.: The Web Service Modeling Framework WSMF. Electronic Commerce Research and Applications 1(2), 113–137 (2002)
36. Gambetta, D.: Can we trust trust? In: Trust: Making and Breaking Cooperative Relations, pp. 213–237. Basil Blackwell, Malden (1988)
37. De Giacomo, G., Lembo, D., Lenzerini, M., Rosati, R.: On reconciling data exchange, data integration, and peer data management. In: PODS, pp. 133–142 (2007)
38. Giunchiglia, F., Shvaiko, P.: Semantic Matching. Knowledge engineering review 18(3), 265–280 (2003)
39. Grandison, T., Sloman, M.: A Survey of Trust in Internet Applications. IEEE Communications Surveys and Tutorials 3(4) (September 2000)
40. Griffiths, N.: Trust: Challenges and Opportunities. AgentLink News, AgentLink 19, 9–11 (2005)
41. Guha, R.V., McCool, R.: Tap: A semantic web testbed. J. Web Sem. 1(1), 81–87 (2003)
42. Halevy, A.Y., Ives, Z.G., Suciu, D., Tatarinov, I.: Schema mediation in peer data management systems. In: ICDE, p. 505 (2003)
43. Hose, K., Klan, D., Sattler, K.-U.: Distributed data summaries for approximate query processing in pdms. In: IDEAS, pp. 37–44 (2006)
44. Jarke, M., Lenzerini, M., Vassiliou, Y., Vassiliadis, P. (eds.): Fundamentals of Data Warehouses. Springer, Heidelberg (1995)
45. Golbeck, J.H.J., Parsia, B.: Trust networks on the semantic web. In: Proceedings of Cooperative Information Agents 2003, August FebruaryJuly-FebruarySeptember (2003)
46. Jøsang, A., Ismail, R., Boyd, C.: A survey of trust and reputation systems for online service provision. Decis. Support Syst. 43(2), 618–644 (2007)
47. Jøsang, A., Knapskog, S.J.: A metric for trusted systems. In: Proceedings of the 21st NIST-NCSC National Information Systems Security Conference, pp. 16–29 (1998)
48. Kashyap, V.: Trust, but verify: Emergence, trust, and quality in intelligent systems. IEEE Intelligent Systems 19(5) (2004)

49. Keller,U., Lausen, H., Roman, D.: Web Service Modeling Ontology (WSMO). WSMO Working Draft (2004), http://www.wsmo.org/2004/d2/v02/20040306/
50. Kementsietsidis, A.: Data sharing and querying for peer-to-peer data management systems. In: EDBT Workshops, pp. 177–186 (2004)
51. Kementsietsidis, A., Arenas, M., Miller, R.J.: Mapping data in peer-to-peer systems: Semantics and algorithmic issues. In: SIGMOD Conference, pp. 325–336 (2003)
52. Madhavan, J., Bernstein, P.A., Domingos, P., Halevy, A.: Representing and Reasoning about Mappings between Domain Models. In: Proc. of the 18th National Conference on Artificial Intelligence and 14th Conference on Innovative Applications of Artificial Intelligence, Edmonton, Alberta, Canada (2002)
53. Manchala, D.W.: Trust metrics, models and protocols for electronic commerce transactions. In: ICDCS 1998: Proceedings of the The 18th International Conference on Distributed Computing Systems, p. 312. IEEE Computer Society, Washington (1998)
54. Mawlood-Yunis, A., Weiss, M., Santoro, N.: Issues for robust consensus building in p2p networks. In: OTM Workshops, vol. 2, pp. 1021–1027 (2006)
55. McKnight, D.H., Chervany, N.L: The meanings of trust. Technical Report WP 96-04, University of Minnesota, Carlson School of Management (1996)
56. Mihaila, G., Raschid, L., Vidal, M.: Using Quality of Data Metadata for Source Selection and Ranking. In: Suciu, D., Vossen, G. (eds.) WebDB 2000. LNCS, vol. 1997, Springer, Heidelberg (2000)
57. Motta, E., Domingue, J., Cabral, L., Gaspari, M.: IRS-II: A Framework and Infrastructure for Semantic Web Services. In: Fensel, D., Sycara, K.P., Mylopoulos, J. (eds.) ISWC 2003. LNCS, vol. 2870, pp. 306–318. Springer, Heidelberg (2003)
58. Motta, E., Domingue, J., Cabral, L., Gaspari, M.: METEOR-S Web Service Annotation Framework. In: Proc. of the 13th International World Wide Web Conference (WWW 2004), New York, NY, USA, pp. 553–562 (2004)
59. Mui, L., Mohtashemi, M., Halberstadt, A.: Notions of reputation in multi-agents systems: A review. In: Falcone, R., Barber, S., Korba, L., Singh, M.P. (eds.) AAMAS 2002. LNCS (LNAI), vol. 2631, pp. 280–287. ACM Press, New York (2002)
60. Olmedilla, D., Rana, O.F., Matthews, B., Nejdl, W.: Security and trust issues in semantic grids. In: Goble, C., Kesselman, C., Sure, Y. (eds.) Semantic Grid: The Convergence of Technologies, number 05271 in Dagstuhl Seminar Proceedings. Internationales Begegnungs- und Forschungszentrum fuer Informatik (IBFI), Schloss Dagstuhl, Germany (2006)
61. Ooi, B.C., Shu, Y., Tan, K.-L.: Relational data sharing in peer-based data management systems. SIGMOD Rec. 32(3), 59–64 (2003)
62. Orlowska, E.: Kripke semantics for knowledge representation logics. Studia Logica, 25 (1990)
63. Rabitti, F., Savino, P.: Querying semantic image databases
64. Rahm, E., Bernstein, P.A.: A survey of approaches to automatic schema matching. VLDB J. 10(4), 334–350 (2001)
65. Ramchurn, S.D., Huynh, D., Jennings, N.R.: Trust in multi-agent systems. Knowl. Eng. Rev. 19(1), 1–25 (2004)
66. Rasmusson, L., Jansson, S.: Simulated social control for secure internet commerce. In: NSPW 1996: Proceedings of the 1996 workshop on New security paradigms, pp. 18–25. ACM Press, New York (1996)
67. Richardson, M., Agrawal, R., Domingos, P.: Trust Management for the Semantic Web. In: Fensel, D., Sycara, K.P., Mylopoulos, J. (eds.) ISWC 2003. LNCS, vol. 2870, Springer, Heidelberg (2003)

68. Sattler, K., Conrad, S., Saake, G.: Interactive Example- Driven Integration and Reconciliation for Accessing Database Integration. Information Systems 28 (2003)
69. Scannapieco, M., Virgillito, A., Marchetti, C., Mecella, M., Baldoni, R.: The DaQuinCIS Architecture: a Platform for Exchanging and Improving Data Quality in Cooperative Information Systems. Information Systems 29(7), 551–582 (2004)
70. Shvaiko, P., Euzenat, J.: A Survey of Schema-based Matching Approaches. Journal on Data Semantics (JoDS) 4, 146–171 (2005)
71. Staab, S.: Emergent semantics. IEEE Intelligent Systems 17(1), 78–86 (2002)
72. Staab, S., Bhargava, B., Lilien, L., Rosenthal, A., Winslett, M., Sloman, M., Dillon, T.S., Chang, E., Hussain, F.K., Nejdl, W., Olmedilla, D., Kashyap, V.: The pudding of trust. IEEE Intelligent Systems 19(5), 74–88 (2004)
73. Wang, R.Y., Kon, H.B., Madnick, S.E.: Data Quality Requirements: Analysis and Modeling. In: Proc. ICDE 1993, Vienna, Austria (1993)
74. Winskel, G.: The formal semantics of programming languages: An introduction. MIT Press, Cambridge (1993)
75. Yan, L.L., Ozsu, T.: Conflict Tolerant Queries in AURORA. In: Proc. CoopIS 1999, Edinburgh, UK (1999)
76. Zeng, L., Benatallah, B., Dumas, M., Kalagnanam, J., Chang, H.: QoS-Aware Middleware for Web Services Composition. IEEE Transactions on Software Engineering 30(5), 311–327 (2004)

Web-Based Bayesian Intelligent Tutoring Systems

C.J. Butz, S. Hua, and R.B. Maguire

Department of Computer Science, University of Regina,
Regina, Saskatchewan, Canada S4S 0A2
{butz,huash111,rbm}@cs.uregina.ca

Abstract. The rapid development of the World Wide Web offers an opportunity to apply a large variety of artificial intelligence technologies in various practical applications. In this chapter, we provide a review of our recent work on developing a Web-based intelligent tutoring system for computer programming. The decision making process conducted in our intelligent system is guided by Bayesian networks, which are a proven framework for uncertainty management in artificial intelligence based on probability theory. Whereas many tutoring systems are static HTML Web pages of a class textbook or lecture notes, our intelligent system can help a student navigate through the online course materials, recommend learning goals, and generate appropriate reading sequences.

10.1 Introduction

Web-based education is currently an area of intense research and development. The benefits of Web-based education are clear: classroom independence, easy accessibility and greater flexibility [5]. Students control their own pace of study and do not depend on rigid classroom schedules. Thousands of Web-based courses and Web-based tutoring systems have been made available over the last five years [5, 63]. Many Web-based tutoring systems, however, are unable to satisfy the heterogeneous needs of users [5, 7]. These tutoring systems are static HTML Web pages, which act simply as copies of regular textbooks. This kind of tutoring system suffers from two major shortcomings, namely, it is neither interactive nor adaptive [5]. Many Web courses present the same learning materials to students with widely differing knowledge levels of the given subject. In fact, Brusilovsky and Maybury [6] explicitly state that an effective system must be robust enough to deal with various types of users. To resolve the traditional "one-size-fits-all" problem, it is necessary to develop systems with an ability to adapt their behavior to the goals, tasks, interests, and other features of individual users and groups of users.

In this chapter, we provide a review of our recent work [9, 10, 11] on developing a Bayesian *Intelligent Tutoring System*, called BITS. BITS can be used on stand-alone computers or as a Web-based application that delivers knowledge through the Internet. BITS is based on *Bayesian networks* [48] - a formal framework for uncertainty management in artificial intelligence and supports student learning. We describe the architecture of BITS and examine the role of each component in the system. In particular, we discuss how to employ Bayesian networks as an inference engine to guide the students' learning

R. Nayak et al. (Eds.): Evolution of the Web in Artificial Intel. Environ., SCI 130, pp. 221–242, 2008.
springerlink.com

processes. Moreover, we describe the features that allow BITS to be accessed via the Web. Beside Web-based tutoring systems, we also provide a comprehensive survey of related work involving Bayesian networks in various Web-based tasks.

There are two favourable features of BITS. Unlike the static tutoring systems mentioned above, BITS can assist a student in navigation through the online materials. More importantly, BITS can recommend learning goals, and generate appropriate reading sequences. For example, a student may want to learn "File I/O" without having to learn every concept discussed in the preceding materials. BITS can determine the minimum prerequisite knowledge needed in order to understand "File I/O" and display the links for these concepts in the correct learning sequence. In this way, one can address the problem of Web-based learners' unproductive navigation, and refocus them on their study objectives by making the tutoring system adaptable to different types of learners. Thus, BITS is very useful to any work applying Bayesian networks as a model for developing adaptive Web-based learning systems for various courses. Our discussion is based on computer programming, since BITS was recently utilized in the initial computer programming course at the University of Regina.

This chapter is organized as follows. In Section 10.2, we review intelligent tutoring systems. We discuss Bayesian networks and probabilistic inference in Section 10.3. In Section 10.4, we describe the architecture of BITS, the role of each module in the system. In particular, we discuss BITS's capability for adaptive guidance by applying the Bayesian network approach. The features that allow BITS to be accessed via the Web are provided in Section 10.5. In Section 10.6, related work is discussed. The conclusion is presented in Section 10.7.

10.2   Background Knowledge

Here, we review Intelligent Tutoring Systems. More specifically, we discuss motivations and desirable features when designing Intelligent Tutoring Systems. We then discuss the general framework of an Intelligent Tutoring System and the function for each component. Finally, we review three popular approaches applied in current Web-based Intelligent Tutoring Systems.

10.2.1   Intelligent Tutoring Systems

Intelligent Tutoring Systems (ITSs) form an advanced generation of traditional education systems instructing via computers, called *Computer Aided Instruction* [8]. In particular, ITSs are computer-based programs that present educational materials in a flexible and personalized way that is similar to one-to-one tutoring [5]. The basic underlying idea of ITSs is to realize that each student is unique. These systems can be used in the traditional educational setting or in distant learning courses, either operating on stand-alone computers or as applications that deliver knowledge through the Internet.

ITSs have been shown to be highly effective in increasing students' performance and motivation levels compared with traditional instructional methods (e.g., [34], [55]). The emergence of the World Wide Web also increased the usefulness of such systems [57]. Their key feature is their ability to provide a user-adapted presentation of the teaching material [2, 19, 59]. This is accomplished by using artificial intelligence methods to represent the pedagogical decisions and the information regarding each student. Bloom [3] demonstrates that individual one-on-one tutoring is the most effective mode of teaching and learning. Carefully designed and individualized tutoring produces the best learning for the majority of people. ITSs uniquely offer a technology that implements computer-assisted one-on-one tutoring.

ITS Architecture. Researchers typically separate an ITS into several different parts, and each part plays an individual function. Fig. 10.1 depicts the basic architecture of an ITS according to [56]. It consists of the following components: (i) the knowledge domain, containing the structure of the domain and the educational content, (ii) the student model, which records information concerning the user, (iii) the pedagogical model, which encompasses knowledge regarding the various pedagogical decisions, and (iv) the user interface.

The key aspect of each component in an ITS is elaborated as follows.

(i) **Knowledge Domain:** The knowledge domain stores the teaching materials that the students are required to study for the topic or curriculum being taught. In a Web-based ITS, the teaching materials are presented to the users as Web pages.

(ii) **Student Model:** The student model is used to record information concerning the user which is vital for the system's user-adapted operation. Usually, this information reflects the system's understanding of one learner's current knowledge level (novice, beginner, intermediate, advanced, etc.) of the subject. Thus, the student model can track a student's understanding and particular need. Without an explicit student model, the pedagogical model is unable to make decisions to adapt instructional content and guidance (see Fig. 10.1) and is forced to treat all students similarly.

Main ITS Component

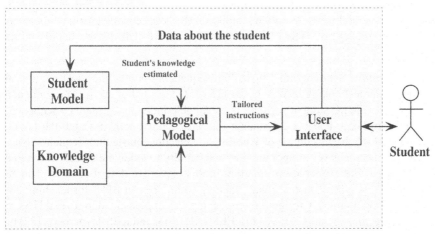

Fig. 10.1. The major components in most Intelligent Tutoring Systems

(iii) **Pedagogical Model:** The pedagogical model provides the knowledge infrastructure in order to tailor presentation of the teaching material according to the student model. In a specific learning session, the pedagogical model decides when to present a new topic, how to provide recommendations and guidance, and which topic to present. As mentioned earlier, the assessment result of the student model is input to this component, so the system's pedagogical decisions reflect differing needs of students. Therefore, this component needs to take appropriate actions to manage one-on-one tutoring, such as selecting a topic based on the user's knowledge of the domain, using a variety of teaching approaches at the appropriate times according to the student's particular needs.

(iv) **User Interface:** The user interface component decides how the system interacts with a user. The dialogue and the screen layouts are controlled by this component. A well-designed interface can enhance the capabilities of an ITS by allowing the system to present instructions and feedback to the student in a clear and direct way. Conversely, if the interface to the ITS is confusing or poorly designed, the effectiveness of the entire instructional session suffers.

Over the past decade, the gradual advances in artificial intelligence methods have been incorporated into ITSs to provide broader and better support for the users of Web-based educational systems [46]. A number of artificial intelligence approaches have been applied in ITSs [1, 39, 40, 45, 49], including rule-based systems, fuzzy logic and neural networks. *Bayesian networks* (BNs) [48] are a formal approach to uncertainty reasoning in artificial intelligence. This technique combines rigorous probability theory as a formal framework with graphical representation and efficient inference mechanisms [62]. In section 10.3, we will discuss details of the Bayesian network approach.

10.2.2   Intelligent Tutoring Technologies

In the reviews of existing ITSs in [7, 43], major intelligent tutoring technologies are identified according to the system's objective: curriculum sequencing, intelligent solution analysis, and problem solving support. All these technologies have been well explored in the field of ITSs [7]. In the following, we will discuss details of each technology.

(i) **Curriculum Sequencing:** Curriculum sequencing is now the most popular and important technology in Web-based ITSs [5]. The goal of curriculum sequencing is to provide the student with the most suitable individually planned sequence of topics to learn. It helps the student find an "optimal path" through the learning material [6]. In the context of Web-based education, curriculum sequencing technology becomes very important due to its ability to guide the student through the hyperspace of available information. In this paper, we describe the implementation of this type of instructional planning technology in BITS for teaching computer programming. Curriculum sequencing was implemented in several ITSs. For instance, the systems CALAT [44], ELM-ART [63] and KBS-Hyperbook [23] provide good curriculum sequencing examples.

(ii) **Intelligent Solution Analysis:** The objective of intelligent solution analysis is to deal with students' solutions of educational problems. Unlike non-intelligent checkers which can only tell whether the solution is correct or not, intelligent solution analysis can tell what is incorrect or incomplete and which missing or wrong knowledge concept may be responsible for the error [7]. Intelligent solution analysis can provide the students with detailed error feedback and update the student model. This technology was implemented early on the Web in the system WITS [47]. The very recent tutoring systems SQL-Tutor [42], German Tutor [22] demonstrate several ways of implementing intelligent solution analysis on the Web.

(iii) **Problem Solving Support:** For many years, problem solving support was considered as the primary duty of an ITS [5]. The purpose for problem solving support is usually to provide the student with intelligent help for each step when resolving a task, such as a project or a problem. When the student is stuck on one step, the system provides a hint showing the next correct solution step for the student, or offering appropriate error feedback. In this setting, the critical problem for the system is to interpret the student's actions and infer the solution plan that the student is currently following based on a partial sequence of observable actions. That is, the system needs to understand the student's plan, and apply this understanding to provide help. Problem solving support technology is not as popular in Web-based systems as in stand-alone intelligent tutoring systems due to its difficult implementation problems [7]. Examples of this type of ITS are [1, 20, 33, 37, 42, 53, 58].

10.3 Bayesian Networks

Bayesian networks [48] are a formal framework for uncertainty management in artificial intelligence. Bayesian networks have been successfully employed by many leading technology companies such as Microsoft [25, 27, 38], NASA [24, 26], Hewlett Packard [4, 50] and Nokia [29]. In this section, we discuss Bayesian networks and probabilistic inference therein.

Let $U = \{v_1, v_2, \ldots, v_n\}$ denote a finite set of discrete random variables. Each variable v_i is associated with a finite domain, denoted $dom(v_i)$, representing the values v_i can take on. For a subset $X \subseteq U$, we write $dom(X)$ for the Cartesian product of the domains of the individual variables in X. Each element $x \in dom(X)$ is called a *configuration* of X. A *joint probability distribution* [54] is a function p on $dom(U)$, such that the following two conditions hold:

(i) $0 \leq p(u) \leq 1.0$, for each configuration $u \in dom(U)$,
(ii) $\sum_{u \in dom(U)} p(u) = 1.0$.

Clearly, it may be impractical to obtain the joint probability distribution on U directly: for example, one would have to specify $2^n - 1$ entries for a distribution over n binary variables.

A *Bayesian network* [48] is a pair $\mathcal{B} = (D, C)$. In this pair, D is a *directed acyclic graph* (DAG) on U. The directed edges in the graph reflect the dependencies that hold among the variables. $C = \{p(v_i|P_i) \mid v_i \in D\}$ is the corresponding set of *conditional probability tables* (CPTs), where P_i denotes the parents of each variable v_i in D. A CPT $p(v_i|P_i)$ has the property that for each configuration of the variables in P_i, the sum of the probabilities of v_i is 1.0.

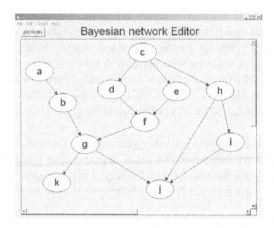

Fig. 10.2. The DAG of the coronary heart disease Bayesian network in Example 1

Example 1. One real-world Bayesian network for *coronary heart disease* [21] is the directed acyclic graph in Fig. 10.2 together with the corresponding CPTs in Fig. 10.3. Note that some minor modifications were made for this Bayesian network. In Fig. 10.3, the missing conditional probabilities can be obtained by the property of a CPT, for instance, $p(a = 0) = 0.504$ and $p(b = 0|a = 0) = 0.948$.

The DAG D graphically encodes *probabilistic conditional independencies* [62]. Based on the probabilistic conditional independencies encoded in D, the product of the CPTs in C is a joint probability distribution $p(U)$. For instance, the independencies encoded in the DAG of Figure 10.2 indicate that the product of the CPTs in Figure 10.3 is a joint probability distribution on $U = \{a, b, c, d, \ldots, k\}$, namely,

$$p(U) \;=\; p(a) \cdot p(b|a) \cdot p(c) \cdot p(d|c) \cdot \ldots \cdot p(k|g).$$

Bayesian networks, thus, provide a semantic and concise modelling tool for modelling uncertainty in complex domains. For example, specifying the joint probability distribution $p(U)$ directly involves stating 2047 prior probabilities ($2^{11} - 1$ for 11 binary variables), whereas by the DAG in Fig. 10.2, only 30 conditional probabilities need be given as shown in Fig. 10.3.

Another important notion we need to discuss is ancestral number. A numbering of the variables in a DAG is called *ancestral* [13], if the number corresponding to any vertex is lower than the numbers corresponding to all of its children. For example, recall the DAG in Fig. 10.2. One ancestral numbering of these variables is $a = 1, b = 2, c = 3, \ldots, k = 11$. We now turn our attention to probabilistic inference in the next discussion.

The main purpose of Bayesian networks is for probabilistic reasoning. Probabilistic reasoning, namely, processing queries, simply means computing $p(V)$ or $p(V|E = e)$, where where V and E are disjoint subsets of U. The *evidence* in the latter query is that the set E of variables is observed to be configuration e, $e \in dom(E)$. When we enter the evidence and use it to update the probabilities, we call it *propagation of evidence*, or

a	$p(a)$	c	$p(c)$	d	e	f	$p(f\|d,e)$	b	f	g	$p(g\|b,f)$	g	h	i	j	$p(j\|g,h,i)$
1	0.496	1	0.577	0	0	1	0.710	0	0	1	0.027	0	0	0	1	0.178
				0	1	1	0.193	0	1	1	0.123	0	0	1	1	0.565

a	b	$p(b\|a)$	c	d	$p(d\|c)$	d	e	f	$p(f\|d,e)$	b	f	g	$p(g\|b,f)$	g	h	i	j	$p(j\|g,h,i)$
0	1	0.052	0	1	0.714	1	0	1	0.485	1	0	1	0.898	0	1	0	1	0.446
1	1	0.358	1	1	0.627	1	1	1	0.602	1	1	1	0.405	0	1	1	1	0.729
														1	0	0	1	0.931
														1	0	1	1	0.582

c	e	$p(e\|c)$	c	h	$p(h\|c)$	h	i	$p(i\|h)$	g	k	$p(k\|g)$	g	h	i	j	$p(j\|g,h,i)$
0	1	0.383	0	1	0.214	0	1	0.104	0	1	0.593	1	1	0	1	0.403
1	1	0.286	1	1	0.651	1	1	0.369	1	1	0.416	1	1	1	1	0.222

Fig. 10.3. The CPTs of the coronary heart disease Bayesian network in Example 1: $p(a)$, $p(b|a)$, $p(c)$, $p(d|c)$, $p(e|c)$, $p(f|d,e)$, $p(g|b,f)$, $p(h|c)$, $p(i|h)$, $p(j|g,h,i)$ and $p(k|g)$.

simply *propagation*. There are many probabilistic inference algorithms [31, 35, 54, 65] for processing queries which seem to work well in practice. There are also numerous implementations of Bayesian network software [28].

10.4   A Bayesian Intelligent Tutoring System

In this section, we provide a review of our recent work on developing a *Bayesian intelligent tutoring system*, called BITS, for computer programming.

10.4.1   General Architecture of BITS

Here, we outline the major components of BITS and describe how they interact with each other.

To simplify the task of developing an intelligent tutoring system, we restrict the scope of the problem as follows: firstly, the system is built to tutor students in the C++ programming language; secondly, only elementary topics are covered, such as those typically found in introductory courses on programming. These topics include concepts such as variables, assignments, and control structures, while more sophisticated topics like pointers and inheritance are excluded.

As BITS is designed, we adapt the common framework of the ITS discussed in Section 10.2.1 and divide the user interface module into two interface sub-modules, the input sub-module and output sub-module. We also separate the adaptive guidance module (this module corresponds with the pedagogical model component in the general framework of an ITS in Section 10.2.1) into three sub-modules: Navigation Support, Prerequisite Recommendations, and Generating Learning Sequences. The student model is implemented by the Bayesian network approach.

Fig. 10.4 shows the overall architecture of BITS. Four modules contained in BITS are: (i) Bayesian networks, (ii) the knowledge base, (iii) the user interface, and (iv) the adaptive guidance module. Details of each component will be explained and examined in the following sections.

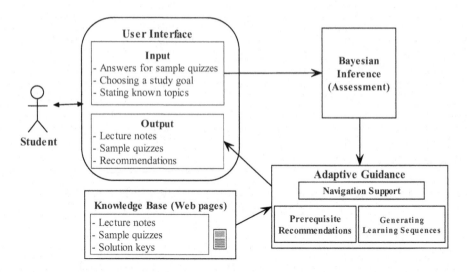

Fig. 10.4. General architecture of BITS

10.4.2   Bayesian Networks in BITS

There are two tasks involved in helping a student navigate in a personalized Web-based learning environment: firstly, the structure of the problem domain must be modelled; secondly, student knowledge regarding each concept in the problem domain should be tracked. Bayesian networks can help us meet both these objectives.

For our purposes, we have identified a set of concepts that are taught in CS110, the introductory computer programming course at the University of Regina. Each concept is represented by a node in the graph. There exist learning dependencies among knowledge concepts, namely, prerequisite relationships. Using a Bayesian Network, the prerequisite relationships among the concepts can be represented directly and clearly. We add a directed edge from one concept (node) to another if knowledge of the former is a prerequisite for understanding the latter. Thus, the DAG can be constructed manually with the aid of the textbook for our CS110 course [16], and it encodes the proper sequence for learning all the concepts in the problem domain.

Example 2. Consider the following instance of the "For Loop" construct in C++:

```
for(i=1; i<=10; i++).
```

To understand the "For Loop" construct, one must first understand the concepts of "Variable Assignment" (i=1), "Relational Operators" (i<=10), and "Increment/ Decrement Operators" (i++). These prerequisite relationships can be modelled as the DAG depicted in Fig. 10.5.

Fig. 10.5 depicts a small portion of the entire DAG implemented in BITS. The entire DAG implemented in BITS consists of 29 nodes and 43 edges. Due to the large size of the DAG, we do not show the whole Bayesian network implemented in BITS in this chapter.

The next task in the construction of the Bayesian network is to specify a CPT for each node given its parents.

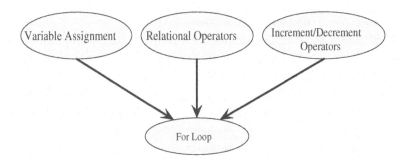

Fig. 10.5. Modelling the prerequisite concepts of the "For Loop" construct

All CPTs for the DAG are obtained from the results of previous CS110 final exams. Firstly, we identify the concepts being tested by each question. Normally, these exam questions consist of multiple choice or filling-in-the-blank. If the student answers the question correctly, then we considered the concept *known*. Similarly, if the student answers the question incorrectly, then the concept is *unknown* (*not known*). The probability of each concept being known, namely $p(v_i = known)$, can then be determined. Moreover, we can compute $p(v_i = known, P_i = known)$ (i.e., the probability that the student correctly answers both the concept v_i and the prerequisite concepts P_i). From $p(v_i = known, P_i = known)$, the desired CPTs $p(v_i = known|P_i = known)$ can be obtained through the following equation:

$$p(v_i = known|P_i = known) \quad = \quad \frac{p(v_i = known, P_i = known)}{p(P_i = known)}.$$

While it is acknowledged that the accuracy of calculating the CPTs in this fashion is arguable [15], this approach to CPT acquisition is the only available way that we currently have. Thereby, by this approach, we calculate every CPT for the entire Bayesian network.

Example 3. Recall the node v_i= For Loop with the parent set P_i={ Variable Assignment, Relational Operators, Increment/Decrement Operators } depicted in Fig. 10.5. A CPT p (For Loop | Variable Assignment, Relational Operators, Increment/Decrement Operators) is shown in Fig. 10.6, where the missing conditional probabilities can be computed by the definition of the CPT.

10.4.3 Knowledge Base

The knowledge base contains the class lecture notes in the form of Web pages, a repository of sample tests (which are in the form of interactive flash multimedia files), and solution keys. The class lecture notes are displayed while the user learns a new concept. On the contrary, a sample quiz is displayed when BITS is trying to determine whether or not a student has understood a particular concept.

All course materials, including both lecture notes and quizzes, are organized by knowledge concepts for C++ programming that correspond to the nodes in the Bayesian

Variable Assignment	Relational Operators	Incre/Decrement Operators	For Loop	$p(F|V,R,I)$
known	known	known	known	0.75
known	known	not known	known	0.39
known	not known	known	known	0.50
not known	known	known	known	0.50
known	not known	not known	known	0.22
not known	known	not known	known	0.29
not known	not known	known	known	0.40
not known	not known	not known	known	0.15

Fig. 10.6. The CPT corresponding to the "For Loop" node in Fig. 10.5

network. This allows the concepts to be indexed and retrieved efficiently. The class lecture notes are displayed while the user is learning a new concept. On the contrary, a sample quiz is displayed when BITS is trying to determine whether or not a student has understood a particular concept.

10.4.4   User Interface Module

A student interacts with BITS through the user interface module. This interaction is partitioned into two sub-modules; an input module for input from a student to BITS, and an output module for output from BITS to a student.

The output module displays the class lecture notes through a Web browser. It uses dialog boxes to display quizzes and offer pedagogical suggestions.

The primary goal of the input module is to update the Bayesian network based on evidence collected from the student. There are two ways in which the student can enter information into BITS:

(i)   A student's direct reply to a BITS query if this student knows a particular concept.
(ii)   A sample quiz result for the corresponding concept to determine whether or not a student has understood a particular concept. Although a quiz question may pertain to several C++ topics, we subjectively associated each question with the most relevant topic.

After the student has finished reading the displayed lecture notes, BITS will ask the student to select one of the following three choices: I understand this concept; I don't understand this concept; I'm not sure (quiz me). These choices are illustrated in the bottom right corner of Fig. 10.7.

If either of the first two choices is selected, then the evidence is obtained under the above-mentioned method (i). The Bayesian network can be immediately updated to reflect the student's knowledge, or his/her lack thereof. On the other hand, the last choice falls under method (ii). BITS will retrieve the appropriate quiz from the knowledge base and present it to the student. The quizzes include: *True/False*, *Multiple Choice* and *Fill-in-the-blank* questions.

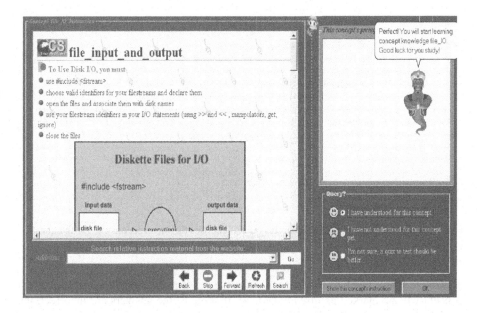

Fig. 10.7. A screen shot of BITS displaying the lecture notes and querying whether this concept is understood for the concept "File I/O"

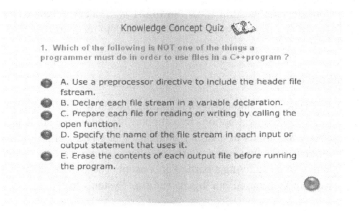

Fig. 10.8. One question on a sample quiz for the concept "File I/O" in Fig. 10.7

Example 4. Suppose the student selects the last choice in Fig. 10.7 to indicate that he/she is not sure whether he/she understands the concept "File I/O." BITS displays the quiz on "File I/O" in Fig. 10.8.

After the student inputs the answer, BITS compares the answer with the solution key. The student is then informed by immediate feedback whether the answer is correct or not. If the student answers all questions correctly in the sample quiz, the Bayesian

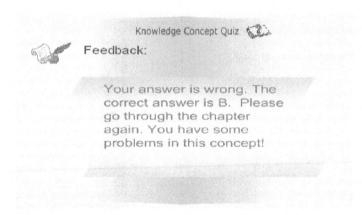

Fig. 10.9. The feedback of BITS when the student chooses the incorrect answer for the question in Fig. 10.8

network is updated and the Navigation Menu is again displayed (but this time the current concept will be marked as *already known*). If some of the answers are incorrect, the correct answer is displayed, and BITS also recommends that the student review the learning material on the current concept again. The Bayesian network is also updated accordingly.

Example 5. Consider the question shown in Fig. 10.8, the correct answer is "*E*". If a student chooses an incorrect answer, say "*A*", the student receives the feedback shown in Fig. 10.9.

10.4.5 Adaptive Guidance

Using the state of the Bayesian network regarding the knowledge of the student, BITS can offer tailored pedagogical options that support the individual student. In this section, we describe three kinds of adaptive guidance that BITS can provide: *navigation support*, *prerequisite recommendations* for problem solving, and *the generation of a learning sequence* when studying a particular concept. These three adaptation methods are current issues in Web-based Intelligent Tutoring Systems [6, 7].

Navigation Support. The navigation menu is used to navigate through the concepts under consideration.

In order to help the student browse the materials, BITS marks each concept with an appropriate traffic light. These traffic lights are computed dynamically from the Bayesian network and indicate the student's knowledge regarding these topics.

In BITS, each concept is marked as belonging to one of the following three categories:

  (i)   already known,
  (ii) ready to learn,
 (iii) not ready to learn.

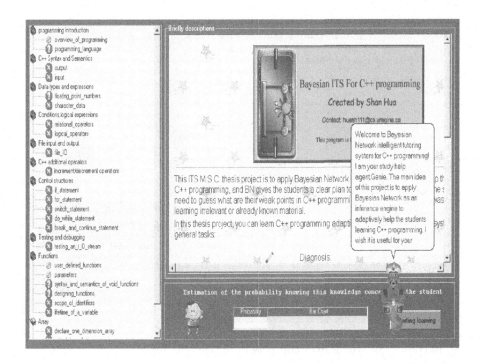

Fig. 10.10. Entry page of BITS with navigation menu (left frame), where green means *ready to learn*, yellow means *already known*, red means *not ready to learn*, and a brief introduction to BITS is provided (right frame)

A concept is considered "already known," if the Bayesian network indicates the probability $p(concept = known|evidence)$ is greater than or equal to 0.70, where evidence is the student's knowledge on previous concepts obtained indirectly from quiz results or directly by the student replying to a query from BITS (see Section 10.4.4). It should be noted that the choice of 0.70 to indicate a concept is already known is subjective. A concept is marked "ready to learn," if the probability $p(concept = known|evidence)$ is less than 0.70 and all of the parent concepts are "already known." Finally, a concept is labelled "not ready to learn," if at least one parent concept is not "already known." Traffic lights are employed as follows: *yellow* (already known), *green* (ready to learn), and *red* (not ready to learn).

When BITS is first started, the concepts are marked with traffic lights based on the initial probabilities obtained from the Bayesian network. The opening screen-shot of BITS is depicted in Fig. 10.10. The navigation menu appears on the left, while a brief introduction to BITS is shown on the right.

A student can highlight a "ready to learn" concept (marked with the green traffic light) and press the start learning button to study this topic. The lecture notes for this topic are retrieved from the database and displayed for the user.

Example 6. Recall the initial state of the navigation menu shown in Figure 10.10. If the student selects the *ready to learn* concept "Floating-point numbers," then BITS displays

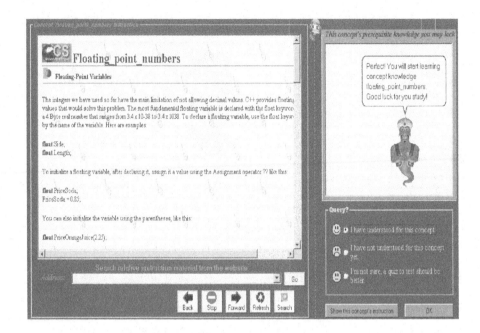

Fig. 10.11. A screen shot of BITS displaying the lecture notes for the concept "Floating-point numbers"

the lecture notes shown in Figure 10.11. After reading the notes, the student selects one of the choices at the bottom right of Figure 10.11. If the student indicates that she understands "Floating-point numbers," then the Bayesian network is updated and the navigation menu is again displayed (this time, however, the concept "Floating-point numbers" is labelled as *already known*).

Prerequisite Recommendations. After reading the lecture notes, the student may indicate that the concept is not understood either by directly answering a query or indirectly through incorrect answers on the corresponding quiz (see Section 10.4.4). In such situations, BITS is designed to present links to prerequisite concepts of this topic, such as the links to each concept in the parent set of the variables in the Bayesian network.

Example 7. Suppose that, after reading the lecture notes for the concept "For Loop," the student indicates that she has not understood. BITS then determines the parent set of the "For Loop" node (i.e., Variable Assignment, Relational Operators, and Increment/Decrement Operators as shown in Fig. 10.5). Finally, BITS displays the links to the lecture notes for these three concepts, as illustrated in the top right corner of Fig. 10.12.

Instead of methods that repeat the problem concept over and over, this approach is useful because it provides the flexibility to revisit the prerequisite concepts and confirm they are indeed understood. The rationale behind our method is that a student may believe that a prerequisite concept is understood when, in fact, it is not, and lacking prerequisites usually affects the student's learning performance.

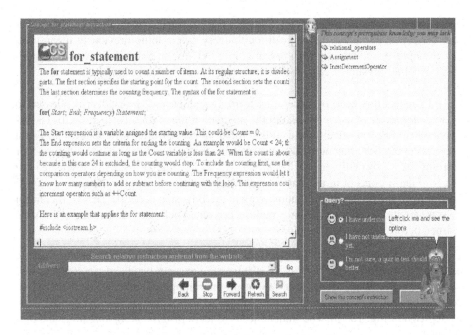

Fig. 10.12. A screen shot of BITS displaying the links to the lecture notes for the prerequisite concepts of "For Loop"

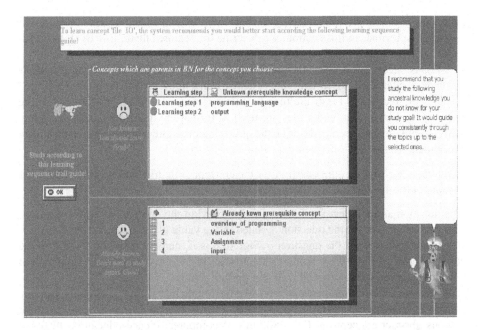

Fig. 10.13. BITS generates a learning sequence for the concept "File I/O" in Fig. 10.10, which is *not ready to learn*

Generating Learning Sequences. Students may sometimes want to learn one particular concept without having to go through every single topic previously mentioned. For example, a student may wish to learn "File I/O" for an impending exam or assignment deadline. In such a case, the student would want to learn a minimum number of concepts.

BITS meets this need by generating learning sequences. The student is allowed to select a *not ready to learn* concept in the navigation menu. In this situation, BITS displays a learning sequence for the chosen topic. In other words, all unknown ancestral concepts in the Bayesian network are revealed to the student in one proper sequence for learning. More formally, let $X = \{x_1, x_2, \ldots, x_n\}$ be the ancestors of a selected *not ready to learn* concept in the Bayesian network. BITS will display the concepts in X according to the fixed ancestral numbering in the Bayesian network. That is, if concept x_i is a parent of concept x_j, then x_i will be presented before x_j, indicating x_i must be learned before x_j can be learned.

Example 8. Suppose the student selects the *not ready to learn* concept "File I/O" in the navigation menu of Fig. 10.10; BITS displays the ancestral concepts in order, grouping by *known* and *unknown* (overview of programming, variable, Assignment, input, marked by *known*, where programming language and output are marked by *unknown*, as depicted in Fig. 10.13). The student needs to learn *programming language*, and *output* first.

10.5   BITS Via the Web

All of the course material, including lecture notes, examples and quizzes, is stored in hypermedia format. In this section, we describe the features that allow BITS to be accessed via the Web.

BITS is interactive because the quizzes used to test the student are not static texts as in other regular Web-based learning systems, rather they are interactions. Each quiz consists of interactive Flash multimedia files together with XML documents. More specifically, Flash multimedia files are used to format the questions displayed, while XML documents are used to describe the quiz contents, store solution keys and validate the student's answer.

Example 9. Recall one quiz question for the concept "File I/O" in Fig. 10.8. The XML document for this question is shown Fig. 10.14.

The question and the answer choices are described in the contents of XML elements. The correct answer for the question is stated in the value of the XML attribute *answer*. The correct answer for the question in Fig. 10.8 is E, namely, $answer = "E"$ in the top left corner of Fig. 10.14. This facility allows BITS to provide dynamic validation of the input answer and to proceed with appropriate action.

BITS uses HTML Web pages to represent the online instructional materials. In some cases, multimedia examples using animated Flash concepts are utilized to illustrate various abstract concepts of C++. Surveys of computer science educators suggest a widespread belief that visualization technology and multimedia play a positive and important role in student learning [51]. Multimedia provides a student with vivid images or

```
<MainElement>
    <Question answer="E"> Which of the following is  NOT one  of the things
        a programmer must do in order to use files in a C++ program?
    <choices>
        <Items> Use a preprocessor directive to include the header file fstream.
        </Items>
        <Items> Declare each file stream in a variable declaration.</Items>
        <Items> Prepare each file for reading or writing  by calling the
                open function.  </Items>
        <Items> Specify the name of the file stream in  each input or output
                statement that uses  it. </Items>
        <Items> Erase the contents of each output file before running the
                program.</Items>
    </choices>
    </Question>
</MainElement>
```

Fig. 10.14. The XML file for the question on the concept "File I/O" in Fig. 10.8

procedures instead of abstract and rigid concepts, and it improves the student's learning efficiency [51].

The lecture notes and quiz questions are displayed using a Web browser embedded in BITS. BITS also provides the student with the ability to access other C++ programming sites on the Web. When studying a new concept, the student can utilize the web browser provided by BITS to search external resources; BITS also recommends a set of URL links to external Websites.

Example 10. As illustrated in the bottom left corner of Fig. 10.11, BITS allows the user to access C++ sites found on the Web.

10.6   Related Work

Recently, there have been several efforts to apply Bayesian networks in many Web-based tasks, including e-commerce [32, 52, 64], information retrieval [12, 17, 18], intelligent agents [36, 61] and student monitoring [14, 15, 33, 37, 40].

Wang et al. [61] proposed a Bayesian network trust model for a file sharing peer-to-peer application on the Web. Since trust is multi-faceted, agents need to develop differentiated trust in different aspects of other agents' behaviors [61]. Thus, an agent needs to consider its trust in a specific aspect of another agent's capability or in a combination of multiple aspects. Bayesian networks provide a flexible method to present differentiated trust and combine different aspects of trust.

In [36], a conversational agent was implemented that utilizes the user model constructed on Bayesian network by considering the users' goal. The agent is applied to the active guide of a Website, which shows that the user modeling based on Bayesian network helps to respond to users' queries appropriately with their goals.

Ji et al. [32] proposed an intelligent electronic shopping system based on Bayesian customer modeling. This system can predict the requirements of customers and provide them with individual product information actively.

In the context of Web-based tutoring systems [14, 15, 33, 37, 40], most of them involve applying Bayesian networks to help a student with problem solving support (refer to Section 10.2.2). For example, Conati et al. [15] developed an intelligent tutoring system, Andes, for physics. Bayesian networks are used to identify which solution the student is working toward. If the student appears stuck in one step, the system provides a hint, telling the student in which direction the next step should go. Furthermore, Liu et al. [37] developed an intelligent system for assisting a user in solving a problem. The primary objective of those systems is to help the student learn how to *problem solve* by providing appropriate error feedback and updating the student model. Unlike [14, 15, 33, 37, 40], the major purpose of BITS is to assist a student in *navigation* through the online materials. Although problem solving is an integral part of computer programming, it is outside the focus of BITS.

Villano [60] first suggested applying Bayesian networks in ITSs. However, Martin and Vanlehn [41] explicitly state that Villano's assessments cannot communicate precisely what a student does not know and cannot identify the components of knowledge that must be taught. BITS, on the other hand, uses yellow traffic lights to indicate known concepts, green traffic lights to indicate ready to learn concepts, and red traffic lights to indicate concepts that the student is not ready to learn.

The *assessment* system proposed by Martin and VanLehn [41] is focused solely on assessing what a student knows. Our *intelligent tutoring system* not only assesses what a student knows, but, in addition, assists the student in navigating the unknown concepts.

It is worth mentioning that Jameson [30] reviewed several frameworks for managing uncertainty in ITSs, including Bayesian networks, the Dempster-Shafer theory of evidence, and fuzzy logic. As Pearl [48] has shown, Bayesian neworks have certain advantages over the other two frameworks. This influenced our decision to use Bayesian networks for uncertainty management in BITS.

10.7   Conclusions

In this chapter, we provide a review of our recent work [9, 10, 11] on developing a Web-based Bayesian ITS, called BITS, for computer programming. Our system takes full advantage of Bayesian networks, which are a proven framework for uncertainty management in artificial intelligence [48, 62]. We describe the architecture of BITS and describe the functionality of each component. In addition, we discuss in detail how to use a Bayesian network in BITS for modelling and inference purposes to guide student study.

Recently, there have been many efforts to employ Bayesian networks in various Web-based applications. For instance, they have been applied in building intelligent agents [36, 61], e-commerce [32, 52, 64], information retrieval [12, 17, 18] and student modelling [14, 15, 33, 37, 40]. In the context of Web-based tutoring systems, Liu et al. [37] developed an intelligent system for assisting a user in solving a problem. Moreover, Conati et al. [15] developed an intelligent tutoring system, Andes, for physics. The primary objective of those systems is to help the student learn how to *problem solve*, while the focus of BITS is quite different; it is to help the student *navigate* the course material. Applying Bayesian networks in Intelligent Tutoring Systems to assess the student's knowledge level was first suggested by Villano in 1992 [60]. However, Villano's assessments cannot communicate precisely what a student does not know and cannot identify the components of knowledge that must be taught [41]. BITS, on the other hand, marks each concept with an appropriate traffic light to indicate the student's knowledge regarding these topics and helps the student select appropriate instructional content to study (see Section 10.4.5). The *assessment* system proposed by Martin and Vanlehn [41] is focused solely on assessing what a student knows (see Section 10.6). Our *intelligent tutoring system* not only assesses what a student knows, but also helps the student navigate unknown concepts (see Section 10.4.5).

Currently, most Web-based learning systems are static Web pages and are simply a copy of regular textbooks. These forms of instructional materials suffer from two major shortcomings [5]: they are not interactive, since students can only passively read the educational materials; they are not adaptive (see Section 10.1). Unlike many traditional tutoring systems which are not interactive nor adaptive [5], BITS is intelligent. It can help a student navigate the online course material using traffic lights. It can recommend learning goals when a particular concept is not understood. Finally, when a student wants to learn a concept without learning all of the previous concepts, BITS can present the minimum prerequisite knowledge needed to understand the desired concept in the proper learning sequence. BITS has been implemented and was recently used in the initial computer programming course at the University of Regina. Empirical studies have shown that individual one-on-one tutoring is the most effective mode of teaching and learning, and intelligent tutoring systems uniquely offer a technology to implement computer-assisted one-on-one tutoring [3]. This research work here, together with [12, 14, 15, 17, 18, 32, 33, 36, 37, 40, 52, 61, 64], demonstrates the practical usefulness of Bayesian networks in Web-based applications.

References

1. Anderson, J.R., Reiser, B.J.: The LISP tutor. Byte 10(4), 159–175 (1985)
2. Angelides, M., Garcia, I.: Towards an intelligent knowledge based tutoring system for foreign language learning. Journal of Computing and Information Technology 1, 15–28 (1993)
3. Bloom, B.: The 2 sigma problem: The search for methods of group instruction as effective as one-to-one tutoring. Educational Researcher 13(6), 4–16 (1984)
4. Bronstein, A., Das, J., Duro, M., Friedrich, R., Kleyner, G., Mueller, M., Singhal, S.: http://www.hpl.hp.com/techreports/2001/HPL-2001-23R1.pdf (April 25, 2005)
5. Brusilovsky, P.: Adaptive and intelligent technologies for Web-based education. Special Issue on Intelligent Systems and Teleteaching 4, 19–25 (1999)
6. Brusilovsky, P., Maybury, M.T.: From adaptive hypermedia to adaptive Web. Communications of the ACM, Special Issue on the Adaptive Web 45(5), 31–33 (2002)
7. Brusilovsky, P., Peylo, C.: Adaptive and intelligent Web-based educational systems. International Journal of Artificial Intelligence in Education, Special Issue on Adaptive and Intelligent Web-based Educational Systems 13(2-4), 159–172 (2003)

8. Burns, H.L., Capps, C.G.: Foundations of intelligent tutoring systems. In: Foundations of Intelligent Tutoring Systems, ch. 3, pp. 55–78. Lawrence Erlbaum Associates, Mahwah (1988)
9. Butz, C.J., Hua, S., Maguire, R.B.: Bits: a Bayesian Intelligent Tutoring System for Computer Programming. In: Proc. of the 9th Western Canadian Conference on Computing Education, pp. 179–186 (2004)
10. Butz, C.J., Hua, S., Maguire, R.B.: A Web-based Intelligent Tutoring System for Computer Programming. In: Proc. of the IEEE/WIC/ACM Conference on Web Intelligence, pp. 159–165 (2004)
11. Butz, C.J., Hua, S., Maguire, R.B.: A Web-based Bayesian Intelligent Tutoring System for Computer Programming. Web Intelligence and Agent Systems: An International Journal 4(1), 77–97 (2006)
12. Calado, P., da Silva, A.S., Laender, A.H.F., Ribeiro-Neto, B.A., Vieira, R.C.: A Bayesian network approach to searching Web databases through keyword-based queries. Information Processing and Management: an International Journal 40(5), 773–790 (2004)
13. Castillo, E., Gutierrez, J., Hadi, A.: Expert systems and Probabilistic Network Models. Springer, Heidelberg (1997)
14. Collins, J.A., Greer, J.E., Huang, S.X.: Adaptive assessment using granularity hierarchies and Bayesian nets. In: Proc. of the 3rd International Conference on Intelligent Tutoring Systems, pp. 569–577 (1996)
15. Conati, C., Gertner, A., Vanlehn, K.: Using Bayesian networks to manage uncertainty in student modeling. User Modeling and User-Adapted Interaction 12(4), 371–417 (2002)
16. Dale, N., Weems, C., Headington, M.: Programming and problem solving with C++. Jones and Bartlett Publishers (2000)
17. de Campos, L.M., Fernández, J.M., Huete, J.F.: Bayesian networks and information retrieval: an introduction to the special issue. Information Processing and Management: an International Journal 40(5), 727–733 (2004)
18. Fung, R., Favero, B.D.: Applying Bayesian networks to information retrieval. Communications of the ACM 38(3), 42–57 (1995)
19. Georgouli, K.: Modelling a versatile mathematical curriculum for low-attainers. In: Proc. of the 8th Panhellenic Conference in Informatics, pp. 463–472 (2001)
20. Gertner, A., Conati, C., Vanlehn, K.: Procedural help in Andes: generating hints using a Bayesian network student model. In: Proc. of 15th National Conference on Artificial Intelligence, pp. 106–111 (1998)
21. Hájek, P., Havránek, T., Jiroušek, R.: Uncertain Information Processing in Expert Systems. CRC Press, Ann Arbor (1992)
22. Heift, T., Nicholson, D.: Web delivery of adaptive and interactive language tutoring. International Journal of Artificial Intelligence in Education 12(4), 310–324 (2001)
23. Henze, N., Nejdl, W.: Adaptation in open corpus hypermedia. International Journal of Artificial Intelligence in Education 12(4), 325–350 (2001)
24. Horvitz, E., Srinivas, S., Rouokangas, C., Barry, M.: A decision-theoretic approach to the display of information for time-critical decisions: The Vista project. In: Proc. of Conference on Space Operations Automation and Research (1992)
25. Horvitz, E.: Agents with beliefs: reflections on Bayesian methods for user modelling. In: Proc. of the 6th International Conference on User Modeling, pp. 441–442 (1997)
26. Horvitz, E., Barry, E.M.: Display of information for time critical decision making. In: Proc. of the 18th Conference on Uncertainty in Artificial Intelligence, pp. 296–305 (1995)
27. Horvitz, E., Breese, J., Heckerman, D., Hovel, D., Rommelse, K.: The Lumiere project: Bayesian user modeling for inferring the goals and needs of software users. In: Proc. of the 14th Conference on Uncertainty in Artificial Intelligence, pp. 256–265 (1998)
28. http://www.ai.mit.edu/~murphyk/Bayes/bnsoft.html (April 28, 2004)

29. http://www.nokia.com/nokia/0,,53720,00.html (April 25, 2005)
30. Jameson, A.: Numerical uncertainty management in user and student modelling: an overview of systems and issues. User Modelling and User-Adapted Interaction 5(3-4), 193–251 (1995)
31. Jensen, F.V., Lauritzen, S.L., Olesen, K.G.: Bayesian updating in causal probabilistic networks by local computations. Computational Statistics Quarterly 4, 269–282 (1990)
32. Ji, J., Zheng, L., Liu, C.: The intelligent electronic shopping system based on Bayesian customer modeling. In: Proc. of the 1st Asia-Pacific Conference on Web Intelligence, pp. 574–578 (2001)
33. Johnson, W.L.: Pedagogical agents for Web-based learning. In: Proc. of 1st Asia-Pacific Conference on Web Intelligence, pp. 43–44 (2001)
34. Koedinger, K.R., Anderson, J.R., Hadley, W., Mark, M.: Intelligent tutoring goes to school in the big city. International Journal of Artificial Intelligence in Education 8, 30–43 (1997)
35. Lauritzen, S.L., Spiegelhalter, D.J.: Local computations with probabilities on graphical structures and their application to expert systems. Journal of Royal Statistical Society, Series B 50(2), 157–224 (1988)
36. Lee, S., Sung, C., Cho, S.: An effective conversational agent with user modeling based on Bayesian network. In: Proc. of the 1st Asia-Pacific Conference on Web Intelligence, pp. 428–432 (2001)
37. Liu, C., Zheng, L., Ji, J., Yang, C., Yang, W.: Electronic homework on the WWW. In: Proc. of the 1st Asia-Pacific Conference on Web Intelligence, pp. 540–547 (2001)
38. Lumière Project: Bayesian Reasoning for Automated Assistance (April 27, 2005),
    http://research.microsoft.com/~horvitz/lum.htm
39. Magoulas, G.D., Papanikolaou, K.A., Grigoriadou, M.: Neuro-fuzzy synergism for planning the content in a Web-based Course. Informatica 25, 39–48 (2001)
40. Martin, J., VanLehn, K.: A Bayesian approach to cognitive assessment, Cognitively Diagnostic Assessment, pp. 141–165 (1995)
41. Martin, J., VanLehn, K.: Student assessment using Bayesian nets. International Journal of Human-Computer Studies 42, 575–591 (1995)
42. Mitrovic, A.: An intelligent SQL tutor on the Web. International Journal of Artificial Intelligence in Education 13(2-4), 171–195 (2003)
43. Murray, T.: Authoring intelligent tutoring systems: an analysis of the state of the art. International Journal of Artificial Intelligence in Education 10, 98–129 (1999)
44. Nakabayashi, K., Koike, Y., Maruyama, M., Touhei, H., Ishiuchi, S., Fukuhara, Y.: An intelligent tutoring system on World-Wide Web: Towards an integrated learning environment on a distributed hypermedia. In: Proc. of the World Conference on Educational Multimedia and Hypermedia, pp. 488–493 (1995)
45. Nkambou, R.: Using fuzzy logic in ITS-course generation. In: Proc. of the 9th IEEE International Conference on Tools With Artificial Intelligence, pp. 190–194 (1997)
46. Nilson, N.: Artificial Intelligence: A New Synthesis. Morgan Kaufmann, San Francisco (1998)
47. Okazaki, Y., Watanabe, K., Kondo, H.: An implementation of an intelligent tutoring system (ITS) on the World-Wide Web (WWW). Educational Technology Research 19(1), 35–44 (1996)
48. Pearl, J.: Probabilistic Reasoning in Intelligent Systems: Networks of Plausible Inference. Morgan Kaufmann, San Francisco (1988)
49. Shiri, M.E., Aimeur, E., Frassen, C.: Student modelling by case-based reasoning. In: Proc. of the 4th International Conference on Intelligent Tutoring Systems. LNCS, pp. 394–404 (1998)
50. Skaanning, C., Jensen, F.V., Kjrulff, U., Parker, L., Pelletier, P., Rostrup-Jensen, L.: (April 25, 2005),
    http://www.cs.auc.dk/research/DSS/papers/skaanning98a.doc

51. Stasko, J., Badre, A., Lewis, C.: Do algorithm animations assist learning? An empirical study and analysis. In: Proc. of the SIGCHI conference on Human factors in computing systems, pp. 61–66 (1993)
52. Robles, V., Lfarrañaga, P., Menasalvas, E., Pérez, M.S., Herves, V.: Improvement of Naive Bayes collaborative filtering using interval estimation. In: Proc. of the 2nd Annual Asia-Pacific Conference on Web Intelligence, pp. 168–174 (2003)
53. Schulze, K.G., Shelby, R.N., Treacy, D.J., Wintersgill, M.C., Vanlehn, K., Gertner, A.: Andes: an intelligent tutor for classical physics. The Journal of Electronic Publishing 6(1) (2000), http://www.press.umich.edu/jep/06-01/schulze.html
54. Shafer, G.: Probabilistic Expert Systems, Society for Industrial and Applied Mathematics (1996)
55. Shute, V.J., Glaser, R.: A large-scale evaluation of an intelligent discovery world: Smithtown. Interactive Learning Environments 1(1), 51–77 (1990)
56. Sleeman, D., Brown, J.S.: Introduction: Intelligent tutoring systems. Intelligent Tutoring Systems, 1–10 (1982)
57. Stern, M., Woolf, B.: Curriculum Sequencing in a Web-based Tutor. In: Proc. of the 4th International Conference on Intelligent Tutoring Systems, pp. 574–583 (1998)
58. Sykes, E.R., Franek, F.: A prototype for an intelligent tutoring system for students learning to program in Java. In: Proc. of the 3rd IEEE International Conference on Advanced Learning Technologies, pp. 485–486 (2003)
59. Vassileva, J.: Dynamic courseware generation. Journal of Computing and Information Technology 5, 87–102 (1997)
60. Villano, M.: Probabilistic student models: Bayesian belief networks and knowledge space theory. In: Proc. of the 2nd International Conference on Intelligence Tutoring System, pp. 491–498 (1992)
61. Wang, Y., Vassileva, J.: Bayesian Network-based trust model. In: Proc. of the 2nd Annual Asia-Pacific Conference on Web Intelligence, pp. 372–378 (2003)
62. Wong, S.K.M., Butz, C.J., Wu, D.: On the implication problem for probabilistic conditional independency. IEEE Transactions on Systems, Man, and Cybernetics, Part A: Systems and Humans 30(6), 785–805 (2000)
63. Weber, G., Brusilovsky, P.: ELM-ART: An adaptive versatile system for Web-based instruction. International Journal of Artificial Intelligence in Education 12(4), 351–384 (2001)
64. Xiang, Y., Ye, C., Stacey, D.A.: Application of Bayesian networks to shopping assistance. In: Proc. of Advances in Artificial Intelligence: 15th Conference of the Canadian Society for Computational Studies of Intelligence, pp. 27–29 (2002)
65. Zhang, N.L., Poole, D.: A simple approach to Bayesian network computations. In: Proc. of the 10th Canadian Conference on Artificial Intelligence, pp. 171–178 (1994)

11

A Survey of Web-Based Collective Decision Making Systems

Jennifer H. Watkins[1] and Marko A. Rodriguez[2]

[1] International and Applied Technology
   Los Alamos National Laboratory
   Los Alamos, New Mexico 87545
   jhw@lanl.gov
[2] Digital Library Research and Prototyping Team
   Los Alamos National Laboratory
   Los Alamos, New Mexico 87545

Abstract. A collective decision making system uses an aggregation mechanism to combine the input of individuals to generate a decision. The decisions generated serve a variety of purposes from governance rulings to forecasts for planning. The Internet hosts a suite of collective decision making systems, some that were inconceivable before the web. In this paper, we present a taxonomy of collective decision making systems into which we place seven principal web-based tools. This taxonomy serves to elucidate the state of the art in web-based collective decision making as well as to highlight opportunities for innovation.

11.1 Introduction

Collective decision making is the aggregation of individuals' information to generate a global solution. There are a variety of reasons that collective decisions are sought. A collective decision may be desirable to represent the opinions of a group, as in a vote. A collective decision may be desirable to collect the best information available, as in expert elicitation. Or a collective decision may be desirable to produce a new combination of ideas held within the group, as in a brainstorm session. The resulting decision may be employed directly or used as decision support for another process. For the purposes of this paper, mechanisms that elicit decisions from a group of people are called collective decision making systems (CDMSs). This designation is used to represent a departure from group decision support systems (a subfield of computer supported collaborative work) as CDMSs are not necessarily collaborative in nature [1]. In addition, this paper refers exclusively to web-based collective decision making systems, often called social software [2]. The unifying purpose of these systems is to structure individual input in such a way as to generate a meaningful aggregate decision, even if that input is implicitly derived or from asynchronous or anonymous contributions.

The human proclivity to decide in a group is long standing. However, web-based tools for collective decision making have advanced this ability and need to a larger scale. In this article, seven types of popular web-based systems are

R. Nayak et al. (Eds.): Evolution of the Web in Artificial Intel. Environ., SCI 130, pp. 243–277, 2008.
springerlink.com

discussed—document ranking, folksonomy, recommender system, vote system, wiki, open source software, and prediction market—within a taxonomy of features. The decision capabilities that determine each type of CDMS are the result of a specific combination of features. These features can be organized into a taxonomy of problem space, implementation, individual features, and collective features. Such a taxonomy serves to distinguish the context under which a particular CDMS can be used and to highlight the similarities between seemingly disparate tools. In addition, this taxonomy reveals combinations within the feature space not utilized by existing systems that could compose a new system and thus a new decision making capability.

The first half of this article describes in detail the history, purpose, and instantiation of the seven types of web-based collective decision making systems in turn. The remainder of the article presents the seven system types within a feature space organized by a taxonomical structure. The aim is to set each system in a broader context while providing a framework to aid in system design. But first, the most fundamental delineation of CDMSs will be outlined—that between the collective and the aggregation mechanism. This dual understanding of CDMSs is the first branch of the taxonomy presented shortly.

All collective decision making systems require a population of participants (i.e., a collective) and a means of aggregating their knowledge into a collective decision (i.e., an aggregation mechanism). For example, deliberation aggregates through conversation; democracy aggregates through voting; a recommender system aggregates through user footprints. The following sections describe these two components.

11.1.1 The Collective

Collective decision making is founded on the belief that people are not flawless decision makers. An individual is a good, but not ideal, complex problem-solver. Collective decision making utilizes a better one, namely the unit of participants. The typical account of decision making involves an expert who applies his or her knowledge to generate a solution. Through collective decision making, however, it is the collective itself that is considered the expert. The collective can be thought of as a meta-individual that possesses, generates, and decides on knowledge in much the same way an individual does. Like an expert, a collective has more knowledge than other individuals through the combination of information held by each member. Collectives are autopoietic, they have continuity in identity despite changes in membership, allowing us to think of them as persistent individuals [3]. Thus, collective decision making is distributed over numerous processes within the collective, as opposed to contained within a single decisive event.

11.1.2 The Aggregation Mechanism

A collective without an aggregator is no more powerful than an individual. An aggregation mechanism serves two purposes in eliciting collective decisions. One, it draws out the pertinent information of each individual in the collective. Two,

Table 11.1. Collective decision making systems and their common aggregation mechanisms

Collective Decision Making System	Aggregation Mechanism
document ranking	PageRank
folksonomy	collaborative tagging
recommender system	collaborative filtering
vote system	plurality
open source software	collaborative development
wiki	collaborative editing
prediction market	market scoring rule

it combines that information in such a way as to make it useful. Every CDMS has a variety of web-based aggregation mechanisms. For example, vote systems may employ approval voting, Borda count, or plurality voting. Table 11.1 lists the CDMSs discussed in this article and their common aggregation mechanisms.

11.2 Web-Based CDMS

We focus exclusively on web-based collective decision making systems, as opposed to, for example, face-to-face decision making. Web-based decision support systems are not only computer mediated but are made powerful through the vast population of individuals that use the Internet. These individuals are utilized by the aggregation mechanisms, either through tracking the combined behavior of many or through scouting for expertise. This online collective provides two potential benefits. One, such a large, dispersed population captures statistical collective intelligence or the generation of knowledge through the weighted averaging of independent, individual judgments [4]. Most of the web-based systems discussed here require these large numbers of self-interested participants to generate an accurate decision. Two, some systems benefit from the ability to amplify expertise. The Condorcet Jury Theorem, from probability theory, states that if each individual in a collective is more likely than not to be correct, then as the size of the group scales, the probability of the collective decision being correct moves toward certainty [5]. Some of the systems discussed discourage participation by those who are not more likely to be correct and thus enjoy the result of this theorem.

Before developing the taxonomy further, the following sections describe the web-based collective decision making systems of interest to this article.

11.2.1 Document Ranking

Document ranking, the system that organizes web-pages for the purpose of document retrieval (the matching of records to queries), uses information inferred from the links between documents. The World Wide Web is a network of web-pages connected through hypertext links. A given web-page becomes embedded in the wider network of web-pages when the person publishing the site creates links to other pages and when the site receives incoming links. Its importance in the web, as determined by its location in the network, produces a ranking. Search engines utilize these rankings to retrieve the most relevant documents in response to keyword queries. Document ranking exploits the aggregate of the individual decisions to link to specific pages, interpreting the resulting network as collective informational content.

Document ranking is the method of information retrieval employed by most popular search engines. PageRank, employed by the Google[1] search engine, is perhaps the most well known of the document ranking algorithms. The PageRank algorithm considers not only the number of incoming links (indegree) to a given web-page, but also the incoming links to the originating web-pages in a recursive manner [6]. Thus, a given page can receive a high ranking through a high indegree or through a single incoming link that itself has a high ranking. The ranking must correspond well with an individual's subjective sense of importance. Search engines achieve this correspondence because the choice to link to a page contains latent human judgement about importance. The structure of the web is determined at the local level when an individual chooses to link to another web-page. Globally, this structure can be interpreted in a variety of ways to inform document retrieval.

PageRank was designed to improve the relevancy of the search results returned. It is superior to text-based ranking functions originally applied to the web that simply relied on a full text keyword search. The web is of such massive scale and amorphous organization that these traditional techniques are infeasible. The harnessing of the collective actions of web-page creators generated an improvement in information retrieval techniques. However, Google uses, as do other search engines, a proprietary combination of criteria to determine document relevancy [7].

While the PageRank algorithm is perhaps the most well-known aggregation mechanism for document ranking, there are a variety of other algorithms that utilize collective decisions for the purpose of information retrieval. For example, the HITS algorithm interprets hypertext links as "conferred authority" [8]. Instead of a single ranking metric, HITS utilizes two measures, hub and authority, along which all web-pages are scored. Due to the semantic understanding individuals encode into network structure through linking pages, a pattern emerges where pages with a high hub score densely link to authoritative, high-quality pages (those with a high authority score) on a given subject. It is this relationship that is exploited by HITS to return precise search results.

Because of the necessity of search engines to locate material on the web, a number of techniques have been developed to falsely inflate the ranking of certain pages. These techniques are known as adversarial information retrieval [9]. The Google bomb is a slang term for the coordinated linking to a particular page with a particular key-word phrase, usually for humorous or political intent. Spamdexing is a complimentary technique used to falsely inflate the ranking of a website in order to increase hits, e.g., for commercial gain. A link farm,

---

[1] URI: http://www.google.com

a specific type of spamdexing involves linking every page in the farm to every other page to increase the rankings of all the pages. The chances for exploitation of search engine results has precipitated search engine optimization, techniques that look to improve the traffic (both volume and quality) to a particular site by orienting it properly for both human and search engine indexing. The result is an algorithmic arms race between search engine companies that strive to maintain relevancy and the search engine optimizers.

Despite the constantly evolving nature of document ranking criteria and algorithms, the essential collective decision core remains the same: individuals create web-pages that link to other web-pages. These links can be perceived as votes of quality. In aggregate, these individual actions sum to an informational content that can be exploited for the purposes of information retrieval. The success of Google's search engine is a testament to the immense utility such unintended footprints can produce. The beauty of document ranking algorithms is that they are able to extract meaning from digitally represented human actions that were made for other purposes. This latent human intent in aggregate forms the data used for our search engines.

11.2.2 Folksonomy

Web-services such as Flickr[2] and Del.icio.us[3] (Del.icio.us is a domain hack for "Delicious" and will hereafter be referred to as such) allow users to label, or tag, resources with descriptive metadata such that the statistical aggregate of all tags creates a collectively designed index, or folksonomy [10]. The folksonomy is used as a tool for information retrieval connecting users to resources via tags. Tagging is the appending of metadata to a resource, most often for the purpose of description. The user tags resources for their own purposes using their own descriptions. Over time, the same resource will be tagged many times and particular tags will be used repeatedly to describe the same resource. This overlap increases the relevance of the tagged resource for retrieval by the tag as a keyword.

The aggregation of many users' tags to create a folksonomy is achieved through a mechanism referred to as collaborative tagging. Through the combination of multiple users' interpretations and thus tags of a particular resource, a folksonomy is generated that indicates the popularity of a particular term to describe a particular resource. Despite the self-interested and uncoordinated actions of the participants, analysis indicates that users' interact with the system through collaborative tagging in a patterned manner to create a coherent tool [11].

The categorization method of folksonomies is in contrast to traditional centralized methods including ontologies, controlled vocabularies, and thesauri [12]. These methods require the careful construction of a world view into which all current and future resources can be placed. Instead of these traditional indexing methods which is an expert-based and time-consuming effort, folksonomies distribute the indexing over a large population of users [13]. In essence, the tagging

---

[2] URI: http://www.flickr.com
[3] URI: http://del.icio.us

of objects by a single person is of less use than the formal classification of those resources by an expert. However, in aggregate the result of many individuals' tags can form a folksonomy that is more robust than traditional methods.

The folksonomy also stands in contrast to newer indexing methods that utilize computer automated crawling of resources, as utilized by search engines [13]. The human indexing provides a semantic understanding of the content of each resource that may not be captured by a web-crawler. Folksonomies more closely resemble traditional human-based classification systems in their ability to understand semantic content, but automated systems in their overhead and cost. Because individuals contribute to the folksonomy primarily for their own benefit, the classification value is merely a by-product of a well-designed system.

In practice, folksonomies are used to describe a variety of resource types, from photos to blog entries. The most common use of folksonomies is to describe the information at a particular URL so that the tagger can find the information again later. This practice is called social bookmarking and is an extension of the bookmarking feature included with most web browsers. Bookmarking began as such with the Mosaic browser. This ability required a hierarchical organization of favorite websites that can quickly become unwieldy with lax management. An increase in the speed and precision of search engines led to dynamic bookmarking where an individual simply searches for a favorite site again. This ability is augmented by social bookmarking which refers to the tagging of a web-page with descriptive metadata for ready retrieval. It is browser-independent and allows users to see how URLs were bookmarked by others and to see the bookmarks of a particular user—thus it is social.

The utility of a folksonomy depends on the duplication of tags. Social bookmarking sites provide a feedback mechanism that encourages the convergence of tags [14]. The Delicious capability to see how other users have tagged a given URL provides feedback that encourages the imitation of others' tags. Thus, early tags of a particular URL are the most popular [15]. The most common depiction of the tags for a particular resource is the tag cloud [16]. A tag cloud depicts an alphabetized list of the tags applied to a given resource. The popularity of the tag, the frequency of its use, is indicated through a relatively larger font size. The presentation of a folksonomy is of primary importance for utility as its clarity aids convergence.

11.2.3   Recommender System

Recommender systems track user behavior, whether implicitly or explicitly, as a means of recommending potentially interesting resources to users in the system. A ranking of the resources not yet seen by a user is produced according to some measure of the user's preferences. The purpose is to filter and organize the overabundance of resources within the system's domain. In other words, recommender systems manage information overload by acting as a search function to provide a personalized subset of the total collection [17]. As one becomes a more finely differentiated individual through interactions with the system, a more individualized filter is developed based on the interactions of other individuals. The purpose is to aid the user in discovering novel and interesting products (i.e., it is primarily a tool implemented for commercial reasons).

The need for a recommender system is based on the typical problem in computer-mediated environments of information overload. Search engines work well when one knows what one is looking for. However, there are situations when this is not the case. Here, a recommender system performs information retrieval without any keyword entry on behalf of the user. Instead, the system infers desires through past interactions with the system.

One class of aggregation mechanism for recommender systems is the suite of collaborative filtering algorithms. Collaborative filtering compares the independent decisions of many users with persistent identity to generate a similarity metric such that users are recommended products they have not accessed but those who are similar to them have [18]. In this sense, the collective decides what will be of interest to the individual. Perhaps the most common technique for establishing similarity is nearest neighbor analysis adapted from pattern recognition research [19]. An alternative content-based algorithm made popular by Amazon.com[4] develops similarity metrics on products instead of users. Products that are similar to a purchased item will be recommended to the user [20].

The persistent use of a particular recommender system is essential to both the individual and the collective as recommendations gain sophistication with more personal- and with more collective-level data. The initial paucity of information with which to infer recommendations is referred to as the cold-start problem [21]. It can become a burden to the user to populate the system with enough information so as to make an accurate recommendation. However, there are a variety of algorithms dedicated to decreasing the impact of this problem so that users will recognize the utility of the system immediately [22, 23]. In addition, a recommender system is a unique web-based application in that it can be implemented so as to work completely unseen to the user. In this effortless instantiation, entry into the collective is automatic when an individual logs in to a site. For example, Amazon.com implicitly tracks user behavior for use in the recommender system. Appropriate recommendations are then inferred from an individual's usage of the site. Other recommender systems require explicit user participation. For example, the Netflix[5] recommender systems requires that the user rate movies they have previously viewed through a simple one through five star interface.

However, the ease of entry into these systems has drawbacks. The use of implicit user tracking technology is seen by some as an invasion of privacy. A user, especially an eclectic user who rates products across many domains, can be identified through the tracking data alone, which can be distributed for use by third parties [24].

---

[4] URI: http://www.amazon.com
[5] URI: http://www.netflix.com

11.2.4   Vote System

The vote system is a time-honored means of gathering individual decisions and aggregating them into a single collective decision. As such, vote systems are the hallmark of democratic governance. The feature space of vote systems has been researched extensively resulting in a large number of aggregation mechanisms. Each mechanism specifies two components — the ballot form and the tally method. The ballot determines the way that an individual can express a decision and the tally method determines how those expressions are aggregated into a collective decision. Common aggregation mechanisms include plurality, Borda count, and approval vote. A plurality vote allows the voter to choose only one option on the ballot. A ballot for a Borda count vote allows all options to be ranked in order of preference by the voter. Points are assigned according to the rank. An approval vote allows the voter to choose as many options as are deemed preferences. For all three mechanisms the majority option wins. Note that in vote systems a majority need not imply more than half of the votes. For multi-winner votes, a tally method other than majority rule may be used, most often for proportional representation.

In addition to the plethora of aggregation mechanisms, there are multiple forms of governance within the democracy designation of which direct democracy and representative democracy may be the most familiar. Regardless of the details of the form of governance, the essential element of a vote system is that it be perceived as fair. Online or offline, vote systems are used to determine collective preference. It is this that sets vote systems apart from other decision systems, as there is no objective measure of accuracy outside of perceived fairness. It is to this end that a wide variety of aggregation mechanisms exist. Each aggregation mechanism elicits votes differently to affect different outcomes to satisfy voters.

The fair transference of individual decisions into a collective decision is studied no more rigorously than in voting systems. The determination of the best systems for aggregating preferences is an important pillar of voting theory literature. To aid in this study, a number of rules have been outlined, all of which are criterion for a fair vote [25]. However, Arrow's General Possibility Theorem proved that there can exist no rank-based vote system between three or more alternatives that will satisfy all fairness requirements [26]. Taken loosely, this implies that in any system individual preferences will fail to aggregate into collective preferences. The result of this theorem is that it is necessary to specify which fairness rules must be met and which can be violated before a vote system is implemented.

While academic interest in vote systems has a long and rich history, the introduction of vote systems to the web is in its infancy. Fields of study such as e-democracy and e-government are increasing the interest in implementing online vote systems. Recognizing the unprecedented potential of the web to facilitate communication on a large and distributed scale, e-democracy embraces the notion of wisdom through collective decisions. This sub-discipline is interested in developing web-based tools that support democratic processes to improve the development of policy [27]. In addition, there are a number of studies proposing the use of a comprehensive system that aids the voter in acquiring information about the candidates, making a decision about the best candidate, and then casting that vote [28, 29]. These studies are designed to facilitate the wider aim of political participation.

E-democracy has met with security and reliability challenges in the development of web-based electronic voting (the casting and/or tallying of votes via the Internet) where it is hoped that more members of society will vote than do in traditional elections. Challenges to developing a secure and reliable system include protecting the secrecy of the vote for each voter, network vulnerabilities, and the appropriate implementation of cryptographic techniques [30]. Nevertheless, direct-recording electronic (DRE) voting machines are used in US elections without paper-based, voter-verfiable copies. While electronic poll site voting and kiosk voting, which are supervised by election officials pose security concerns, the concerns deepen for remote voting via the Internet. It is difficult to guarantee in such situations that the voter is who they say, is not being coerced to vote in a particular manner, or is not selling their vote [31].

Despite the interest in moving political elections online, current online vote systems remain confined almost entirely to polling interfaces where no actual decision is affected. There are some interesting exceptions. Estonia became the first country to implement an online electronic voting system for a national election in 2007, where 30,275 Estonians voted through the system.[6] After the Pentagon ruled the system insufficiently secure to implement for soldiers living overseas, the 2004 Michigan democratic caucus elections had the option of using an electronic voting system to register votes.

While the most obvious application of voting systems is to the execution of political government, a vote system need not be limited by this restriction. With the ease of communication and tally functions, web-based voting systems have the possibility to establish new algorithms and methods for producing fair collective decisions. Smartocracy,[7] a social software voting site, utilizes an online social network to spread voting power to those the voter trusts as proxy, not those elected to represent him or her [32].

11.2.5   Wiki

A wiki is a highly distributed way to gather, create, and share knowledge. A wiki is server software that allows users to freely create and edit web-page content using any web browser [33]. The purpose of a wiki is to capture the collective knowledge held by participants such that the resulting documents transcend the abilities of individual contributors. In a wiki system, any individual can use a simple markup language to create pages and link them to other internal pages. These pages allow content and organizational contributions and edits at any time. Every facet of a wiki web-page, the content, its organization, the links,

---

[6] "Estonia scores world first with web poll," The Age, March 1, 2007 http://www.theage.com.au/news/web/estonia-scores-world-first-with-web-poll/2007/03/01/1172338771317.html

[7] URI: http://smartocracy.net

even its very existence, is alterable by any member of the collective, regardless of original authorship. The result is a network of collaboratively generated documents that contains the authorial wisdom of all its contributors.

A wiki aggregates decisions through the mechanism of collaborative editing. Collaborative editing simply refers to the ability to alter and contribute freely regardless of original authorship. Here a collective produces a document through the melding of asynchronous and independently made individual decisions. Collaborative editing falls under the research of computer-supported cooperative work, a sub-discipline for which wikis are merely one solution [34]. In the past, this sort of group work has required small groups that interact face-to-face. However, through wikis the creation of content takes place on a large-scale and has become a distributed process that often involves strangers. Wikipedia[8] is an online encyclopedia that uses wiki software to allow anyone to contribute. It is a multilingual collaboration to capture the collective's knowledge in encyclopedic format. There are articles written in over 200 languages[9] with the largest collection, the English language version, approaching its 2 millionth article at an exponential rate [35]. Wikipedia has become the de facto source of encyclopedic information on the web with over 50 million queries a day. It is powered using MediaWiki, open source software distributed by the Wikimedia Foundation that is used to host numerous other wikis. Thus, many of the features in Wikipedia are also standard features in other wikis. Despite the predominance of Wikipedia, the wiki system should not be conflated with its most popular example. Wikis are knowledge management and content creation tools that serve projects at multiple scales both public and internal to institutions [36].

Wikipedia, like other social software encourages a sense of community amongst members of the collective. Like most communities, that of Wikipedia divides labor both explicitly through levels of permissions and organically through users identifying a given task and choosing to complete it. This bare bones infrastructure supports over 200,000 edits per day. The community is also governed by a number of rules and guidelines for participation, mostly guiding the type of content that is appropriate. To handle debate and disputes there are talk pages for articles kept separate from the encyclopedic content.

However, the scale and decentralization of Wikipedia leads to inaccuracies, sabotage, vandalism, and exploitation. In order to maintain the open nature of the project, most of these problems are handled by the vigilance of other Wikipedians, a collective-based solution. Other problems require a change in the structure of the wiki itself, an aggregation mechanism solution. For example, some contributors add erroneous links in Wikipedia to another site so as to falsely inflate that site's PageRank through the prestige of Wikipedia. To combat this spam problem, Wikipedia software applies a NOFOLLOW rule to every link on the site. Essentially, this prohibits web-crawling robots from following the links on Wikipedia and inflating the adjacent sites' PageRanks. Despite indicators that Wikipedia would be overcome by malicious and unintentionally poor content, it

---

[8] URI: http://en.wikipedia.org
[9] statistics from http://meta.wikipedia.org/wiki/Statistics accessed July 14, 2007.

remains a viable source of information on the web. In fact, due to the success of the Wikipedia paradigm, it is only one of a multitude of wiki-based projects hosted by the Wikimedia Foundation.

The use of wikis dramatically changes the ownership practices involved in publishing. Thus, the Wikimedia Foundation embraces the copyright licenses developed to account for the new business practices associated with the web. Creative Commons[10] has developed a number of licensing options that differ from traditional copyright in that only some proprietary rights are reserved. The goal is to protect the rights in which a given producer is interested while making access to the work as available as possible. All written material on Wikipedia is under a creative commons license. In addition, Wikimedia contributors have the option to choose the degree of copyright under which their work is protected.

11.2.6 Open Source Software

Open source software is an online collaborative development method employed to create computer software [37]. The name refers to the free availability of the source code that composes a piece of software. A programmer can contribute code to, alter, or delete the source code to produce changes in the software. Through the democratic inclusion of a large collective of interested participants, innovative solutions are contributed and bugs in the code are efficiently found and fixed.

As with wikis, open source software aggregates through the process of collaborative editing [38]. Here participants asynchronously construct, maintain, and improve upon a software project. However, code is a precise and complex dependency-based representation that can be fragile to change. Unlike text that can maintain coherency through various perturbations, code can be "broken". Thus, open source projects use a number of strictures and conventions to maintain the integrity of the existing product. One of the most fundamental of these is versioning whereby once the software reaches a certain state it is named and delineated from other code. Software as developed through a cyclical process of writing and testing and versioning allows the use of revision control to track incremental changes. Other practices include the use of a hierarchical permissions structure whereby only a subset of the total number of programmers can move code to a testing phase, incorporate changes into new versions, and declare a version ready for release.

The Linux operating system[11] is a prototypical example of open source software. This operating system was originally written by Linus Torvalds in 1991 as a processor-independent version of Unix, a proprietary program. Torvalds successfully received help to development problems he posted to a programming-oriented list-serv. This spurred him to provide the entire source code to the programming community for further contributions. The community proved to be a willing and productive collective for software development. Since Linux, the

[10] URI: http://creativecommons.org
[11] URI: http://www.linux.org

development of software through open source techniques has blossomed. Source-Forge[12] is a repository of open source software available for use, improvement, and critique. It hosts over 100,000 projects and over one million registered users, larger than any other resource of its kind.

Open source software has become a successful CDMS on a number of counts. It is lauded for its low cost, flexibility, reliability, and robustness when compared with proprietary equivalents. This is typified in that companies, perhaps most notably the formerly staunch licensor IBM, that rely on proprietary software for revenue are now incorporating open source business models [39]. Open development is also an exemplar of a new online and collaborative political economy that can fuel innovation [40]. For example, the LAMP quad is the powerhouse behind innumerable online services. This stack of technologies includes an operating system, server software, a database program, and a scripting language interpreter. LAMP stands for Linux, Apache, MySQL, and Perl (alternatively PHP or Python)—all open source software available for free. In addition to providing the tools online applications need to innovate, the open source approach is spreading to other arenas as well, notably to educational materials [41].

The licensing of software for free distribution and modification of the source code is an essential component of the open source paradigm. The Free Software Foundation initiated the GNU General Public License (GPL) for the free modification and distribution of software. This collection of licenses is sometimes referred to as copyleft, a play on the term copyright, as it grants rights to the user as opposed to reserving rights of the producer. The GNU GPL is referred to pejoratively as "viral" as all subsequent uses of the code must also be under a GPL license. In addition, the Open Source Initiative[13] maintains the integrity of the term open source through an industry-accepted definition. The definition allows the commercial use of open source software for those individuals and companies that wish to harness the economics of open source [42].

11.2.7 The Prediction Market

The prediction market is a forecasting tool where the pertinent information held by each trader is revealed and aggregated through a game-like exchange [43]. As in traditional financial markets, this exchange refers to the buying and selling of contracts (stocks) by participants who choose the price at which they are willing to trade. The contracts in a prediction market represent not shares in a company, but forecasts of specific event outcomes (such as the winner of a political election) and their price reflects the probability that the outcome will take place [44]. The market value of each contract fluctuates according the price at which traders buy and sell it. Thus, a market value is the collective's estimate of the probability of a future event. Contracts in a given event are sold with a value between 0 and 100 (interpretable as a probability). Thus, a market value of 100 for a given contract suggests the event is certain to happen. As in traditional

---

[12] URI: http://sourceforge.net
[13] URI: http://www.opensource.org

markets, traders who buy low and sell high earn the difference in prices, while those who sell low and buy high lose the difference. Unlike traditional financial markets, traders may also earn money by owning contracts whose forecast of the event outcome was correct. In the most simple payout scheme, this contract would have a value of 100 and the value of contracts in any other outcome would drop to zero.

There are a number of aggregation mechanisms for prediction markets. While in all of them traders leverage their privately held knowledge competitively to out-predict others, the manner in which trades are elicited and affect prices differ. Perhaps the most familiar mechanism of aggregation in traditional financial markets is continuous double auction (CDA) [45]. Here bids (the price for which a trader is willing to buy a given contract) are matched to asks (the price for which another trader is willing to sell the given contract) to complete a trade and update the market price. However, unlike traditional markets, prediction markets have a terminus, usually just before the forecasted event is to take place. This feature encourages other aggregation mechanisms that would not be feasible in traditional markets. A notable option is Hanson's suite of market scoring rules that allow a trader to update the market price at any time without engaging another trader in an auction. The trader can be thought of as updating the market price by compensating the trader whose price was replaced [46]. Another alternative is Pennock's dynamic pari-mutuel market adapting features from pari-mutuel betting into a CDA [47].

Prediction markets have garnered attention for their ability to accurately predict the future. They perform as well or better than traditional prediction techniques such as polling. For example, the Iowa Electronic Markets[14] (IEM) in the 2004 presidential election correctly predicted the number of electoral votes by which Bush would win [48]. The IEM out-predicts polls 75% of the time. The Hollywood Stock Exchange[15] (HSX) in 2007 correctly identified seven out of the eight winners in the most popular Oscar categories as they did in 2006.[16] In 2005, all eight winners were predicted correctly.

Prediction markets are powerful decision tools because they generate an exact probability for each forecast and that probability varies through time as more information is revealed. In 2001, the Defense Advanced Research Projects Agency (DARPA) funded two grants in electronic market-based decision support that came to be known as the Policy Analysis Market. The purpose of the grants was to develop a system that could capture the forecasting accuracy of prediction markets for problems of governmental interest. Problems included political instability, how US policy would affect the instability, and how instability would affect US interests [49]. The area of focus was the Middle East. Intended for public use, these markets would have explored how a combination of events would lead to a future event. However, the project was cancelled before large-scale human subjects testing took place, so little data is available. The markets were

[14] URI: http://www.biz.uiowa.edu/iem/
[15] URI: http://www.hsx.com
[16] HSX Press Release http://www.hsx.com/about/press/070226.htm

closed due to political outrage over the use of gambling for devastating events. The markets were later instantiated by a private company.

Prediction markets, despite their impressive success stories, are not a panacea for all future uncertainties. Typically, prediction markets have only been successful on topics that are of interest to a large number of people (e.g., politics, sports, Hollywood) and thus have a large pool of possible traders from which to draw. A large trading population is important to draw out accurate information by making it worthwhile for informed traders to participate. To some extent the effect of thin participation can be mitigated by specially designed market algorithms, such as market scoring rules, that encourage trade in such a situation [50]. In addition, legality remains a problem for prediction markets in the United States. A prediction market can be construed as a form of gambling where traders place real-money bets on essentially valueless contracts. To avoid legal entanglement, many prediction markets use play-money (and performance rankings) instead of real-money. While money is a universal motivator, it has been found that play-money markets can provide performance comparable to real-money counterparts [51].

A taxonomy of features necessary to define these systems will be detailed next.

11.3   Taxonomy of Collective Decision Making Systems

All collective decision making systems can be placed within a taxonomy of features that both distinguish systems from each other and that highlight system similarities. The presented taxonomy of CDMSs is organized into four primary classes—problem space, implementation, individual features, and collective features. Each of the seven web-based CDMSs maintain unique signatures within the feature space circumscribed by the taxonomy. The features of each system will be described in the sections that follow. The virtue of the taxonomy is to make apparent places for innovation. While many systems have common features, it is the sum of features that makes a system unique. Essentially, a change in one feature could potentially create a unique system with benefits and drawbacks dissimilar from the original system, thus filling a new niche. Note that Table 11.2 summarizes what follows in tabular form and that Fig. 11.1 displays the information as a dendrogram.

11.3.1   Problem Space

Every collective decision making system is designed to generate a decision for a particular type of problem. This class characterizes that problem space.

Decision Type. Decision type is a primary delineation for classifying decision making systems. While, for example, folksonomies and document rankings do not appear to be similar systems, the purpose of both as defined by decision type is information retrieval. The seven collective decision making systems provide decisions for only four types of problems: information retrieval, governance, content creation, and forecast. None of the four decision types inherently require a collective; however, through a well-designed system the power of a multitude

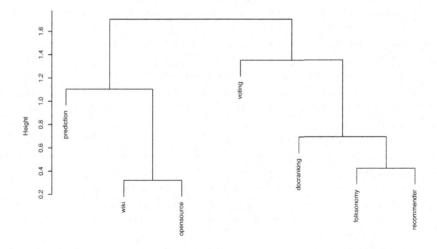

Fig. 11.1. Taxonomy-based comparison of CDMSs

of thinkers can be harnessed to produce powerful results. These four decision types do not completely fill the problem space, nor do the seven systems yield the complete set of systems that can generate these decision types.

Information retrieval is an interdisciplinary area of research encompassing the science behind the search for resources, whether text, documents, or records in a database. A primary aim is to control information overload, a common occurrence on the web. One way to manage the immense amount of information available online is to rank options via a pertinent algorithm to provide a list of search results. Search engines, such as Google, employ a variety of document ranking algorithms to this end; however, search engines are just one example of information retrieval on the web. Like search engines, folksonomies also follow a query-resource format where keyword queries connect users to applicable resources. In addition, folksonomies aggregate resources at the user level. Who tagged a resource can be as important as the tag itself as one resource in common suggests the possibility of the discovery of additional interesting resources. For example, CiteULike,[17] a social bookmarking site for academic papers, organizes a user's favorite papers into a personal library that any other user can peruse. Thus, every user's library serves as that user's bookmarks as well as an impersonal recommendation list for other users who have liked one or more resources in that library. To facilitate consumerism, most recommender systems use information retrieval techniques to anticipate the resource desires of a user. This anticipation is also an attempt to control information overload through the ranking of products.

Governance, a fond topic among political scientists and philosophers, is the administration of power over a population. This includes both the allocation of decision making rights and the aggregation of those decisions and is therefore

---

[17] URI: http://www.citeulike.org

a prime application for collective decision making techniques. Since its advent, democratic rule has been executed through the vote. While a multitude of implemented and theoretical voting derivations exist, the form always follows the casting of a vote by a specified population during a predetermined time followed by a tally. The indication of one's wishes through this general form is one of the most fundamental collective decision makings systems. The vote is different from a poll in that at the conclusion of a poll no decision is reached, and thus no governance takes place.

Content creation is a self explanatory term to refer to all works created. While teamwork resulted in content long before the advent of the Internet, web-based systems such as the wiki and open source software has enabled distributed collaboration across time and geography. Anonymous and asynchronous collaboration is the norm online. In addition, this collaboration is taking place on a massive scale; at the time of this writing Wikipedia reports more than 4.8 million registered contributors.

Prediction or forecasting is the estimation of the state of future events. The generation of formal predictions to minimize risk has historically been entrusted to haruspices, augurs, oracles, chartists, and other prophecy experts. However, online systems with the power to aggregate the opinions of many individuals uses collective decision making to reveal what is not readily apparent to individuals. Prediction markets are clearly forecasting tools. Recommender systems, to a lesser extent, also involve prediction, as the system attempts to anticipate the desires of individuals. However, until these systems routinely generate accurate recommendations not attributable to the mere power of suggestion as well as surprising recommendations, they will be excluded from the prediction categorization.

Decision Principle. The decision principle of a system refers to the manner in which one decision is chosen over another, regardless of the algorithm implemented. This is a primary distinguishing factor between systems. The decision principle may differ within a common decision type, thus it is a prime place for innovation. The application of a new decision principle for a decision type could yield a more effective CDMS.

All three information retrieval systems utilize a different decision principle. Document ranking requires the graph theoretic principle of centrality. Centrality measures the importance of a web-page relative to its position in the network. There are multiple measures of centrality and thus multiple algorithms for its determination.

Folksonomies develop through a measurement of frequency. The frequency of a given tag for a particular item represents how well that item is described by the tag. The tag cloud, a means of displaying the relative frequency of words, is often substituted for a ranked list. In addition to indexing items, the frequency of the use of a tag throughout the entire system, as opposed to for a single item, produces a zeitgeist of the user community at a given time. Many social bookmarking sites show a tag cloud of the most popular recent tags aggregating frequency overall users and all URLs.

Recommender systems that utilize collective decision making operate through the decision principle of similarity. Similarity metrics are used to determine the amount of coherence between two people or two items. Specifically, many recommender systems use matrix similarity. Through collaborative filtering, a similarity metric is determined between pairs of individuals in the system. Note that there are many ways to instantiate similarity metrics within the collaborative filtering paradigm.

Like folksonomies, vote systems aggregate through frequency, or a tally of the votes. Just as the higher frequency tags in a folksonomy suggest its popularity, so does the frequency of votes for a given candidate. In political elections, each vote is unweighted; however, online vote systems can implement a weighting system where it is not just the vote that is considered but the context of the vote as well in the form of a weighting [52].

Consensus plays a role in content creation systems as the content remains stable only as long as all participants individually believe it is satisfactory. This definition is more subtle than that of face-to-face meetings where all members of the collective explicitly give their assent. Here, the members may not have seen the most recent document so their assent is implicit in their not having looked at or changed it. Also, a member may join the collective at any time and alter the stable version. In order to keep contributors informed of changes, Wikipedia implements a "watch list" feature on which users can add pages in which they are interested. These pages are monitored automatically and alert the user when changes are made. The two-minute correction time for some types of vandalism in Wikipedia is attributable to the number of people looking out for that page at any given time [53]. Simple features like the watch list help CDMSs to perform optimally.

Trade, the decision principle powering prediction markets, is the most formalized instantiation of the consensus principle. In markets, the traders independently choose when and in what to participate. There are none but the most basic rules to guide trader behavior. Adam Smith's "invisible hand" is a metaphor for decentralization where markets are driven by the forces of supply and demand. In other words, self-interested individuals in a market produce global effects reflected in the prices of contracts. Prediction markets elicit the foreknowledge of individuals as it develops over time and weights and aggregates it. Here, a stable market value suggests that all participants individually believe that the valuation is correct. Thus, the last contribution (the last trade) stands as the current valuation.

Goal. The goal of a decision support system refers to the decision output that will be produced if the system is performing optimally. It is through the statement of a goal that a system's performance can be evaluated. The goal of a system is directly tied to the decision type (or purpose) of the system thus, as there were four decision types, there are four goals we will discuss.

The goal of all information retrieval systems is to perform a quality retrieval of the available and pertinent information. A vote system is unique in that the ultimate goal is a subjective feeling of satisfaction. While in spirit a vote system may be charged with producing a solution that, for example, maximizes public utility, the actual goal is the satisfaction of the population as to its execution. Most vote systems have no recourse for poor alignment between votes and utility, but do have recourse in the form of recounts for tallies that were perceived as unfair. The goal of content creation systems is also a subjective quality best categorized as utility. If the system can generate useful documents or source code then it has reached its goal. Often this utility is tied to comparisons with commercial counterparts. Thus, the accuracy of Wikipedia is compared to Encyclopaedia Britannica and the speed and cost of production of open source versus proprietary software is debated. The goal of a prediction market is to generate an accurate prediction about the state of the future.

Accuracy Metric. All decision making systems can be evaluated and improved if there is a metric by which to judge their accuracy. Again, accuracy is tied to the other features of problem space and thus only four metrics will be discussed.

Information retrieval systems are typically evaluated according to precision and recall. Precision measures the ratio of relevant results to the total number of retrieved results. High recall means that the retrieved results are a comprehensive sample of the relevant results available in the collection. Both metrics are necessary to describe a good information retrieval system as achieving high recall simply by retrieving all documents is not useful in reducing information overload. However, precision and recall are inversely proportional, thus with an improvement in one comes a decline in the other [54]. As search engine users tend to review only the first score of results, document ranking values precision over recall. Although recall is difficult to test in search engines as the space of all relevant documents for a given query is not obvious, numerous studies have compared search engines, and thus various document ranking techniques, in terms of precision and recall [55, 56].

Folksonomies use keyword matching to connect a queried tag to resources that have been labeled with that tag. This is information retrieval and thus precision and recall are important metrics. However, folksonomies serve as an alternative to professionally generated indices and are thus rated for accuracy by comparison. A major criticism of folksonomies is that imprecision and inconsistency in the use of tags produces an index that lacks rigor [57]. This is certainly something that a controlled vocabulary accounts for. However, the robustness of the system to change exceeds that of traditional taxonomies. For example, the Dewey Decimal System is an often used example of a system that has ossified due to the crystallization of the predominant worldview at the time of its inception. Its development by one man, and thus one perspective, stands in stark contrast to the fuzzy categorizations that develop through the myriad contributions of a folksonomy's collective. Thus, the conclusion of one system's superiority over another's is counterproductive as the strengths of one system are the weaknesses of the other. Each system type, whether folksonomy or traditional taxonomy, has an appropriate application based on the given problem and constraints.

Research on collaborative filtering algorithms is well-established. Their importance for commercial applications has made their refinement a priority. To illustrate the potential benefit of such technology to online business, Netflix is offering a million dollar reward for improving upon their current collaborative filtering algorithm by ten percent. It appears, however, that there may be a maximal accuracy bound due to the vagaries of individual ratings [58]. In addition, system improvement is a multi-faceted problem that extends beyond accuracy metrics [59]. The goal is to create a system that reduces the onus of participation on the user while providing unexpected recommendations, known as serendipity in the literature. This notion of serendipity complicates the evaluation of recommender systems. There are a variety of features on which accuracy metrics can be tabulated easily. The most common method is the leave-n-out method whereby a portion of the dataset is removed from view of the system to determine if the system will recommend the missing data. However, this method fails to account for user satisfaction. Users are often more pleased to be recommended novel products than accurate but obvious ones. The disjoint between user satisfaction and accuracy precludes the dismissal of systems that produce serendipitous recommendations but fail accuracy metrics. An additional mechanism to aid user satisfaction is the ability to succinctly explain why a particular recommendation was made to the user so as to decrease user skepticism [58].

The goal of a vote system is more subjective than other systems; their accuracy metric is perceived fairness. Although not a rigorous metric, there is a rich literature centered around the fairness of a particular algorithm for a vote. These arguments center on the desired outcome for the vote based on a number of factors. For example, [60, 61] explore the utility of majority rule compared to other electoral systems while [62] explores the superiority of the Borda count aggregation mechanism. Although, this is a well studied field of social choice theory, there is not yet consensus on which type of algorithm is best for a given situation. Indeed, it depends highly on historical process and public opinion.

Systems that develop content are held to the quality standards of each individual. For example, Wikipedia supports the recent changes patrol. These contributors use the watchlist to review edits to entries to maintain quality and monitor vandalism. In addition, tags can be added to the top of entries to indicate that a dispute needs to be resolved. As with folksonomies, the accuracy of documents and software can be compared to their proprietary counterparts. The journal *Nature* conducted an inquiry into the accuracy of Wikipedia compared with a resource generated in the traditional model—the Encyclopaedia Britannica. Of the science-oriented articles studied, Wikipedia contained 162 errors and Britannica 123 [63]. While this study has been formally and vehemently opposed by Britannica for not taking into account the nature of the errors, the real discrepancy is that the two formats excel under different constraints. Wikipedia is not limited by size as the hosting of a website is significantly cheaper than printed copies. In addition, the army of contributors to Wikipedia are required less for their authority than for their robust response to new information and the diversity of information they possess.

There is no objective utility function to rate the decision output of an open source software system. Any system that satisfies its contributors is a success.

Open source software is an excellent example of the power of collective decision making in that the distributed nature of the collaborative development process exceeds that of commercially developed software in a number of ways. Open source software is in some cases preferred to commercial software when compared by cost of production, time to release, and quality [64].

The vanguards of prediction market research, the Iowa Electronic Markets, demonstrate the superior predictive abilities of their markets to polling organizations using standard error of forecast in their accuracy analysis [65]. This compares the ability of a statistically representative sample to the self-selecting traders. An alternative compares the estimates of experts to that of the market [44]. Businesses are beginning to augment traditional methods of prediction, representative samples and expert elicitation, with prediction markets. Internal prediction markets aid corporations in gathering information that is not typically expressed by the corporate hierarchy and face-to-face meetings. Hewlett-Packard, Google, Yahoo!, Microsoft, and Intel all experiment with prediction markets. The results are mixed but encouraging. Hewlett-Packard designed a proprietary form of market called BRAIN that weights the trades of employees based on past trade successes. They report that price estimates went from 4% error using traditional methods to a 2.5% error with prediction markets using much less time and effort [66].

11.3.2   Implementation

Implementation refers to the characteristics prescribed by the problem space and system design. This class outlines the specialized skills required of the collective to participate. It is of particular utility when describing web-based systems.

Solution Space. The solution space is the set of all solutions that could be chosen as a decision. As with the problem space class, solution space is a primary defining characteristic of collective decision making systems. Here we discuss the four solution spaces applicable to each decision type.

Information retrieval systems are limited only by the total number of relevant and irrelevant results available in the system—the system's collection. For search engines like Google, the collection encompasses all of the artifacts on the web that are linked to other artifacts. The indexable web was assessed in 2005 at 11.5 billion pages, 8 billion of which Google had indexed [67]. Folksonomies are also concerned with the tagging of this collection. The Netflix collection utilized by their recommender system includes over 81,000 movie titles. It is precisely for these massive solution spaces that information retrieval systems are engineered.

Originally, ballots were blank papers used to write-in the names of candidates running for political office. Most often in vote systems today ballots restrict the solution space by pre-specifying the options. Ballots became necessary to specify the precise option that received the vote, as, for example, multiple people with the same name could claim a written-in vote. Therefore, a pre-printed ballot is used to designate all available options. An additional feature of the pre-printed ballot is secrecy, as eliminating handwriting deters connecting a voter to their vote. Thus, the Australian ballot specifies not only privacy to vote, but a pre-printed ballot as well.

The solution space for content creation documents is not limited in any functional sense. Wikipedia entries are not limited in length the way many offline resources are as the cost is less than in printed counterparts. In addition, entries are not limited to text. The Wikimedia Commons[18] is a database of freely distributable images, sound bytes, and video clips that are combinable with Wikipedia articles to enrich the entries. Open source software is similarly free from length constraints in the solution space where fast, efficient algorithms are the primary concern. However, the need for the compatibility of software does serve to define a solution space for some applications. Perhaps the largest constraint on content creation systems is that the content must be original, un-copyrighted, or properly attributed to be legal.

Prediction markets require the most rigidly defined solution space. A market must have a disjoint set of contracts where the fulfillment of one contract necessarily negates the fulfillment of the others. In addition, the contracts must exhaust the solution space. Every possible future outcome must be accounted for. For example, it is common in election-based prediction markets to see a question with two contracts—1) a Republican wins the election 2) another party wins the election. By not naming the Democratic Party in the second contract, the possibility of a win by a third party is left open and thus it covers the solution space.

A prediction market must be built around a question that has an objective answer once the contracts expire and this answer must clearly refer to a single contract. Otherwise, the results will be nullified and traders will require compensation. For example, TradeSports[19] encountered controversy when the outcome of their North Korea Missile market failed to completely satisfy either outcome. While a test missile was launched in accordance with the prediction of one contract, the launch was not verified by the Department of Defense which suggested that the other contract was more accurate.[20]

Interface Complexity. Interface complexity is particularly important for understanding the population from which decisions are originating and applies most acutely to web-based collective decision making systems. The interface complexity is not restrictive if only standard computer skills are necessary. However, some systems require special skills or a unique context to operate and thus restrict some potential members of the collective from participating. While most online decision systems do not require a specific representative population as do statistical polls, it is worth considering the segments of the population that are excluded due to the demands of the interface.

Interfaces that are not restrictive require only standard computing techniques and web navigation skills. Folksonomy interfaces are not restrictive as they

---

[18] URI: http://commons.wikimedia.org

[19] URI: http://www.tradesports.com

[20] TradeSports Press Release
http://www.tradesports.com/aav2/news/news_58.html

require simply entering personal tags through the keyboard. Users of browser-based bookmarking tools should feel comfortable using social bookmarking. Recommender systems also have a very low complexity, with participation in some occurring automatically and most only requiring a simple rating system. For e-voting to take hold, the interface must be as non-restrictive as possible. User interfaces that support accurate decision entries is of prime importance to vote systems for political elections and has been the subject of much federally-sponsored and independent research [68, 69]. Electronic vote systems in use today employ an extremely simple point-and-click or touchscreen interface. While the accurate and accessible functioning of the interface is important, the perception of such an interface is essential as vote systems are satisfaction based [70].

Restrictive interfaces require skills that are not yet part of the standard repertoire of general computer users. For example, wikis employ user-friendly interfaces similar to non-restrictive systems. However, MediaWiki powered editing sites such as Wikipedia use a wiki markup language called wikitext that, while very simplified, is more complex than the use of word processing programs. Prediction markets also have an interface that is moderately complex. Unfamiliarity with trade processes makes the prediction market a specialized decision tool that may alienate some potential contributors. There are a number of systems attempting to overcome this hurdle in commercial ventures. Inkling markets,[21] for instance, focuses on ease of use by simplifying the burden on the user to interpret price movement. While the interface itself is simplistic, the market was simplified as well through the implementation of an alternative market design. Here, traders use a scale (e.g., the market price is slightly low, low, or way too low) to input their decision instead of placing a bid or ask.

Some systems are highly restrictive. Document ranking is a unique system type in that the interface to contribute (by linking a web-page) is not controlled by document ranking systems. While every user of the Google search engine has benefited from document ranking, only a subset of the web-using population has contributed to the system. Document ranking requires the control of a web-page so as to link it to other web-pages. While the actual linking of pages is simple (especially through website creation software such as VCOM's Web Easy Pro and Apple's iWeb), the occasion to do so is more restrictive. Open source software is perhaps the most specialized of the systems discussed as they require skill in software development for their evaluation and contribution. While not strictly interface-based, the high complexity designation is due to the specialization of skills needed to work in open source software as it restricts the population eligible for membership in the collective.

Skill Set. To describe the reasoning behind the interface complexity rating, the skill set distinction describes the online actions required. All web-based CDMS require a minimum level of computer and Internet competency to operate. However, none of the systems discussed are intended to require training.

---

[21] URI: http://inklingmarkets.com

The Scottish Qualifications Authority[22] outlines four areas of knowledge that encompass basic computing skills. These are computer skills (mouse and keyboard operation, opening and closing files, locating files), e-mail skills, word processing skills, and web skills. Folksonomies, recommender systems, and vote systems all have a low interface complexity rating because they require no more than these basic skills. Wikis have a restrictive designation as they require more sophistication in operation than the aforementioned systems. The markup language is not What You See Is What You Get (WYSIWYG) as word processors are. Prediction markets require only basic computer skills, but demand of the interested user understanding of market trading. Sites that host one of these systems usually include tutorials to familiarize the new user with the system. Document ranking is a highly restrictive system as the contributor must have the skills and means to publish a website to the web. Open source software is restrictive in that intelligent participation requires specialized knowledge in computer programming, software debugging, documentation authoring, etc.

Contributor/User. The contributor/user distinction refers to whether the individuals interacting with the system are the only ones to benefit from the decision or if the benefits reach both the contributors and web users in general. While contributor and user have been used interchangeably throughout this article, in this section contributor refers to the participants entering their decisions into the system (the collective) and user refers to others who use the collective decision. Note that, as will be discussed shortly, all systems are of benefit to the contributor, thus no external enticements are usually required for recruitment into the collective. The only question is whether general users benefit as well.

Both document ranking and folksonomies are of utility to general users as well as contributors. Anyone who has used the Google search engine but has never linked a web-page is evidence of this. Recommender systems become more worthwhile to the contributor as they contribute and build up a pattern of behavior. However, they generate little overall value to a first-time user. In addition, recommender systems that use collaborative filtering to find similarity between users provide no utility to those who are not participating in the system. However, content-based recommender systems that compare the similarity between products could be useful to a first-time user.

Vote systems typically generate decisions that affect solely the populace that votes on them. This is a defining characteristics of direct democracy. However, there is a range of other arrangements that could be envisioned—one person deciding for all (dictatorship), a representative group deciding for one (jury), etc. Despite the potential benefits and detriments an individual may receive without participating in a vote as a result of that vote, there is no sense of a general user in a vote system. Therefore, a vote system is of benefit only to the contributors who are given the opportunity to express their views.

Both wikis and open source software are also of utility to both contributors and general users. Both systems utilize a collective to generate a product of

[22] URI: http://www.sqa.org.uk

wide interest. There are over 50 million page requests on Wikipedia everyday, but only 200,000 edits. Prediction markets provide a game-like environment for contributors to elicit information about the future for others. Unlike vote systems, where there is no sense in which a general user can participate in the system, prediction markets are used similarly to polls. Their prices are tracked through time as forecasts for particular events by individuals who do not trade. This service is even sellable; the Hollywood Stock Exchange was the first to produce a commercialization plan where the information generated by those playing in the markets was sold to interested buyers in the entertainment industry.

11.3.3   Individual Features

The preceding two classes of the taxonomy have dealt with features of the aggregation mechanism. This class and the following pertain to the composition and statistics of the collective. Individuals that compose the collective maintain independent choice in web-based collective decision making systems. Thus, individuals are important to consider when examining the role of the collective.

Motivation. The use of these systems by a large user-base is in many ways inexplicable. The notion of the most highly consulted online encyclopedia,[23] Wikipedia, being written by unpaid volunteers is in complete paradox to standard economic motivational theories. In addition, low voter turnout in national elections suggests that simply being asked for your opinion is not a sufficient motivation for many. Because of the necessity of large collectives to activate the problem solving potential of these systems, engaging motivating factors is an essential feature of every CDMS.

The need for affiliation is a primary motivational factor in human behavior [71, 72]. This need motivates individuals to make connections with those they want to be associated with. The wild popularity of social networking sites demonstrates that this need for affiliation and ability to connect transfers to the web. The structure of the web is set by a similar desire for connectedness, where a hyperlink serves as an affiliative bond. Through linking, the individual is prescribing where the published page fits in the network of web-pages. Therefore, the PageRank algorithm characterizes an incoming link as a vote of quality for that site as the originator of the link chose to associate with it.

Folksonomies are of particular utility for those who wish to organize and index information. Delicious, for example, replaces browser-based bookmarking with online bookmarking accessible from any computer. A contributor generally tags websites they wish to find again with a word that is meaningful to them without regard for others. The result is a personalized sample of the web. Recommender systems provide personalized advice out of an overwhelming number of options to facilitate browsing and purchasing online. A recommender system is able to best choose similar users if each user has a rich history of behavior in the system allowing the systems to "get to know" the user.

---

[23] According to Alexa global top 500 URI: http://www.alexa.com
accessed July 11, 2007.

Vote systems, as a method to elicit the desires of the populace, function by allowing each voter to express their beliefs. A vote system is cooperative in that an individual hopes that others are deciding in the same way they are, thus increasing the likelihood that their desire will be chosen. On the other hand, prediction markets are competitive in that a trader makes the most money if they express a view that most others do not have and they are correct in their prediction. These distinctions affect the way information is shared in the system. For different reasons, participants in each system may be unwilling to share the decision they registered.

Users of both wikis and open source software systems, as forms of content creation, are fundamentally motivated by a desire to impart knowledge to create valuable tools. Each edit in a content creation system is motivated by a criticism of the work in its present form. As in document ranking, contributors are motivated by the existing content.

Expertise. Expertise is the knowledge an individual must have for the system to generate an accurate decision. This is not to be confused with the skill needed to operate the system interface. It is of fundamental importance to distinguish systems that are for experts from ones that operate on more general principles. Systems that do not require experts work simply from a statistical collective intelligence perspective where the more people who participate, the more likely a satisfactory result will emerge. On the other hand, expertise-based systems must elicit information from those specially knowledgeable to generate satisfactory results.

None of the three information retrieval systems require expertise. They function through a process of averaging public opinion, although the algorithm varies in each system. However, all three systems have counterparts that do use experts instead of a collective. Mahalo[24] is a social search engine that ranks web-pages by hand. Individuals contribute the best web-pages for a set of popular search terms. These individuals are selected through an online application based on their frequent and high quality participation in other social software sites, rendering them experts. Folksonomies are often compared to their taxonomic counterparts generated by professional taxonomists. Before Amazon.com, librarians served as the experts in connecting people with their media whims and needs. It is through system design that expert knowledge-keepers can be replaced by an amorphous collective of fallible and untrained individuals.

Like information retrieval systems, vote systems replace the judgment of a single individual with the opinions of the collective. Vote systems do not require expertise as they are held to no accuracy metric other than satisfaction. For this reason, campaigns such as "get out the vote" continue. Every person allowed to vote is so encouraged regardless of their knowledgeability.

On the other hand, content creation systems and prediction markets require expertise. In order to create a work, a participant must be able to provide a useful and unique contribution. Prediction markets will identify the inexpert

[24] URI: http://www.mahalo.com

through his or her dropping portfolio value; however, knowledge of the future state must be present to be amplified in the market.

Membership. Membership refers to the method by which participants become a part of the collective. Almost all of the systems discussed rely on the principle of self-selection. In other words, individuals provide the initial impetus to participate and are not selected upon by the system for fitness in the collective. Document ranking is the only system that does not rely on self-selection. While it is up to each individual to link to whatever web-pages they please, they do not choose to lend this decision to document ranking systems. Instead, the decisions of the collective are co-opted by robots that traverse the web by following these links to determine the structure of the web. It is worth noting that a next generation of search engine designed as social software and typified by Sproose[25] ranks pages based on contributors' explicit votes.

The systems that have a self-selecting collective also require the use of a consistent user name to maintain a persistent identity through time. The log-in serves to organize anonymous and asynchronous interactions with the system into a coherent entity. It also enables the discrete tracking of user behavior for automatic membership. For example, Amazon.com exploits the tracking of a logged-in user to automatically enroll the user in their recommender system. The desires of the user is inferred from past purchases. The fundamental difference between the co-opting of decisions made by individuals for the purposes of document ranking and that of Amazon.com is that the individual on Amazon.com has explicitly engaged in a user relationship with the website by logging in. The confirmation of identity is also important in vote systems where only one vote is allowed by every eligible participant. Problems with the verification of identity is a major impediment to the establishment of online voting systems [73].

Folksonomies, recommender systems, and vote systems take very little care to maintain the quality of their collective. As these systems are not expert-based this is not surprising. However, wikis, open source software, and prediction markets are all systems requiring expertise and thus contain interesting features to ameliorate the impact of unhelpful members. Wikipedia posts a "wanted list" of contributors and IP addresses that have engaged in vandalism to identify those whose edits should be monitored. Persistent vandals are permanently blocked from participating. Open source software provides a hierarchical arrangement where contributions are reviewed before being incorporated into the code. In most cases, packages are signed to provide accountability for poor contributions and to look out for malicious content. To encourage traders to play only if they are reasonably assured of their decision, prediction markets offer incentives based on participants' performance. The monetary and prestige-based incentives encourage one to participate if they desire the reward or not to participate if the consequences are too great. Traders form a self-selecting population where each individual chooses if, when, and the extent of their participation. To participate

---

[25] URI: http://www.sproose.com

without knowledge may lead to financial losses for the trader, which hinders future participation.

11.3.4 Collective Features

Statistics regarding the collective in a CDMS may be difficult to interpret as the collective itself is an amorphous and changing collection of individuals. However, any collective decision must consider its aggregation mechanism in conjunction with the facts of its collective.

Size. Size refers to the number of individuals needed in a collective to produce a collective decision of quality. This extremely relative measure is designated either variable or large for our systems of interest. Systems that require a large collective suggests that statistical collective intelligence plays a role in generating quality results. In other words, it is through high participation levels that accuracy develops. Conversely, systems that can handle a variable population size suggests that expertise is required. The only system (of the seven) where this does not hold true is the vote system. A vote system allows a variable population size but does not require expertise. As vote systems are based on the principle of fairness, the vote need only satisfy this requirement. An exception is the requirement of a quorum adopted by some voting bodies. A quorum is the minimum number of people needed to be present to participate in a vote to make it legitimate. In web-based votes where an individual's "presence" during a vote is difficult to guarantee, the institution may require a per-option quorum[26] where an *option* must receive the number of votes equal to the quorum before it can be considered a winner. The per-option quorum protects against a non-monotic situation where the vote cast to reach quorum allows another option to win. Full participation in voting systems can alternatively be simulated when presence to vote is infeasible. For example, the trust-based social network algorithm dynamically distributed democracy (DDD) simulates complete participation in a direct democracy as user participation wanes [74].

Information retrieval systems work best with a large number of contributions. This is because of the reliance on statistical collective intelligence to provide a complete and rich description of the solution space. As more information is contributed through individual interaction with the system a cleaner probability distribution is generated.

Content creation tools allow a variable collective size. The size necessary depends on the complexity of the decision and the distribution of knowledge on the topic. If there are three foremost experts in an area, then others may not be necessary. Open source software systems echo the sentiment of more is better with Torvald's famous quote, "Given enough eyes, all bugs are shallow" [37]. The size of the population of prediction markets necessary to generate an accurate solution is not a well-researched subject. While traditional financial markets operate with thousands of participants a day, prediction markets can handle, but do not

---

[26] Implemented by Debian URI: http://www.debian.org

require this amount of traffic [75]. Ostensibly, this is an expert-based system, so if the knowledge to predict the future is held between a few, then those are the only ones that need participate. However, the noise trader in traditional markets induces experts to participate by moving prices away from a correct value [76]. In other words, the poor contributions of noise traders allow experts to include relevant information and thus earn money by moving a price back in line. The Iowa Electronic Markets (IEM) advise that 20 to 30 participants can generate accurate predictions.[27]

Diversity. The role of diversity is a well-studied area of collective dynamics [77, 78]. Diversity is the fundamental mechanism behind the emergence of collective decision making. A collective is necessarily diverse, although the ways in which the individuals differ are of importance. Some systems benefit by utilizing a population that represents different pieces of information because the diverse contributions help to cover the solution space. For these systems, it is through diversity and a large collective size that optimal solutions are generated. In other systems, diversity allows an individual to improve upon the contributions of another [79]. A collective is used precisely because only through a large distribution do patterns of consensus become apparent. Thus, all systems balance the exploitation of diversity with the capturing of similarity. The seven systems are classified according to their most prominent use of diversity—coverage of the solution space or incremental improvement upon the current solution.

Information retrieval systems rely on a comprehensive index of the collection that makes up the solution space. Thus, a collective is used to gather information about this space. Large numbers are required in the collective to incorporate enough diversity to cover the solution space. If all users were totally homogenous no general distribution would be required. Recommender systems require participants to have similar preferences, but a diversity of experiences leading to differences in the items they have accessed.

Diversity is not always a desired characteristic in collective systems. For example, in a vote system, it would be best if every participants' views were in total accord. As long as the vote system properly delegates the favorable position, the system will be regarded as universally fair. While debate is a cornerstone of democracy, consensus is ideal for the vote system. In such a case, all votes would return a unanimous decision. Thus, neither type of diversity is desired in a vote system.

Both content creation systems and prediction markets require diversity to produce incremental improvements in the system. To generate a collective decision in these systems, it is important that each person has a different skill set, element of knowledge, or critique to contribute. In prediction markets, diversity is the impetus for trade. It is the individually different valuations of contract prices that initiate trades. The market aggregates the incremental movements of contracts toward an accurate prediction. The competitiveness of prediction markets, where a trader succeeds at anothers failure, encourages the contribution of

[27] IEM FAQ http://fluprediction.uiowa.edu/fluhome/FAQ.html

diverse prediction-relevant information. Before each participant chooses to trade in a market, they must evaluate the uniqueness of their information. A trader has an opportunity to perform the best if they have unique information. In other words, if the market price does not already reflect a trader's information he or she can earn money by buying or selling shares to bring the actual price closer to their estimation.

Interaction. Interaction is a property of the collective that refers to the amount of feedback experienced by the contributors from other members of the collective. For our purposes, this feature is broken down into three types of interaction. Imitative refers to a level of interaction that urges a normative response in the user. Strategic refers to the expression of decisions based on a strategic analysis of options. Stigmergic refers to the indirect communication left by individuals in a shared space [80].

Document ranking is not inherently interaction-based. Contributors simply choose to link to other web-pages and in aggregate this produces a connected network. Recommender systems do no require direct interaction between others in the system. In face, the lack of transparency connecting past preferences to recommendations leads some users to "test" the system to try to reveal why a given recommendation was made. To counteract this behavior, some sites now explain their recommendations [58]. For example, Amazon.com explains that a given item was recommended based on a specific item that was either viewed or purchased by the user. Folksonomies also do not require interaction; however, the convergence of tags to produce a coherent system depends upon individuals choosing to tag as others have. The popularity of tags that were originally used for a document and other patterns in tagging behavior suggest imitative interaction [15, 11].

A vote system may require strategic interaction with others in the system. In nearly contemporaneous papers, Gibbard and Satterthwaite presented a theorem of broad circumstances in which voters have an incentive to strategically vote in a manner that does not reflect their true preferences [81, 82]. For example, in some systems, if an individual votes on an option that is not in serious contention, a third party vote for example, that is considered a wasted vote. The voter would have better expressed their desires, if they knew that there would not be strong support for their first choice, by choosing a more likely contender. The best strategy is dictated by the aggregation algorithm employed. Prediction markets, like all financial markets, also involve strategic interaction with the system as they are game-like. Specific strategies for each aggregation mechanism of prediction markets have been researched both to aid in strategy implementation and understand their effects on system accuracy [83, 84].

Content creation systems have a high level of stigmergic interaction as the work itself functions as the feedback within which users interact. The large number of contributors that participate in these systems extends our pre-Internet notions of the size of collaboration. The scale of collaboration in wikis and open source software is reminiscent of insect colonies. Thus, it is apt that the tools used to facilitate this collaboration are similar to those of insect colonies [85].

Table 11.2. Collective decision making systems in the feature space

	Document Ranking	Folksonomy	Recommender	Vote	Wiki	Open Source	Prediction Market
Problem Space							
Decision Type	information retrieval	information retrieval	information retrieval	governance	content creation	content creation	prediction
Decision Principle	centrality	frequency	similarity	frequency	consensus	consensus	trade
Goal	quality retrieval	quality retrieval	quality retrieval	satisfaction	document utility	code utility	predictive accuracy
Accuracy Metric	precision recall	precision recall	precision recall	fairness	usability	usability	forecast standard error
Implementation							
Solution Space	number of artifacts	number of artifacts	number of artifacts	ballot	creative output	creative output	disjoint + exhaustive
Interface Complexity	very restrictive	not restrictive	not restrictive	not restrictive	restrictive	very restrictive	restrictive
Skill Set	web-page design	basic skills	basic skills	basic skills	wikitext syntax	programming	market trading
Contributor/User	both	both	contributors	contributors	both	both	both
Individual Features							
Motivation	connectedness	organization	personalized advice	cooperative	critical	critical	competitive
Expertise	unnecessary	unnecessary	unnecessary	unnecessary	necessary	necessary	necessary
Membership	co-opted	self-selecting	auto/self-selecting	self-selecting	self-selecting	self-selecting	self-selecting
Collective Features							
Size	large	large	large	variable	variable	variable	variable
Diversity	coverage	coverage	coverage	none	improvement	improvement	coverage + improvement
Interaction	none	imitative	none	strategic	stigmergic	stigmergic	strategic

Both use the environment to leave information that communicates to others. Wikis improve efficiency in this communication process by assembling a list of pages that need to be written (essentially the links looking for articles) and open source software often employs postings of known problems to focus the efforts of myriad contributors [86]. In sum, these features facilitate the high interaction levels of these productive systems that might otherwise be overwhelmed by the chaos of so many contributors.

11.4 Conclusion

The move to web-based collective decision making systems has precipitated an enhanced ability to gather useful information from individuals as well as aggregate this information using scalable techniques for a variety of outcomes. The taxonomy presented defines each system by the unique combination of their features and highlights similarities between the systems. Unexplored combinations suggest a potential for the development of additional systems to meet our decision-making needs. It is left to future work to examine the feature space of web-based collective decision making systems to determine the unexploited options and the unexplored combinations to design new tools. Each system has its own particular benefits, specific applications in the problem space, and disadvantages. If the variations between the systems are explored and the best system for a particular problem is determined, then CDMSs will have reached the extent of their abilities to facilitate decisions.

References

1. Desanctis, G., Gallupe, R.B.: A foundation for the study of group decision support systems. Management Science 33(5), 589–609 (1987)
2. Tepper, M.: The rise of social software. netWorker 7(3), 18–23 (2003)
3. Luhmann, N.: Soziale Systeme. Suhrkamp Verlag, Frankfurt am Main (1984)
4. Atlee, T.: Forms of collective intelligence (May 2006), Retrieved January 20, 2007 from, http://www.community-intelligence.com
5. Condorcet, M.d.: Essai sur l'application de l'analyse a la probabilite des decisions rendues a la pluralite des voix, Imprimerie Royale, Paris (1785)
6. Page, L., Brin, S., Motwani, R., Winograd, T.: The PageRank citation ranking: Bringing order to the web. Technical report, Stanford Digital Library Technologies Project (1998)
7. Brin, S., Page, L.: The anatomy of a large-scale hypertextual web search engine. Computer Networks 30(1–7), 107–117 (1998)
8. Kleinberg, J.M.: Authoritative sources in a hyperlinked environment. In: Proceedings of the ACM-SIAM Symposium on Discrete Algorithms, pp. 668–677. ACM Press, New York (1998)
9. Castillo, C., Chellapilla, K., Davison, B.D. (eds.): Proceedings of the 3rd International Workshop on Adversarial Information Retrieval on the Web. ACM Press, New York (2007)

10. Mathes, A.: Folksonomies - cooperative classification and communication through shared metadata. Computer Mediated Communication - LIS590CMC (graduate course) (December 2004)
11. Cattuto, C., Loreto, V., Pietronero, L.: Semiotic dynamics and collaborative tagging. Proceedings of the National Academy of Science 104(5), 1461–1464 (2007)
12. Hammond, T., Hannay, T., Lund, B., Scott, J.: Social bookmarking tools (1). D-Lib Magazine 11(4) (April 2005)
13. Voss, J.: Tagging, folksonomy, and co.—renaissance of manual indexing. In: 10th International Symposium for Information Science, Cologne (January 2007)
14. Udell, J.: Collaborative knowledge gardening. InfoWorld (August 2004)
15. Golder, S.A., Huberman, B.A.: Usage patterns of collaborative tagging systems. Journal of Information Science 32(2), 198–208 (2006)
16. Steinbock, D., Pea, R., Reeves, B.: Wearable tag clouds: Visualizations to support new collaborations. In: Conference on Computer Supported Collaborative Learning, Camden, New Jersey (2007)
17. Resnick, P., Varian, H.: Recommender systems. Communications of the ACM 40(3), 56–58 (1997)
18. Adomavicius, G., Tuzhilin, A.: Toward the next generation of recommender systems: A survey of the state-of-the-art and possible extensions. IEEE Transactions on Knowledge and Data Engineering 17(6), 734–749 (2005)
19. Samet, H.: Similarity searching: Indexing, nearest neighbor finding dimensionality reduction, and embedding methods for applications in multimedia databases. In: International Conference on Pattern Recognition, Cambridge, UK (August 2004)
20. Vucetic, S., Obradovic, Z.: A regression-based approach for scaling-up personalized recommender systems in e-commerce. In: Workshop on Web Mining for E-Commerce at 6th ACM-SIGKDD, Boston, MA (2000)
21. Maltz, D., Ehrlich, K.: Pointing the way: Active collaborative filtering. In: Proceedings of the CHI 1995 Human Factors in Computing Systems, ACM, New York (1995)
22. Ahn, H.J.: A hybrid collaborative filtering recommender system using a new similarity measure. In: Proceedings of the 6th WSEAS International Conference on Applied Computer Science, Hangzhou, China (April 2007)
23. Middleton, S.E., Shadbolt, N.R., De Roure, D.C.: Ontological user profiling in recommender systems. ACM Transactions on Information Systems 22(1), 54–88 (2004)
24. Ramakrishnan, N., Keller, B.J., Mirza, B.J., Grama, A.Y., Karypis, G.: Privacy risks in recommender systems. IEEE Internet Computing, 54–62 (2001)
25. Straffin Jr., P.D.: Topics in the Theory of Voting. Birkhauser, Boston (1980)
26. Arrow, K.: Social Choice and Individual Values, 2nd edn. John Wiley and Sons, Chichester (1963)
27. Thornton, A.: Does internet create democracy? Ecquid Novi 22(2) (2001)
28. Robertson, S.P., Wania, C.E., Park, S.J.: An observational study of voters on the Internet. In: 40th Annual Hawaii International Conference on System Sciences (HICSS 2007), Waikoloa, Hawaii (January 2007)
29. Robertson, S.P.: Voter-centered design: Toward a voter decision support system. ACM Transactions on Computer-Human Interaction 12(2), 263–292 (2005)
30. Kohno, T., Stubblefield, A., Rubin, A.D., Wallach, D.S.: Analysis of an electronic voting system. In: IEEE Symposium on Security and Privacy, pp. 27–40 (May 2004)
31. Schyren, G.: How security problems can compromise remote Internet voting systems. Electronic Voting in Europe, 121–131 (2004)

32. Rodriguez, M.A., Steinbock, D.J., Watkins, J.H., Gershenson, C., Bollen, J., Grey, V., deGraf, B.: Smartocracy: Social networks for collective decision making. In: 40th Annual Hawaii International Conference on Systems Science (HICSS 2007), Waikoloa, Hawaii (2007)

33. Leuf, B., Cunningham, W.: The Wiki Way: Quick Collaboration on the Web. Addison-Wesley, Reading (2001)

34. Oster, G., Urso, P., Molli, P., Imine, A.: Data consistency for P2P collaborative editing. In: Proceedings of the 2006 Conference on Computer Supported Cooperative Work, pp. 259–268 (2006)

35. Almeida, R.B., Mozafari, B., Cho, J.: On the evolution of Wikipedia. In: Proceedings of ICWSM, Boulder, CO (2007)

36. Ebersbach, A., Glaser, M., Heigl, R., Dueck, G., Adelung, A.: Wiki: Web Collaboration. Springer, New York (2005)

37. Raymond, E.S.: The Cathedral and the Bazaar: Musings on Linux and Open Source by an Accidental Revolutionary. O'Reilly, Sebastopol (1999)

38. Yamauchi, Y., Yokozawa, M., Shinohara, T., Ishida, T.: Collaboration with Lean Media: How open-source software succeeds. In: Proceedings of the 2000 ACM Conference on Computer Supported Cooperative Work, pp. 329–338 (2000)

39. Samuelson, P.: IBM's pragmatic embrace of open source. Communications of the ACM 49, 21–25 (2006)

40. Weber, S.: The Success of Open Source. Harvard University Press, Cambridge, MA (2004)

41. Baldi, S., Heier, H., Mehler-Bicher, A.: Open courseware and open source software. Communications of the ACM 46, 105–107 (2003)

42. Lerner, J., Tirole, J.: Some simple economics of open source. Journal of Industrial Economics 52, 197–234 (2002)

43. Berg, J., Rietz, T.A.: Prediction markets as decision support systems. Information Systems Frontiers 5(1), 79–93 (2003)

44. Wolfers, J., Zitzewitz, E.: Prediction markets. Journal of Economic Perspectives 18(2), 107–126 (2004)

45. Smith, E., Farmer, D.J., Gillemot, L., Krishnamurthy, S.: Statistical theory of the continuous double auction. Quantitative Finance 3(6), 1469–7688 (2003)

46. Hanson, R.: Logarithmic market scoring rules for modular combinatorial information aggregation. Journal of Prediction Markets 1(1), 3–15 (2007)

47. Pennock, D.M.: A dynamic pari-mutuel market for hedging, wagering, and information aggregation. In: ACM Conference on Electronic Commerce, New York (2004)

48. McCrory, G.: Iowa electronic markets forecasted bush win in presidential election. University of Iowa News Service (November 2004) Retrieved January 29, 2007, http://www.news-releases.uiowa.edu

49. Hanson, R.: Policy analysis market archive. George Mason University (2003) Retrieved January 29, 2007, http://hanson.gmu.edu/policyanalysismarket.html

50. Abramowicz, M.B.: The hidden beauty of the quadratic market scoring rule: A uniform liquidity market maker, with variations. Technical report, The George Washington Law School (March 2007)

51. Servan-Schrieber, E., Wolfers, J., Pennock, D.M., Galebach, B.: Prediction markets: Does money matter? Electronic Markets 14(3) (2004)

52. Rodriguez, M.A.: Social decision making with multi-relational networks and grammar-based particle swarms. In: 40th Annual Hawaii International Conference on System Sciences (HICSS 2007), Waikoloa, Hawaii (2007)

53. Viegas, F.B., Wattenberg, M., Dave, K.: Studying cooperation and conflict between authors with history flow visualizations. In: Proceedings of SIGCHI Conference on Human Factors in Computing Systems, Vienna, Austria, pp. 575–582 (2004)
54. Cleverdon, C.: On the inverse relationship of recall and precision. Journal of Documentation 28, 195–201 (1972)
55. Shafi, S.M., Rather, R.A.: Precision and recall of five search engines for retrieval of scholarly infomation in the field of biotechnology. Webology 2(2) (August 2005)
56. Clark, S., Willett, P.: Estimating the recall performance of search engines. ASLIB Proceedings 49(7), 184–189 (1997)
57. Guy, M., Tonkin, E.: Folksonomies: Tidying up tags? D-Lib Magazine 12(1) (January 2006)
58. Herlocker, J.L., Konstan, J.A., Terveen, L.G., Riedl, J.T.: Evaluating collaborative filtering recommender systems. ACM Transactions on Information Systems 22(1), 5–53 (2004)
59. McNee, S.M., Riedl, J., Konstan, J.A.: Being accurate is not enough: How accuracy metrics have hurt recommender systems. In: Extended Abstracts of the 2006 ACM Conference on Human Factors in Computing Systems, Montreal, Canada (April 2006)
60. Blais, A.: The debate over electoral systems. International Political Science Review 12, 239–260 (1991)
61. Norris, P.: Choosing electoral systems: Proportional, majoritarian and mixed systems. International Political Science Review 18, 297–312 (1997)
62. Reilly, B.: Social choice in the south seas: Electoral innovation and the Borda count in the pacific island countries. International Political Science Review 23(4), 355–372 (2002)
63. Giles, J.: Internet encyclopaedias go head to head. Nature, 900–901 (December 2005)
64. Fitzgerald, B.: A critical look at open source. Computer 37, 92–94 (2004)
65. Berg, J., Nelson, F., Rietz, T.A.: Accuracy and forecast standard error of prediction markets. Technical report, University of Iowa, College of Business Administration (July 2003)
66. Fine, L.: HP Prediction Markets. Confab.Yahoo! (December 2006)
67. Gulli, A., Signorini, A.: The indexable web is more than 11.5 billion pages. In: 14th International World Wide Web Conference, Chiba, Japan (May 2005)
68. Wiklund, M.: Procuring a user-centered voting system. Technical report, United States of America Federal Election Commission (October 2003)
69. Bederson, B.B., Lee, B., Sherman, R.M., Herrnson, P.S., Niemi, R.G.: Electronic voting system usability issues. In: Proceedings of SIGCHI Conference on Human Factors in Computing Systems. ACM Press, New York (2003)
70. Nass, C.: Perceptions of the voting process: Measurement and relationship to interface design. In: Proceedings of the Workshop on Election Standards and Technology, American Association for the Advancement of Science, Washington, D.C. (January 2002)
71. Murray, H.A.: Explorations in Personality. Oxford University Press, New York (1938)
72. McClelland, D.C.: The Achieving Society. Van Nostrand, Princeton, NJ (1961)
73. Jefferson, D., Rubin, A.D., Simons, B., Wagner, D.: A security analysis of the secure electronic registration and voting experiment (SERVE) (January 2004), Retrieved June 15, 2007 from, www.servesecurityreport.org

74. Rodriguez, M.A., Steinbock, D.J.: A social network for societal-scale decision-making systems. In: Proceedings of the North American Association for Computational Social and Organizational Science Conference (NAACSOS 2004), Pittsburgh, PA, USA (2004)

75. Hanson, R.D.: Decision markets. IEEE Intelligent Systems 14(3), 16–19 (1999)

76. Wolfers, J., Zitzewitz, E.: Five open questions about prediction markets. Technical report, National Bureau of Economic Research (February 2006)

77. Johnson, N.L.: Diversity in decentralized systems: Enabling self-organizing solutions. Technical report, Los Alamos National Laboratory (November 1999)

78. Page, S.E.: The Difference: How the Power of Diversity Creates Better Groups, Teams, Schools, and Societies. Princeton University Press, Princeton (February 2007)

79. Hong, L., Page, S.E.: Problem solving by heterogeneous agents. Journal of Economic Theory 97(1), 123–163 (2001)

80. Grasse, P.P.: La reconstruction du nid et les coordinations interindividuelles, la theorie de la stigmergie. Insectes Sociaux 6, 41–84 (1959)

81. Gibbard, A.: Manipulation of voting schemes: A general result. Econometrica 41(4), 587–601 (1973)

82. Satterthwaite, M.A.: Strategy-proofness and Arrow's conditions: Existence and correspondence theorems for voting procedures and social welfare functions. Journal of Economic Theory 10, 187–217 (1975)

83. Dimitrov, S., Sami, R.: Non-myopic strategies in prediction markets. In: ACM Conference on Electronic Commerce, San Diego (June 2007)

84. Nikolova, E., Sami, R.: A strategic model for information markets. In: ACM Conference on Electronic Commerce, San Diego (June 2007)

85. Elliot, M.: Stigmergic collaboration: The evolution of group work. M/C Journal 9(2) (May 2006)

86. Heylighen, F.: Why is open access development so successful?: Stigmergic organization and the economics of information. In: Lutterbeck, B., Barwolff, M., Gehring, R.A. (eds.) Open Source JahrBuch, Lehmanns Media, Berlin (2007)

Author Index

Editors

Richi Nayak received PhD in computer science in 2001 from the Queensland University of Technology, Brisbane, Australia and masters from the Indian Institute of Technology, Roorkee, India. She is senior lecturer at the Faculty of Information Technology, Queensland University of Technology. Her research interests are data mining and knowledge discovery. In recent years she has focused her research on Web intelligence,

Web service discovery and XML data management. She has been successful in applying theories of data mining in a variety of practical settings. These application domains are active ageing, structural health monitoring, asset management, software engineering, e-commerce, m-commerce and information retrieval. Her publication include 1 edited workshop proceeding, 7 Book Chapters, 10 Refereed Journal Articles, 35+ Refereed Conference Articles.

Nikhil Shripal Ichalkaranje is Senior Technology Advisor in the Australian Department of Broadband, Communications and the Digital Economy. He is also an adjunct

Senior Research Fellow at the University of South Australia. His research interests include artificial intelligence, computer communications/networking and robotics. Nikhil holds a Masters Degree in Computer Network Systems from Swinburne University of Technology Melbourne and a PhD in Computer Systems Engineering from the University of South Australia Adelaide. Nikhil has co-edited 4 books along with several publications in the area of artificial intelligence and their applications.

Lakhmi C. Jain is a Director/Founder of the Knowledge-Based Intelligent Engineering Systems (KES) Centre, located in the University of South Australia.

His interests focus on the applications of novel techniques such as knowledge-based systems, virtual intelligent systems, defence systems, intelligence-based medical systems, e-Education and intelligent agents.